T0240831

Monthly Problem Gems

Monthly Problem Gems

Hongwei Chen

Christopher Newport University

CRC Press
Taylor & Francis Group
Boca Raton London New York

CRC Press is an imprint of the
Taylor & Francis Group, an **informa** business

First edition published [2021]
by CRC Press
6000 Broken Sound Parkway NW, Suite 300, Boca Raton, FL 33487-2742

and by CRC Press
2 Park Square, Milton Park, Abingdon, Oxon, OX14 4RN

© 2021 Hongwei Chen

CRC Press is an imprint of Taylor & Francis Group, LLC

Library of Congress Control Number: 2020951721

ISBN: 978-0-367-76678-8] (hbk)
ISBN: 978-0-367-76677-1 (pbk)
ISBN: 978-1-003-16806-5 (ebk)

Typeset in CMR10 font
by KnowledgeWorks Global Ltd.

Contents

Preface

Problems and their solutions lie at the heart of mathematics. The problem section in *The American Mathematical Monthly* (AMM) provides one of the most challenging and interesting problem sections among the various journals and online sources currently available.

According to the statistics on JSTOR, not only is the AMM the most widely-read mathematical journal in the world, but its problem section is the most often downloaded article. The published problems and solutions have become a treasure trove rife with mathematical gems. This book is an outgrowth of a collection of 62 monthly problems that the author has worked in the last two decades. Each selected problem has a central theme, contains gems of sophisticated ideas connected to important current research, and opens new vistas in the understanding of mathematics. This collection is intended to encourage the reader to move away from routine exercises and toward creative solutions, as well as offering the reader a systematic illustration of how to organize the transition from problem-solving to exploring, investigating, and discovering new results. This book seeks to elucidate the principles underlying these ideas and immerse the readers in the processes of problem-solving and research. For example, our solution to Problem 10949 (a weak version of the Riemann hypothesis, see Chapter 4.10) provides an elementary gateway to the Riemann hypothesis, which has appeal to anyone with rudimentary exposure to number theory.

This book also emphasizes historical connections, provides interesting anecdotes, and adds the details of existing results that have been relegated to widely scattered books and journals. To fulfill the above goals, this book follows a very structured format:

1. State the numbered Monthly problem along with its proposer(s);

2. Address the heuristic consideration of the solution — a careful discussion of the challenges and related mathematical ideas needed to solve the problem, focusing on mathematical insight and intuition;

3. Present one or more solutions based on the heuristics; most are either my published solution in the Monthly or a solution I submitted to the Monthly (which is different from the published one), aimed at presenting and developing problem-solving techniques;

4. Reveal possible extensions of the problem to promote further research. The materials vary in their degree of rigor and sophistication

to accommodate readers with diverse interests. This often refers the reader to additional related Monthly problems (more than 180), but some extensions may serve as a springboard to further research and lead to publications.

This book is organized into five chapters as follows:

Chapter 1. Limits. This chapter contains 15 Monthly problems involving the various ways in which the notion of limit may appear. In particular, it shows how to use the Stolz-Cesàro theorem to determine the growth rate of various Pólya-type sequences.

Chapter 2. Infinite Series. This chapter covers 12 Monthly problems related to series, including those involving Euler sums and Riemann zeta values, and presents deeper techniques like special integrals, Abel's summation formula, and Fourier series to test the behavior of the series and to determine their values.

Chapter 3. Integrations. This chapter presents a combination of approaches to evaluate 15 irresistible Monthly integral problems. These evaluations provide nice applications of infinite series, gamma and beta functions, parametric differentiation and integration, Laplace transforms, and contour integration.

Chapter 4. Inequalities. This chapter collects 10 Monthly problems involving inequalities. Their solutions encounter many familiar characters, among them: Cauchy's inequality, L'Hôpital's monotone rule, majorization and convexity, Hilbert's identity, Hardy's inequality, and differential equations. What makes our trip worthwhile is the realization that these old chestnuts still have something new to tell us. At any rate, they all appear here in novel and surprising ways. Add them to your inequality tool box.

Chapter 5. Monthly Miniatures. This chapter introduces 10 Monthly problems related to the mean value theorem, eigenvalues and determinants of some special matrices, weighted trigonometric sums and Dirichlet series, infinite sum-product identity, and polynomial zero identities. In addition to developing valuable techniques for solving these problems, it also includes some interesting historical accounts of ideas and contexts. Among them, we will encounter various extensions of the mean value theorem, Chebyshev polynomials, $(0, 1)$-matrices, the Cauchy matrix, Dodgson condensation, Dirichlet convolution, the Woods-Robbbins identity, and the alternating sign matrix conjecture. This chapter is intended to facilitate a natural transition from problem-solving to independent exploration of new results.

Most of the ideas involved in the solutions will be comprehensible to the good undergraduate math student. A rich glossary of important theorems and formulas is included for easy reference. For some definitely challenging problems, I have endeavored to provide the related background so that they can be followed and enjoyed without frustration. As problems in each new issue of the AMM arrive, the reader may regard this book as a starter set, acquire a jumping of point to new ideas, and extend one's personal profile of problems and solutions.

I wish to take this opportunity to express my thanks to Dr. Brian Bradie, who has shared many years of enjoyment of problem-solving, and carefully read the earlier draft of the manuscript, and corrected many errors. I am also grateful to my chair, Dr. Christopher Kennedy, and dean, Dr. Nicole Guajardo, for their constant encouragement and support. Dr. Kennedy also read the most of manuscript and offered his help in making it better. Special thanks to Christopher Newport University for granting me a sabbatical to write this manuscript. Thanks to Steven Kennedy, Gail Nelson, and the anonymous reviewers for their many valuable suggestions for improving the book. Thanks also to Robert Ross and the CRC Press for their help in making this book a reality and for all assistance they offered. Finally, thanks to all individuals who contributed problems and solutions, from whom I have learned much.

Naturally all possible errors are my own responsibility. Comments, corrections, and suggestions from readers are always welcome: hchen@cnu.edu. Thank you in advance.

1

Limits

We begin with 15 limit problems with the intent of providing an implementation organizing the natural transition from problem solving toward exploring, investigating, and discovering new results.

1.1 A rational recurrence

Problem 11559 (Proposed by M. Bataille, 118(3), 2011). For positive p and $x \in (0, 1)$, define the sequence $\{x_n\}$ by $x_0 = 1, x_1 = x$, and, for $n \geq 1$,

$$x_{n+1} = \frac{px_{n-1}x_n + (1-p)x_n^2}{(1+p)x_{n-1} - px_n}.$$

Find positive real numbers α, β such that $\lim_{n \to \infty} n^\alpha x_n = \beta$.

Discussion.
Although a rational recurrence is simple in its form, it is usually difficult to find an exact formula for the general term. Here we try to reformulate the problem into an equivalent but simpler form. First we observe the ratio $y_{n+1} := x_{n+1}/x_n$ satisfies a first-order rational recurrence, and that the limit to be proved implies $y_{n+1} \to 1$. We then recall that the solution to the special case

$$z_{n+1} = \frac{z_n}{1 + az_n}$$

has a closed form. Taking into account these facts, we are ready to start.

Solution.
Let $y_{n+1} := x_{n+1}/x_n$ for $n \geq 0$. It is easily verified that

$$y_{n+1} = \frac{p + (1-p)y_n}{(1+p) - py_n}.$$

This implies that

$$1 - y_{n+1} = \frac{1 - y_n}{1 + p(1 - y_n)}.$$

1

Setting $z_{n+1} = 1/(1 - y_{n+1})$ yields

$$z_{n+1} = z_n + p.$$

Repeatedly applying this recurrence leads to

$$z_{n+1} = z_1 + np = \frac{1}{1 - a_1} + np = \frac{1}{1 - x} + np.$$

Thus

$$y_{n+1} = 1 - \frac{1}{z_{n+1}} = \frac{p(1 - x)n + x}{p(1 - x)n + 1},$$

and

$$x_n = \prod_{k=0}^{n-1} \frac{x_{k+1}}{x_k} = \prod_{k=0}^{n-1} y_{k+1}$$

$$= \prod_{k=0}^{n-1} \left(\frac{p(1 - x)k + x}{p(1 - x)k + 1} \right) = \prod_{k=0}^{n-1} \left(\frac{k + a}{k + b} \right), \tag{1.1}$$

where $a = x/p(1 - x), b = 1/p(1 - x)$. Since $\Gamma(x + 1) = x\Gamma(x)$, we have

$$\Gamma(n + x) = (n - 1 + x)(n - 2 + x) \cdots (1 + x)(x) \, \Gamma(x),$$

and so

$$\prod_{k=0}^{n-1} (k + x) = \frac{\Gamma(n + x)}{\Gamma(x)}.$$

Thus

$$x_n = \frac{\Gamma(n + a)\Gamma(b)}{\Gamma(n + b)\Gamma(a)}. \tag{1.2}$$

Using Stirling's formula for the gamma function,

$$\Gamma(x) \sim \sqrt{2\pi} \, x^{x - 1/2} e^{-x},$$

we readily deduce that

$$\frac{\Gamma(n + a)}{\Gamma(n + b)} \sim \frac{\sqrt{2\pi} \, (n + a)^{n + a - 1/2} e^{-(n + a)}}{\sqrt{2\pi} \, (n + b)^{n + b - 1/2} e^{-(n + b)}}$$

$$\sim \frac{n^{n + a - 1/2} \left(1 + \frac{a}{n}\right)^{n + a - 1/2} e^{-a}}{n^{n + b - 1/2} \left(1 + \frac{b}{n}\right)^{n + b - 1/2} e^{-b}}$$

$$\sim \frac{n^a e^a e^{-a}}{n^b e^b e^{-b}} = n^{a - b} = n^{-1/p}. \tag{1.3}$$

From (1.2) it follows that

$$\lim_{n \to \infty} n^{1/p} x_n = \lim_{n \to \infty} n^{1/p} \frac{\Gamma(n + a)\Gamma(b)}{\Gamma(n + b)\Gamma(a)} = \frac{\Gamma(b)}{\Gamma(a)},$$

giving

$$\alpha = \frac{1}{p}, \quad \beta = \frac{\Gamma(b)}{\Gamma(a)}.$$

□

Remark. The asymptotic ratio (1.3) can also be obtained by the formula (see [49], 8.328.2, p. 945)

$$\lim_{x \to \infty} \frac{\Gamma(x+a)}{x^a \Gamma(x)} = 1.$$

In particular, if $0 < a < 1$, Wendel [93] established the following elegant inequality:

$$\left(\frac{x}{x+a}\right)^{1-a} \leq \frac{\Gamma(x+a)}{x^a \Gamma(x)} \leq 1, \quad \text{for } x > 0.$$

A full expansion of the asymptotic ratio due to Tricomi and Erdélyi [91] is given by

$$\frac{\Gamma(z+a)}{\Gamma(z+b)} \sim \sum_{n=0}^{\infty} \frac{(-1)^n}{n!} B_n^\sigma(a) \frac{\Gamma(b-a+n)}{\Gamma(b-a)} \frac{1}{z^{b-a+n}},$$

with $|\arg z| \leq \pi/2 - \delta, \delta > 0, \operatorname{Re}(b-a) > 0$ and the $B_n^\sigma(z)$ (often called as *generalized Bernoulli polynomials*) satisfy

$$\frac{t^\sigma e^{zt}}{(e^t - 1)^\sigma} = \sum_{n=0}^{\infty} B_n^\sigma(z) \frac{t^n}{n!}, \quad |t| < 2\pi.$$

Mellin's formula ([49], 8.325.2, p. 945),

$$\prod_{k=1}^{\infty} \left[\left(1 - \frac{y}{z+k}\right) e^{y/k}\right] = \frac{e^{\gamma y} \Gamma(z+1)}{\Gamma(z-y+1)}, \quad (z > y > 0), \quad (1.4)$$

where γ is Euler's constant, provides another approach to solve this problem. First, we have

$$\lim_{n \to \infty} n^y \prod_{k=1}^{n-1} e^{-y/k} = \exp\left[\lim_{n \to \infty} y\left(\ln n - \sum_{k=1}^{n-1} 1/k\right)\right] = e^{-\gamma y}, \quad (1.5)$$

where we have used the well-known limit

$$\lim_{n \to \infty} \left(\sum_{k=1}^{n} 1/k - \ln n\right) = \gamma.$$

Next, we rewrite (1.1) as

$$x_n = \prod_{k=0}^{n-1} \left(\frac{k+a}{k+b}\right) = \frac{a}{b} \prod_{k=1}^{n-1} \left(1 - \frac{b-a}{k+b}\right).$$

Because

$$n^{b-a}x_n = \frac{a}{b}\left(n^{b-a}\prod_{k=1}^{n-1}e^{-(b-a)/k}\right)\left(\prod_{k=1}^{n-1}\left(1-\frac{b-a}{k+b}\right)e^{(b-a)/k}\right),$$

Using (1.4) and (1.5), again, we find that

$$\lim_{n\to\infty}n^{b-a}x_n = \frac{a}{b}e^{-\gamma(b-a)}\frac{e^{\gamma(b-a)}\Gamma(b+1)}{\Gamma(a+1)} = \frac{\Gamma(b)}{\Gamma(a)}.$$

We now end this section with three more Monthly problems for additional practice:

1. **Problem E3356** (Proposed by W. W. Chao, 98(4), 1991). Suppose $a_1 = 1$ and $a_{n+1} = n/a_n$ for $n = 1, 2, \ldots$. Evaluate

$$\lim_{n\to\infty}n^{-1/2}\left(\frac{1}{a_1}+\frac{1}{a_2}+\cdots+\frac{1}{a_n}\right).$$

2. **Problem 11528** (Proposed by A. Sîntămărian, 117(9), 2010). Let p, a, and b be positive integers with $a < b$. Consider a sequence $\langle x_n \rangle$ defined by $nx_{n+1} = (n + 1/p)x_n$ and an initial condition $x_1 \neq 0$. Evaluate

$$\lim_{n\to\infty}\frac{x_{an}+x_{an+1}+\cdots+x_{bn}}{nx_{an}}.$$

   Hint: Let $y_n = (n-1)x_n/(1+1/p)$. Then show that $x_n = y_{n+1}-y_n$.

3. **Problem 11659** (Proposed by A. Stadler, 119(7), 2012). Let x be real with $0 < x < 1$, and consider the sequence $\{a_n\}$ given by $a_0 = 1, a_1 = 1$, and, for $n > 1$,

$$a_n = \frac{a_{n-1}^2}{xa_{n-2}+(1-x)a_{n-1}}.$$

   Show that

$$\lim_{n\to\infty}\frac{1}{a_n} = \sum_{k=-\infty}^{\infty}(-1)^k x^{k(3k-1)/2}.$$

1.2   Asymptotic behavior for a Pólya type recurrence

Problem 11995 (Proposed by D. S. Marinescu and M. Monea, 124(7), 2017). Suppose $0 < x_0 < \pi$, and for $n \geq 1$ define $x_n = \frac{1}{n}\sum_{k=0}^{n-1}\sin x_k$. Find $\lim_{n\to\infty}x_n\sqrt{\ln n}$.

Discussion.
Before proceeding to this problem, it is interesting to recall Pólya's classical recurrence ([75], Problem 173, p. 38): Suppose $0 < x_0 < \pi$, and for $n \geq 1$ define $x_n = \sin(x_{n-1})$. Then

$$\lim_{n \to \infty} \sqrt{\frac{n}{3}} \, x_n = 1.$$

There are a number of ingenious proofs. Based on the Stolz-Cesàro theorem, Chen gave a short proof as follows: Since $\sin x < x$ for all $x > 0$, it follows that $0 < x_n < x_{n-1}$ whenever $0 < x_0 < \pi$. Thus, x_n converges as $n \to \infty$. Let $\lim_{n \to \infty} x_n = L$. Taking the limit in $x_n = \sin(x_{n-1})$ yields $L = \sin L$, which implies that $L = 0$. Define

$$a_n = n, \quad b_n = \frac{1}{x_n^2}.$$

Using the Stolz-Cesàro theorem, we have

$$\lim_{n \to \infty} n x_n^2 = \lim_{n \to \infty} \frac{n}{1/x_n^2} = \lim_{n \to \infty} \frac{a_{n+1} - a_n}{b_{n+1} - b_n}.$$

Setting $t = x_n$, we have

$$\lim_{n \to \infty} \frac{a_{n+1} - a_n}{b_{n+1} - b_n} = \lim_{n \to \infty} \frac{1}{\frac{1}{\sin^2(x_n)} - \frac{1}{\sin^2(x_{n-1})}}$$

$$= \lim_{t \to 0} \frac{1}{\frac{1}{\sin^2 t} - \frac{1}{t^2}}$$

$$= \lim_{t \to 0} \frac{t^2 \sin^2 t}{t^2 - \sin^2 t}$$

$$= 3,$$

thereby proving the expected limit

$$\lim_{n \to \infty} \sqrt{\frac{n}{3}} \, x_n = \sqrt{\lim_{n \to \infty} n x_n^2 / 3} = 1.$$

We can solve this Monthly problem similarly.

Solution.
By the assumption, we have

$$x_{n+1} = \frac{\sum_{k=0}^{n-1} \sin(x_k) + \sin x_n}{n+1} = \frac{n x_n + \sin x_n}{n+1}.$$

This implies that

$$0 < x_{n+1} \leq \frac{n x_n + x_n}{n+1} = x_n$$

whenever $0 < x_0 < \pi$. From the monotone convergence theorem, we conclude that x_n converges. Let $\lim_{n\to\infty} x_n = L$. Applying the Cauchy theorem yields

$$L = \lim_{n\to\infty} \frac{1}{n} \sum_{k=0}^{n-1} \sin(x_k) = \lim_{n\to\infty} \sin(x_n) = \sin(L),$$

which implies that $L = 0$. Next, using the Taylor series approximation of $\sin x$, we have

$$x_{n+1} = \frac{nx_n + x_n - x_n^3/6 + o(x_n^3)}{n+1} = x_n\left(1 - \frac{x_n^2}{6(n+1)} + \frac{o(x_n^2)}{n+1}\right).$$

Note that $(1-x)^{-2} = 1 + 2x + o(x)$ for small x. We find that

$$\frac{1}{x_{n+1}^2} = \frac{1}{x_n^2}\left(1 + \frac{x_n^2}{3(n+1)} + \frac{o(x_n^2)}{n+1}\right).$$

Using the Stolz-Cesàro theorem again, we obtain

$$\lim_{n\to\infty} x_n^2 \ln n = \lim_{n\to\infty} \frac{\ln n}{\frac{1}{x_n^2}} = \lim_{n\to\infty} \frac{\ln(n+1) - \ln n}{\frac{1}{x_{n+1}^2} - \frac{1}{x_n^2}}$$

$$= \lim_{n\to\infty} \frac{\ln(n+1) - \ln n}{\frac{1}{3(n+1)} + o(\frac{1}{n+1})} = 3.$$

Hence,

$$\lim_{n\to\infty} x_n\sqrt{\ln n} = \sqrt{3}.$$

$\square$

Remark. To study the properties of a sequence without an explicit formula, we often tend to determine the asymptotic behavior of the sequence. For example, Pólya's classical case asserts that

$$x_n = \sqrt{\frac{3}{n}} + o\left(\frac{1}{\sqrt{n}}\right).$$

Modifying the technique used in the above solution, we can improve this asymptotic formula to

$$x_n = \sqrt{\frac{3}{n}} - \frac{3\sqrt{3}\ln n}{10n\sqrt{n}} + o\left(\frac{\ln n}{n\sqrt{n}}\right).$$

This implies that

$$\lim_{n\to\infty} \frac{n}{\ln n}\left(1 - \sqrt{\frac{n}{3}}x_n\right) = \frac{3}{10}.$$

Another famous example is the normalized *logistic sequence* $\{x_n\}$ defined below.

Problem E3034 (Proposed by D. Cox, 91(4), 1984). Let $0 < x_0 < 1$ and $x_{n+1} = x_n(1 - x_n)$ for $n = 0, 1, \ldots$. Prove that

$$\lim_{n \to \infty} \frac{n}{\ln n} (1 - nx_n) = 1.$$

This is a refinement of the well-known results

$$\lim_{n \to \infty} x_n = 0 \quad \text{and} \quad \lim_{n \to \infty} nx_n = 1,$$

the latter of which appeared as **Putnam Problem 1966-A3** (see [4], p. 5.)

Motivated by the above results, we offer a generalization as an exercise. Let

$$f(x) = x - ax^\alpha + bx^{2\alpha - 1} + o(x^{2\alpha - 1}) \quad (\text{as } x \to 0),$$

where $\alpha > 1$ and $a > 0$. For $x_0 \to 0$, define $x_{n+1} = f(x_n)$ for $n = 0, 1, \ldots$. Prove that

$$x_n = \frac{1}{[a(\alpha - 1)]^{1/(\alpha - 1)}} \left(\frac{1}{n} \right)^{1/(\alpha - 1)}$$
$$+ \frac{b - \alpha a^2/2}{[a(\alpha - 1)]^{(2\alpha - 1)/(\alpha - 1)}} \frac{\ln n}{n^{\alpha/(\alpha - 1)}} + o \left(\frac{\ln n}{n^{\alpha/(\alpha - 1)}} \right).$$

As a consequence, this yields another solution to the following Monthly problem.

Problem 11773 (Proposed by M. Omarjee, 121(4), 2014). Given a positive real number a_0, let

$$a_{n+1} = \exp \left(-\sum_{k=1}^{n} a_k \right) \quad \text{for } n \geq 0.$$

For which values of b does $\sum_{n=0}^{\infty} (a_n)^b$ converge?

Hint: Note $a_{n+1} = \exp(-\sum_{k=1}^{n-1} a_k)e^{-a_n} = a_n e^{-a_n}$. Thus $f(x) = xe^{-x}$ and

$$a_n = \frac{1}{n} - \frac{\ln n}{2n^2} + o \left(\frac{\ln n}{n^2} \right).$$

Furthermore, for sequences that converge to zero faster, we have the following better estimate: If x_n converges to zero and for $k > 1$

$$\lim_{n \to \infty} n^k (x_n - x_{n+1}) = A,$$

then

$$\lim_{n \to \infty} n^{k-1} x_n = \frac{A}{k - 1}.$$

Indeed, for $\epsilon > 0$, there is $N \in \mathbb{N}$ such that for $n > N$

$$A - \epsilon \leq n^k (x_n - x_{n+1}) \leq A + \epsilon.$$

By adding the inequalities

$$(A - \epsilon)\frac{1}{n^k} \leq x_n - x_{n+1} \leq (A + \epsilon)\frac{1}{n^k},$$

for $n > N, m \geq 2$, we have

$$(A - \epsilon)\sum_{j=n}^{n+m-1}\frac{1}{j^k} \leq x_n - x_{n+m} \leq (A + \epsilon)\sum_{j=n}^{n+m-1}\frac{1}{j^k}.$$

Letting $m \to \infty$, then multiplying by n^{k-1}, and using $\zeta(k) = \sum_{j=1}^{\infty} 1/j^k$, we find that

$$(A - \epsilon)n^{k-1}\left(\zeta(k) - \sum_{j=1}^{n-1}\frac{1}{j^k}\right) \leq n^{k-1}x_n \leq (A + \epsilon)n^{k-1}\left(\zeta(k) - \sum_{j=1}^{n-1}\frac{1}{j^k}\right).$$

Now, the claimed limit follows from the rate of convergence

$$\lim_{n \to \infty} n^{k-1}\left(\zeta(k) - \sum_{j=1}^{n-1}\frac{1}{j^k}\right) = \frac{1}{k-1}.$$

We have seen that the Stolz-Cesàro theorem plays a central role in determining the growth rate of a sequence. This approach can be extended to other classes of recurrent sequences. Here are five problems for additional practice:

1. **Problem E1557** (Proposed by D. R. Hayes, 70(9), 1963). If $S_n = (1/n^2)\sum_{k=0}^{n} \ln\binom{n}{k}$, evaluate $S = \lim_{n \to \infty} S_n$.

2. **Problem 6376** (Proposed by I. J. Schoenberg, 89(1), 1982). Let a_n $(n = 1, 2, \ldots)$ be reals. Show that

$$\lim_{n \to \infty} a_n \sum_{k=1}^{n} a_k^2 = 1 \quad \text{implies that} \quad \lim_{n \to \infty} (3n)^{1/3}a_n = 1.$$

3. **Problem 12079** (Proposed by M. Omarjee, 125(10), 2018). Choose x_1 in $(0, 1)$, and let

$$x_{n+1} = \frac{1}{n}\sum_{k=1}^{n} \ln(1 + x_k) \quad \text{for } n \geq 1.$$

Compute $\lim_{n \to \infty} x_n \ln n$.

4. **Mathematics Magazine Problem 2087** (Proposed by F. Stanescu, 93(1), 2020). Consider the sequence defined by $x_1 = a > 0$ and

$$x_n = \ln\left(1 + \frac{x_1 + x_2 + \cdots + x_{n-1}}{n - 1}\right) \quad \text{for } n \geq 2.$$

Compute $\lim_{n \to \infty} x_n \ln n$.

5. Let $\{a_n\}$ be a sequence of real numbers such that $e^{a_n} + na_n = 2$ for all $n \in \mathbb{N}$. Show that

$$\lim_{n\to\infty} n(1 - na_n) = 1 \quad \text{and} \quad \lim_{n\to\infty} n[n(1 - na_n) - 1] = \frac{1}{2}.$$

1.3 A nonlinear recurrence with Fibonacci exponents

Problem 11976 (Proposed by R. Bosch, 124(4), 2017). Given a positive real number s, consider the sequence $\{u_n\}$ defined by $u_1 = 1, u_2 = s$, and $u_{n+2} = u_n u_{n+1}/n$ for $n \geq 1$.

(a) Show that there is a constant C such that $\lim_{n\to\infty} u_n = \infty$ when $s > C$ and $\lim_{n\to\infty} u_n = 0$ when $s < C$.

(b) Calculate $\lim_{n\to\infty} u_n$ when $s = C$.

Discussion.
To search for patterns in this sequence, we compute the first few terms. Based on the definition of u_n, taking $n = 1, 2, \ldots, 6$, we obtain

$$u_3 = s, \; u_4 = \frac{s^2}{2}, u_5 = \frac{s^3}{2 \cdot 3},$$

$$u_6 = \frac{s^5}{2^2 \cdot 3 \cdot 4}, \; u_7 = \frac{s^8}{2^3 \cdot 3^2 \cdot 4 \cdot 5}, u_8 = \frac{s^{13}}{2^5 \cdot 3^3 \cdot 4^2 \cdot 5 \cdot 6}.$$

In the above forms, we see that

- The exponents of s are $1, 2, 3, 5, 8, 13$, respectively, the Fibonacci sequence.

- The denominators are products of the form $k^{F_{n-k}}$, where F_n is the nth Fibonacci number.

Thus, the pattern of u_n emerges and can be established by induction.

Solution.
(a) Let F_n be the nth Fibonacci number with $F_1 = F_2 = 1$, and $F_n = F_{n-1} + F_{n-2}$ for $n > 2$. It follows by induction that

$$u_{n+2} = \frac{s^{F_{n+1}}}{2^{F_{n-1}} \cdot 3^{F_{n-2}} \cdots (n-1)^{F_2} \cdot n^{F_1}}.$$

Let

$$b_n = 2^{F_{n-1}} \cdot 3^{F_{n-2}} \cdots (n-1)^{F_2} \cdot n^{F_1}.$$

and $\phi = (1 + \sqrt{5})/2$, the golden ratio. We now show that

$$b_n \sim C^{F_{n+1}} \qquad \text{as } n \to \infty,$$

where $C = e^L = 3.20096\ldots$ and

$$L := \sum_{k=1}^{\infty} \frac{\ln k}{\phi^k} = 1.16345\ldots.$$

To see this, taking the logarithm of b_n yields

$$\ln b_n = \sum_{k=1}^{n} F_{n+1-k} \ln k.$$

Recall Binet's formula

$$F_n = \frac{1}{\sqrt{5}} \left(\phi^n - \bar{\phi}^n \right),$$

where $\bar{\phi} = -1/\phi = (1 - \sqrt{5})/2$. We compute

$$
\begin{aligned}
\ln b_n &= \frac{1}{\sqrt{5}} \left(\sum_{k=1}^{n} (\phi^{n+1-k} - \bar{\phi}^{n+1-k}) \ln k \right) \\
&= \frac{1}{\sqrt{5}} \phi^{n+1} \sum_{k=1}^{n} \phi^{-k} \ln k - \frac{1}{\sqrt{5}} \sum_{k=1}^{n} \bar{\phi}^{n+1-k} \ln k.
\end{aligned}
$$

Therefore,

$$\frac{\sqrt{5} \ln b_n}{\phi^{n+1}} = \sum_{n=1}^{\infty} \frac{\ln k}{\phi^k} - \frac{1}{\phi^{n+1}} \sum_{k=1}^{n} \bar{\phi}^{n+1-k} \ln k.$$

Since $|\bar{\phi}| < 1$, using Stirling's formula for $n!$ yields

$$\left| \sum_{k=1}^{n} \bar{\phi}^{n+1-k} \ln k \right| \leq \sum_{k=1}^{n} \ln k = \ln n! = n \ln n - n + O(\ln n)$$

and so

$$\frac{1}{\phi^{n+1}} \left| \sum_{k=1}^{n} \bar{\phi}^{n+1-k} \ln k \right| \to 0, \qquad \text{as } n \to \infty.$$

Thus, as $n \to \infty$, we have

$$\frac{\sqrt{5} \ln b_n}{\phi^{n+1}} \sim L$$

and

$$b_n \sim e^{L \cdot \phi^{n+1}/\sqrt{5}} = (e^L)^{\phi^{n+1}/\sqrt{5}} \sim C^{F_{n+1}}.$$

When $\ln s \neq L$, i.e., $s \neq e^L = C$, we have

$$\ln u_{n+2} = F_{n+1} \ln s - \ln b_n \sim F_{n+1}(\ln s - l) = F_{n+1} \ln \frac{s}{C}.$$

Therefore

$$\lim_{n \to \infty} u_n = \begin{cases} 0 & \text{if } s < C, \\ \infty & \text{if } s > C. \end{cases}$$

(b) When $s = C$, we have

$$\begin{aligned}
\ln u_{n+2} &= F_{n+1}l - \ln b_n \\
&= \frac{1}{\sqrt{5}} \phi^{n+1} \sum_{k=n+1}^{\infty} \frac{\ln k}{\phi^k} - \frac{1}{\sqrt{5}} \bar{\phi}^{n+1}l + \frac{1}{\sqrt{5}} \sum_{k=1}^{n} \bar{\phi}^{n+1-k} \ln k \\
&\sim \frac{1}{\sqrt{5}} \phi^{n+1} \sum_{k=n+1}^{\infty} \frac{\ln k}{\phi^k} \geq \frac{1}{\sqrt{5}} \ln(n+1),
\end{aligned}$$

and so $\lim_{n \to \infty} u_n = +\infty$. $\qquad\square$

Remark. Tyler provided a generalization by relaxing the initial condition to $u_1 = t > 0$, then proving that $u_n \to \infty$ when $st^{\phi-1} \geq C$ and $u_n \to 0$ when $st^{\phi-1} < C$.

Let $U_n = 1/u_n$. The recurrence becomes $U_{n+2} = nU_nU_{n+1}$, which then can be viewed as a variant of *derangements* in combinatorics. A derangement is a permutation of the elements of a set such that no element appears in its original position. In other words, a derangement is a permutation that has no fixed points. If D_n is the number of derangements of a set of size n, then it is well-known that D_n satisfies

$$D_{n+1} = n(D_n + D_{n-1}) \quad \text{and} \quad D_n = n! \sum_{k=0}^{n} (-1)^k \frac{1}{k!}.$$

In particular, $D_n/n! \to e^{-1} = 0.367879\ldots$ as $n \to \infty$. So for large n, more than one third of all permutations are derangements.

In contrast to what we know about derangements, there are many interesting open questions related to the u_n. For example,

1. Is there a combinatorics interpretation for u_n? Does u_n count something?

2. Is it possible to find the exact value of $L = \sum_{k=1}^{\infty} \frac{\ln k}{\phi^k}$?

3. Is there a better asymptotic formula for b_n? For example, is there an α such that

$$b_n = \frac{C^{F_{n+1}}}{n} \left(1 + \frac{\alpha}{n} + o(1/n)\right)?$$

4. More generally, is there a closed form for the generating function of $\{\ln n\}$?

For this last question, using the Laplace transform

$$\mathcal{L}(\ln t) = -\frac{\gamma + \ln s}{s},$$

we have

$$\sum_{n=1}^{\infty} \ln n \, x^n = -\frac{\gamma x}{1-x} - \int_0^1 \frac{xe^{-t}\ln t}{(1-xe^{-t})^2}\, dt, \qquad \text{for } x \in (0,1).$$

But, it is currently unknown whether this integral can be represented by elementary functions or not.

1.4   Rate of convergence for an integral

Problem 11941 (Proposed by O. Furdui, 123(9), 2016). Let

$$L = \lim_{n\to\infty} \int_0^1 \sqrt[n]{x^n + (1-x)^n}\, dx.$$

(a) Find L.

(b) Find

$$\lim_{n\to\infty} n^2 \left(\int_0^1 \sqrt[n]{x^n + (1-x)^n}\, dx - L \right).$$

Discussion.
When an integral sequence is hopelessly incomputable, a potential approach is to find two simpler sequences that "squeeze" it and use the squeeze theorem. For part **(b)**, observe that the substitution $x = t/(1+t)$ transforms the integral

$$\int_0^{1/2} \sqrt[n]{x^n + (1-x)^n}\, dx = \int_0^1 \frac{\sqrt[n]{1+t^n}}{(1+t)^3}\, dt$$

and

$$\frac{\sqrt[n]{1+t^n}}{(1+t)^3} \to \frac{1}{(1+t)^3}, \qquad \text{as } n \to \infty.$$

Thus once finding the value of L, the desired result follows upon determining the limit under the integral sign.

Solution.
(a) We claim $L = 3/4$. To see this, let $I_n = \int_0^1 \sqrt[n]{x^n + (1-x)^n}\, dx$. We have

$$\begin{aligned}
I_n &= \int_0^{1/2} \sqrt[n]{x^n + (1-x)^n}\, dx + \int_{1/2}^1 \sqrt[n]{x^n + (1-x)^n}\, dx \\
&\geq \int_0^{1/2} (1-x)\, dx + \int_{1/2}^1 x\, dx = \frac{3}{4}.
\end{aligned}$$

On the other hand, since $x \leq 1-x$ for $x \in [0, 1/2]$ and $1-x \leq x$ for $x \in [1/2, 1]$, we have

$$
\begin{aligned}
I_n &= \int_0^{1/2} \sqrt[n]{x^n + (1-x)^n}\, dx + \int_{1/2}^1 \sqrt[n]{x^n + (1-x)^n}\, dx \\
&\leq \int_0^{1/2} \sqrt[n]{2}\,(1-x)\, dx + \int_{1/2}^1 \sqrt[n]{2}\, x\, dx = \frac{3}{4}\,\sqrt[n]{2}.
\end{aligned}
$$

The squeeze theorem implies that $L = \lim_{n\to\infty} I_n = 3/4$ as claimed.

(b) The limit is $\pi^2/48$. Notice that $\int_0^{1/2}(1-x) = \int_{1/2}^1 x\, dx = 3/8$. We claim

$$
\lim_{n\to\infty} n^2 \left(\int_0^{1/2} \sqrt[n]{x^n + (1-x)^n}\, dx - \frac{3}{8} \right) = \frac{\pi^2}{96}, \tag{1.6}
$$

and

$$
\lim_{n\to\infty} n^2 \left(\int_{1/2}^1 \sqrt[n]{x^n + (1-x)^n}\, dx - \frac{3}{8} \right) = \frac{\pi^2}{96}, \tag{1.7}
$$

from which the required limit follows. To prove (1.6), we compute

$$
\begin{aligned}
&\lim_{n\to\infty} n^2 \left(\int_0^{1/2} \left(\sqrt[n]{x^n + (1-x)^n} - (1-x) \right) dx \right) \\
&= \lim_{n\to\infty} n^2 \left(\int_0^{1/2} (1-x) \left(\sqrt[n]{1 + \left(\frac{x}{1-x}\right)^n} - 1 \right) dx \right) \\
&\quad (\text{let } t = x/(1-x)) \\
&= \lim_{n\to\infty} n^2 \left(\int_0^1 \frac{1}{(1+t)^3} \left(\sqrt[n]{1 + t^n} - 1 \right) dt \right) \quad (\text{let } u = t^n) \\
&= \lim_{n\to\infty} n \left(\int_0^1 \frac{1}{(1+u^{1/n})^3} \left(\sqrt[n]{1+u} - 1 \right) u^{1/n-1} du \right) \\
&= \int_0^1 \left(\lim_{n\to\infty} \frac{1}{(1+u^{1/n})^3} \cdot n \left(\sqrt[n]{1+u} - 1 \right) \cdot u^{1/n-1} \right) du \\
&= \frac{1}{8} \int_0^1 \frac{\ln(1+u)}{u}\, du = \frac{\pi^2}{96},
\end{aligned}
$$

where we have used the well-known facts:

$$
\lim_{n\to\infty} n \left(\sqrt[n]{1+u} - 1 \right) = \ln(1+u) \quad \text{and} \quad \int_0^1 \frac{\ln(1+u)}{u}\, du = \sum_{n=1}^{\infty} \frac{(-1)^{n+1}}{n^2} = \frac{\pi^2}{12}.
$$

Equation (1.7) follows from (1.6) upon substituting $1 - x$ for x. $\qquad\square$

Remark. The above results can be generalized as follows. We leave the proof to the reader. For part **(a)**, if f and g are nonnegative and integrable on $[a, b]$, then

$$
\lim_{n\to\infty} \int_a^b \sqrt[n]{f^n(x) + g^n(x)}\, dx = \int_a^b \max\{f(x), g(x)\}\, dx.
$$

For part **(b)**, if f is positive continuous function on $[0,1]$ with $f(0) = 1$ and $g(x)$ is continuous on $[0,1]$, then

$$\lim_{n\to\infty} n^2 \left(\int_0^1 \sqrt[n]{f(x^n)} g(x)\, dx - \int_0^1 g(x)\, dx \right) = g(1) \int_0^1 \frac{\ln f(x)}{x}\, dx.$$

In particular, letting $f(x) = 1 + x$ and $g(x) = 1/(1+x)^3$ yields the result in part **(b)**.

Finally, we offer seven more problems for additional practice.

1. **Problem E1245** (Proposed by M. S. Klamkin, 63(10), 1956). If

    $$b_{n+1} = \int_0^1 \min(x, a_n)\, dx, \quad a_{n+1} = \int_0^1 \max(x, b_n)\, dx.$$

    Prove that the sequences $\{a_n\}$ and $\{b_n\}$ both converge and find their limits.

2. **Problem 4828** (Proposed by M. S. Klamkin, 66(2), 1959). Do the sequences $\{a_n\}, \{b_n\}, \{c_n\}$ converge, where

    $$a_{n+1} = \int_0^1 \min(x, b_n, c_n)\, dx, \quad b_{n+1}$$
    $$= \int_0^1 \mathrm{mid}(x, c_n, a_n)\, dx, \quad c_{n+1} = \int_0^1 \max(x, a_n, b_n)\, dx$$

    and $\mathrm{mid}(a, b, c) = b$ if $a \ge b \ge c$. If so, can you find their limits?

3. **Problem 11225** (Proposed by J. L. Díaz-Barrero, 113(5), 2006). Find

    $$\lim_{n\to\infty} \frac{1}{n} \left(\int_0^n \frac{x \ln(1 + x/n)}{1 + x}\, dx \right).$$

4. **Problem 11611** (Proposed by O. Furdui, 118(10), 2011). Let f be a continuous function from $[0,1]$ into $[0, \infty)$. Find

    $$\lim_{n\to\infty} n \int_0^1 \left(\sum_{k=n}^\infty \frac{x^k}{k} \right)^2 f(x)\, dx.$$

5. **Problem 12120** (Proposed by M. Bataille, 126(6), 2019). For positive integers n and k with $n \ge k$, let $a(n, k) = \sum_{j=0}^{k-1} \binom{n}{j} 3^j$.

    (a) Evaluate

    $$\lim_{n\to\infty} \frac{1}{4^n} \sum_{k=1}^n \frac{a(n, k)}{k}.$$

(b) Evaluate

$$\lim_{n\to\infty} n\left(4^n L - \sum_{k=1}^{n} \frac{a(n,k)}{k}\right),$$

where L is the limit in part (a).

6. **Problem 12153** (Proposed by O. Kouba, 127(1), 2020). For a real number x whose fractional part is not $1/2$, let $\langle x \rangle$ denote the nearest integer to x. For a positive integer n, let

$$a_n = \left(\sum_{k=1}^{n} \frac{1}{\langle\sqrt{k}\rangle}\right) - 2\sqrt{n}.$$

(a) Prove that the sequence $a_1, a_2, \ldots$ is convergent, and find its limit L.

(b) Prove that the set $\{\sqrt{n}(a_n - L) : n \geq 1\}$ is dense subset of $[0, 1/4]$.

7. **Mathematics Magazine Problem 2097** (Proposed by O. Kouba, 93(3), 2020). For a real number $x \notin 1/2 + \mathbb{Z}$, denote the nearest integer to x by $\langle x \rangle$. For any real number x, denote the largest smaller than or equal to x and the smallest integer larger than or equal to x by $\lfloor x \rfloor$ and $\lceil x \rceil$, respectively. For a positive integer n let

$$a_n = \frac{2}{\langle\sqrt{n}\rangle} - \frac{1}{\lfloor\sqrt{n}\rfloor} - \frac{1}{\lceil\sqrt{n}\rceil}.$$

(a) Prove that the series $\sum_{n=1}^{\infty} a_n$ is convergent and find its sum L.

(b) Prove that the set

$$\left\{\sqrt{n}\left(\sum_{k=1}^{n} a_k - L\right) : n \geq 1\right\}$$

is dense in $[0, 1]$.

1.5   How closely does this sum approximate the integral?

Problem 11535 (Proposed by M. Tetiva, 117(9), 2010). Let f be continuously differentiable function on $[0, 1]$. Let $A = f(1)$ and let $B = \int_0^1 x^{-1/2} f(x)dx$.

Evaluate

$$\lim_{n \to \infty} n \left(\int_0^1 f(x)dx - \sum_{k=1}^n \left(\frac{k^2}{n^2} - \frac{(k-1)^2}{n^2} \right) f \left(\frac{(k-1)^2}{n^2} \right) \right)$$

in terms of A and B.

Discussion.
The statement to be proved suggests we work backward. Indeed, if we split the summation in the proposed limit into

$$\frac{1}{n} \sum_{k=1}^n \frac{2(k-1)}{n} f \left(\frac{(k-1)^2}{n^2} \right) \quad \text{and} \quad \frac{1}{n} \sum_{k=1}^n f \left(\frac{(k-1)^2}{n^2} \right),$$

then they are exactly the equidistant Riemann sums of $2xf(x^2)$ and $f(x^2)$ associated with the left-endpoint rule over $[0,1]$, respectively. On the other hand, by substitution, we have

$$B = \int_0^1 x^{-1/2} f(x) \, dx = 2 \int_0^1 f(x^2)dx, \quad \int_0^1 2xf(x^2) \, dx = \int_0^1 f(x) \, dx.$$

Thus, the problem reduces to determining the convergence rate for the equidistant Riemann sum associated with the left-endpoint rule.

Solution.
It is well-known that the error in approximating $\int_0^1 f(x)dx$ by the Riemann sums with left-endpoint rule is $O(1/n)$. More precisely, this result can be rephrased as follows: If $f(x)$ is continuously differentiable on $[0,1]$, then

$$\lim_{n \to \infty} n \left(\int_0^1 f(x) \, dx - \frac{1}{n} \sum_{k=1}^n f \left(\frac{k-1}{n} \right) \right) = \frac{1}{2} (f(1) - f(0)). \qquad (1.8)$$

For the sake of completeness, we give a proof. Let F be an antiderivative of f. We have

$$\int_0^1 f(x) \, dx - \frac{1}{n} \sum_{k=1}^n f \left(\frac{k-1}{n} \right)$$

$$= \sum_{k=1}^n \left(\int_{(k-1)/n}^{k/n} f(x) \, dx - \frac{1}{n} f \left(\frac{k-1}{n} \right) \right)$$

$$= \sum_{k=1}^n \left(F \left(\frac{k}{n} \right) - F \left(\frac{k-1}{n} \right) - \frac{1}{n} F' \left(\frac{k-1}{n} \right) \right)$$

$$= \sum_{k=1}^n \frac{1}{2n^2} F''(\xi_k), \qquad \text{(using Taylor's theorem)}$$

for some $\xi_k \in ((k-1)/n, k/n)$ and $1 \le k \le n$. Thus

$$\lim_{n\to\infty} n \left(\int_0^1 f(x)dx - \frac{1}{n} \sum_{k=1}^n f\left(\frac{k-1}{n}\right) \right)$$

$$= \frac{1}{2} \lim_{n\to\infty} \frac{1}{n} \sum_{k=1}^n F''(\xi_k) = \frac{1}{2} \int_0^1 F''(x)\,dx$$

$$= \frac{1}{2} \int_0^1 f'(x)\,dx = \frac{1}{2}(f(1) - f(0)).$$

This proves (1.8). Using $\int_0^1 f(x)\,dx = \int_0^1 2xf(x^2)\,dx$, we have

$$\lim_{n\to\infty} n \left(\int_0^1 f(x)\,dx - \sum_{k=1}^n \left(\frac{k^2}{n^2} - \frac{(k-1)^2}{n^2} \right) f\left(\frac{(k-1)^2}{n^2}\right) \right)$$

$$= \lim_{n\to\infty} n \left(\int_0^1 2xf(x^2)\,dx - \sum_{k=1}^n \frac{2(k-1)+1}{n^2} f\left(\frac{(k-1)^2}{n^2}\right) \right)$$

$$= \lim_{n\to\infty} \left[n \left(\int_0^1 2xf(x^2)\,dx - \frac{1}{n} \sum_{k=1}^n 2\frac{(k-1)}{n} f\left(\frac{(k-1)^2}{n^2}\right) \right) \right.$$

$$\left. - \frac{1}{n} \sum_{k=1}^n f\left(\frac{(k-1)^2}{n^2}\right) \right].$$

Replacing $f(x)$ by $2xf(x^2)$ in (1.8), we find that

$$\lim_{n\to\infty} n \left(\int_0^1 f(x)\,dx - \sum_{k=1}^n \left(\frac{k^2}{n^2} - \frac{(k-1)^2}{n^2} \right) f\left(\frac{(k-1)^2}{n^2}\right) \right)$$

$$= \frac{1}{2} (2xf(x^2))|_0^1 - \int_0^1 f(x^2)\,dx = f(1) - \frac{1}{2} \int_0^1 x^{-1/2} f(x)\,dx = A - \frac{1}{2} B.$$

$\square$

Remark. We can generalize (1.8) with a similar proof: If $f(x)$ is continuously differentiable on $[0,1]$ and $\alpha \in [0,1]$, then

$$\lim_{n\to\infty} n \left(\int_0^1 f(x)\,dx - \frac{1}{n} \sum_{k=1}^n f\left(\frac{k-\alpha}{n}\right) \right) = \frac{1}{2}(2\alpha - 1)(f(1) - f(0)). \quad (1.9)$$

For the midpoint Riemann sum, along the same lines, we see the convergence rate is second order: If f is twice continuously differentiable on $[0,1]$, then

$$\lim_{n\to\infty} n^2 \left(\int_0^1 f(x)\,dx - \frac{1}{n} \sum_{k=1}^n f\left(\frac{2k-1}{2n}\right) \right) = \frac{1}{24} (f'(1) - f'(0)). \quad (1.10)$$

However, there is no general result on the convergence rates of non-equidistant Riemann sums. Observe that the summation in the proposed problem

$$\sum_{k=1}^{n} \left(\frac{k^2}{n^2} - \frac{(k-1)^2}{n^2} \right) f\left(\frac{(k-1)^2}{n^2} \right)$$

is indeed a left-endpoint Riemann sum of $\int_0^1 f(x)dx$ associated with the non-equidistant partition

$$0 < \frac{1}{n^2} < \frac{2^2}{n^2} < \frac{3^2}{n^2} < \cdots < \frac{(n-1)^2}{n^2} < 1.$$

The reader may directly prove

$$\lim_{n \to \infty} n \left(\int_0^1 f(x)dx - \sum_{k=1}^{n} \left(\frac{k^2}{n^2} - \frac{(k-1)^2}{n^2} \right) f\left(\frac{(k-1)^2}{n^2} \right) \right)$$
$$= \int_0^1 x^{-1/2} f(x)\, dx.$$

Another related topic is to find conditions ensuring the monotonicity of right and left Riemann sums with equidistant partitions. Recently, Borwein et. al. experimentally found that if f is decreasing on $[0, 1]$ and its symmetrization, $F(x) := (f(x) + f(1-x))/2$ is concave then its right Riemann sums increase monotonically with partition size. The story they have told highlights the many accessible ways that the computer and the Internet can enrich mathematical research and instruction. We refer the reader to the paper [17] (The online version is available at `https://carma.newcastle.edu.au/jon/riemann.pdf`).

We challenge reader directly to verify the monotonicity for the special case $f(x) = \frac{1}{1+x^2}$. You may encounter the application of Descartes' rule of signs.

1. Let

$$R_n = \sum_{k=1}^{n} \frac{n}{n^2 + k^2} \qquad \text{for all } n \geq 1.$$

   Show that R_n is increasing.

2. Let

$$L_n = \sum_{k=0}^{n-1} \frac{n}{n^2 + k^2} \qquad \text{for all } n \geq 1.$$

   Show that L_n is increasing.

1.6   An arctangent series

Problem 11438 (Proposed by D. Bailey, J. Borwein and J. Waldvogel, 116(5), 2009). Let

$$P(x) = \sum_{k=1}^{\infty} \arctan\left(\frac{x-1}{(k+x+1)\sqrt{k+1}+(k+2)\sqrt{k+x}}\right).$$

(a) Find a closed form expression for $P(n)$ when n is a nonnegative integer.

(b) Show that $\lim_{x \to -1+} P(x)$ exists, and find a closed-form expression for it.

Discussion.
Since there is no trigonometric identity available for the n-partial sums of the series directly, one attempt is to determine whether the general term admits representation as a difference. If so, we can evaluate the series by "telescoping."

Solution.
(a) Notice that

$$\frac{\frac{1}{\sqrt{k+1}} - \frac{1}{\sqrt{k+n}}}{1 + \frac{1}{\sqrt{k+1}} \cdot \frac{1}{\sqrt{k+n}}} = \frac{\sqrt{k+n} - \sqrt{k+1}}{1 + \sqrt{k+1} \cdot \sqrt{k+n}}.$$

Rationalizing the numerator gives

$$\frac{\frac{1}{\sqrt{k+1}} - \frac{1}{\sqrt{k+n}}}{1 + \frac{1}{\sqrt{k+1}} \cdot \frac{1}{\sqrt{k+n}}} = \frac{n-1}{(k+n+1)\sqrt{k+1}+(k+2)\sqrt{k+n}}.$$

Using the identity

$$\arctan \alpha - \arctan \beta = \arctan\left(\frac{\alpha - \beta}{1 + \alpha\beta}\right),$$

we obtain

$$\begin{aligned}
P(n) &= \sum_{k=1}^{\infty} \arctan\left(\frac{n-1}{(k+n+1)\sqrt{k+1}+(k+2)\sqrt{k+n}}\right) \\
&= \sum_{k=1}^{\infty} \left(\arctan \frac{1}{\sqrt{k+1}} - \arctan \frac{1}{\sqrt{k+n}}\right).
\end{aligned}$$

Clearly, $P(1) = 0$. The series for $P(0)$ telescopes to give

$$P(0) = -\arctan 1 + \lim_{k \to \infty} \arctan \frac{1}{\sqrt{k+1}} = -\frac{\pi}{4}.$$

In general, for $n \geq 2$, the series telescopes into the form

$$P(n) = \sum_{k=2}^{n} \arctan \frac{1}{\sqrt{k}}.$$

(b) Now use the inequality $\arctan t < t$ for $t > 0$. If $k \geq 2, x > -1$, then

$$\arctan \left(\frac{x-1}{(k+x+1)\sqrt{k+1}+(k+2)\sqrt{k+x}} \right) \leq \frac{|x|+1}{k\sqrt{k+1}+(k+2)\sqrt{k-1}}.$$

By the Weierstrass M-test, the series $P(x)$ converges uniformly, and therefore it is continuous for $x > -1$. As in (a), we have

$$P(x) = \sum_{k=1}^{\infty} \left(\arctan \frac{1}{\sqrt{k+1}} - \arctan \frac{1}{\sqrt{k+x}} \right)$$

so

$$P(x+1) = P(x) + \arctan \frac{1}{\sqrt{1+x}}. \tag{1.11}$$

Thus

$$\lim_{x \to -1^+} P(x) = \lim_{x \to -1^+} \left(P(x+1) - \arctan \frac{1}{\sqrt{1+x}} \right) = P(0) - \frac{\pi}{2} = -\frac{3\pi}{4}.$$

$\square$

Remark. The proposers revealed that they discovered the value $-3\pi/4$ experimentally: They calculated a high accuracy numerical value, then used this value as input to the *Inverse Symbolic Calculator* (ISC, available at `http://wayback.cecm.sfu.ca/projects/ISC/ISCmain.html`) to identify the exact value.

It is interesting to know whether there are more general closed forms for P, say at half-integers. In this case, from (1.11), we only need to figure out the exact value of

$$P(1/2) = \sum_{k=1}^{\infty} \left(\arctan \frac{1}{\sqrt{k+1}} - \arctan \frac{1}{\sqrt{k+1/2}} \right),$$

which seems still an open problem. Here are six Monthly problems for additional practice.

1. **Problem 11853** (Proposed by H. Ohtsuka, 122(7), 2015). Find

$$\sum_{n=1}^{\infty} \frac{1}{\sinh 2^n}.$$

2. **Problem 11930** (Proposed by C. I. Vălean, 123(8), 2016). Find

$$\sum_{n=1}^{\infty} \sinh^{-1} \left(\frac{1}{\sqrt{2^{n+2}+2}+\sqrt{2^{n+1}+2}} \right).$$

3. **Problem 12090** (Proposed by H. Ohtsuka, 126(2), 2019). The Pell-Lucas numbers Q_n satisfy
$Q_0 = 2$, $Q_1 = 2$, and $Q_n = 2Q_{n-1} + Q_{n-2}$ for $n \geq 2$. Prove

$$\sum_{n=1}^{\infty} \arctan\left(\frac{1}{Q_n}\right) \arctan\left(\frac{1}{Q_{n+1}}\right) = \frac{\pi^2}{32}.$$

4. **Problem 12101** (Proposed by H. Lee, 126(3), 2019). Find the least upper bound of

$$\sum_{n=1}^{\infty} \frac{\sqrt{x_{n+1}} - \sqrt{x_n}}{\sqrt{(1 + x_{n+1})(1 + x_n)}}$$

over all increasing sequences $x_1, x_2, \ldots$ of positive real numbers.

5. **Problem 12118** (Proposed by H. Ohtsuka, 126(6), 2019). Compute

$$\sum_{n=0}^{\infty} \frac{1}{F_{2mn} + iF_m}$$

where m is an odd integer and F_n is the nth Fibonacci number.

6. **Problem 11505** (Proposed by B. Burdick, 117(5), 2010). Define a_n to be the periodic sequence given by $a_1 = a_3 = 1, a_2 = 2, a_4 = a_6 = -1, a_5 = -2$, and $a_n = a_{n-6}$ for $n \geq 7$. Let F_n be the Fibonacci sequence with $F_1 = F_2 = 1$. Show that

$$\sum_{k=1}^{\infty} \frac{a_k F_k F_{2k-1}}{2k-1} \sum_{n=0}^{\infty} \frac{(-1)^{kn}}{F_{kn+2k-1}F_{kn+3k-1}} = \frac{\pi}{4}.$$

Hint: Show that for $|x| < 1$,

$$\arctan\left(\frac{2x\cos\theta}{1-x^2}\right) = 2\sum_{k=1}^{\infty} \frac{(-1)^{k-1}\cos(2k-1)\theta}{2k-1} x^{2k-1}.$$

1.7 Geometric mean rates

Problem 11935 (Proposed by D. M. Bătinetu-Giurgiu and N. Stanciu, 123(8), 2016). Let f be a function from $\mathbb{Z}^+$ to $\mathbb{R}^+$ such that $\lim_{n\to\infty} f(n)/n = a$, where $a > 0$. Find

$$\lim_{n\to\infty} \left(\sqrt[n+1]{\prod_{k=1}^{n+1} f(k)} - \sqrt[n]{\prod_{k=1}^{n} f(k)} \right).$$

Discussion.

Searching for solutions sometimes can be done by examining the special cases. The solutions to the special cases may shed light on the general solutions. Notice that when $f(k) = k$ the problem becomes the well-known result ([77], p. 9):

$$\lim_{n \to \infty} \left(\sqrt[n+1]{(n+1)!} - \sqrt[n]{n!} \right) = \frac{1}{e}.$$

It is fortunate here that the same argument goes through for the general problem.

Solution.

We show that the limit is a/e. To see this, we first establish three facts:

F1 $\lim_{n \to \infty} \frac{n}{\sqrt[n]{\prod_{k=1}^{n} f(k)}} = \frac{e}{a}$. To prove this, we rewrite

$$\frac{n}{\sqrt[n]{\prod_{k=1}^{n} f(k)}} = \sqrt[n]{\frac{n!}{\prod_{k=1}^{n} f(k)}} \cdot \frac{n}{\sqrt[n]{n!}} = \sqrt[n]{\prod_{k=1}^{n} \frac{k}{f(k)}} \cdot \frac{n}{\sqrt[n]{n!}}.$$

Stirling's formula implies that

$$\lim_{n \to \infty} \frac{n}{\sqrt[n]{n!}} = e.$$

The Cauchy theorem yields

$$\lim_{n \to \infty} \frac{\ln \frac{1}{f(1)} + \ln \frac{2}{f(2)} + \cdots + \ln \frac{n}{f(n)}}{n} = \lim_{n \to \infty} \ln \frac{n}{f(n)} = \ln \frac{1}{a}.$$

Exponentiating both sides gives

$$\lim_{n \to \infty} \sqrt[n]{\prod_{k=1}^{n} \frac{k}{f(k)}} = \frac{1}{a}.$$

These limits confirm $\mathbf{F_1}$ immediately.

F2 $\lim_{n \to \infty} \left(\frac{\sqrt[n+1]{\prod_{k=1}^{n+1} f(k)}}{\sqrt[n]{\prod_{k=1}^{n} f(k)}} \right)^n = e$. This follows from

$$\left(\frac{\sqrt[n+1]{\prod_{k=1}^{n+1} f(k)}}{\sqrt[n]{\prod_{k=1}^{n} f(k)}} \right)^n = \left(\sqrt[n(n+1)]{\frac{f^n(n+1)}{\prod_{k=1}^{n} f(k)}} \right)^n = \sqrt[n+1]{\frac{f^n(n+1)}{\prod_{k=1}^{n} f(k)}}$$

$$= \sqrt[n+1]{\frac{f^n(n+1)}{(n+1)^n}} \cdot \frac{n+1}{\sqrt[n+1]{\prod_{k=1}^{n+1} f(k)}} \cdot \sqrt[n+1]{\frac{f(n+1)}{n+1}} \to a \cdot \frac{e}{a} \cdot a^0 = e.$$

F3 Let $a_n = \sqrt[n+1]{\prod_{k=1}^{n+1} f(k)} - \sqrt[n]{\prod_{k=1}^{n} f(k)}$. Then

$$\lim_{n \to \infty} \left(1 + \frac{a_n}{\sqrt[n]{\prod_{k=1}^{n} f(k)}}\right)^{\sqrt[n]{\prod_{k=1}^{n} f(k)}/a_n} = e.$$

Recalling the limit definition of e, it suffices to show that

$$\lim_{n \to \infty} \frac{a_n}{\sqrt[n]{\prod_{k=1}^{n} f(k)}} = 0.$$

This can be done by

$$\frac{a_n}{\sqrt[n]{\prod_{k=1}^{n} f(k)}} = \frac{\sqrt[n+1]{\prod_{k=1}^{n+1} f(k)}}{\sqrt[n]{\prod_{k=1}^{n} f(k)}} - 1 \sim e^{1/n} - 1 \to 0,$$

where we have used **F2**.

Since

$$\left(\frac{\sqrt[n+1]{\prod_{k=1}^{n+1} f(k)}}{\sqrt[n]{\prod_{k=1}^{n} f(k)}}\right)^n = \left(1 + \frac{a_n}{\sqrt[n]{\prod_{k=1}^{n} f(k)}}\right)^n$$

$$= \left[\left(1 + \frac{a_n}{\sqrt[n]{\prod_{k=1}^{n} f(k)}}\right)^{\sqrt[n]{\prod_{k=1}^{n} f(k)}/a_n}\right]^{n a_n / \sqrt[n]{\prod_{k=1}^{n} f(k)}},$$

taking logarithm gives

$$\ln\left(\frac{\sqrt[n+1]{\prod_{k=1}^{n+1} f(k)}}{\sqrt[n]{\prod_{k=1}^{n} f(k)}}\right)^n = a_n \frac{n}{\sqrt[n]{\prod_{k=1}^{n} f(k)}} \ln\left(1 + \frac{a_n}{\sqrt[n]{\prod_{k=1}^{n} f(k)}}\right)^{\sqrt[n]{\prod_{k=1}^{n} f(k)}/a_n}.$$

Finally, letting $n \to \infty$ and using **F1–F3**, we find that

$$\lim_{n \to \infty} a_n = \frac{a}{e}$$

as claimed. □

Remark. Here are some similar Monthly problems for your practice.

1. **Problem 11771** (Proposed by D. M. Bătinetu-Giurgiu, 121(4), 2014). Find

$$\lim_{n \to \infty} \sqrt[n]{(2n-1)!!} \left(\tan\left(\frac{\pi \sqrt[n+1]{(n+1)!}}{4 \sqrt[n]{n!}}\right) - 1\right).$$

2. **Problem 11338** (Proposed by O. Furdui, 115(1), 2008). Let Γ
   denote the classical gamma function, and let $G(n) = \prod_{k=1}^{n} \Gamma(1/k)$.
   Find
   $$\lim_{n\to\infty} \left(G(n+1)^{1/(n+1)} - G(n)^{1/n} \right).$$

3. **Problem 11724** (Proposed by A. Cusumano, 120(7), 2013). Let
   $f(n) = \sum_{k=1}^{n} k^k$ and let
   $g(n) = \sum_{k=1}^{n} f(k)$. Find
   $$\lim_{n\to\infty} \left(\frac{g(n+2)}{g(n+1)} - \frac{g(n+1)}{g(n)} \right).$$

4. **Problem 11808** (Proposed by D. M. Bătinetu-Giurgiu, 121(10),
   2014). Compute
   $$\lim_{n\to\infty} n^2 \int_{((n+1)!)^{-1/(n+1)}}^{(n!)^{-1/n}} \Gamma(nx)\, dx.$$
   Hint: Use the mean value theorem for the integrals.

5. **Problem 11875** (Proposed by D. M. Bătinetu-Giurgiu and N.
   Stanciu, 122(10), 2015).
   Let $f_n = (1 + 1/n)^n ((2n-1)!! L_n)^{1/n}$. Find
   $$\lim_{n\to\infty} (f_{n+1} - f_n).$$
   Here L_n denotes the nth Lucas number, given by $L_0 = 2, L_1 = 1$,
   and $L_n = L_{n-1} + L_{n-2}$ for $n \geq 2$.

1.8   Limits of the weighted mean ratio

Problem 11811 (Proposed by V. Mikayelyan, 122(1), 2015). Let $\{a\}$ and
$\{b\}$ be infinite sequences of positive numbers. Let $\{x\}$ be the infinite sequence
given for $n \geq 1$ by
$$x_n = \frac{a_1^{b_1} a_2^{b_2} \cdots a_n^{b_n}}{\left(\frac{a_1 b_1 + a_2 b_2 + \cdots + a_n b_n}{b_1 + b_2 + \cdots + b_n} \right)^{b_1 + \cdots + b_n}}.$$

(a) Prove that $\lim_{n\to\infty} x_n$ exists.

(b) Find the set of all c that can occur as this limit, for suitably chosen $\{a\}$
and $\{b\}$.

Discussion.

As usual, we search for ideas from a special case: Let $b_n = 1$ for all $n \geq 1$. Then

$$x_n = \frac{a_1 a_2 \cdots a_n}{\left(\frac{a_1 + a_2 + \cdots + a_n}{n}\right)^n} = \left(\frac{\sqrt[n]{a_1 a_2 \cdots a_n}}{\frac{a_1 + a_2 + \cdots + a_n}{n}}\right)^n.$$

The AM-GM inequality implies that $x_n \leq 1$ for all $n \geq 1$. Moreover, if $a_n = n$ we have

$$x_n = \frac{n!}{\left(\frac{1 + 2 + \cdots + n}{n}\right)^n} = \frac{n!}{\left(\frac{n+1}{2}\right)^n},$$

which is monotonically decreasing. This suggests that x_n is decreasing in general. Consider

$$\frac{x_{n+1}}{x_n} = \frac{a_{n+1} \left(\frac{a_1 + a_2 + \cdots + a_n}{n}\right)^n}{\left(\frac{a_1 + \cdots + a_n + a_{n+1}}{n+1}\right)^{n+1}}.$$

Notice that x_n is decreasing if and only if $x_{n+1}/x_n \leq 1$, which is equivalent to

$$a_{n+1}\left(\frac{a_1 + a_2 + \cdots + a_n}{n}\right)^n \leq \left(\frac{a_1 + \cdots + a_n + a_{n+1}}{n+1}\right)^{n+1}.$$

Taking logarithms on both sides, then dividing by $(n+1)$ we obtain

$$\frac{1}{n+1}\ln a_{n+1} + \frac{n}{n+1}\ln\left(\frac{a_1 + a_2 + \cdots + a_n}{n}\right) \leq \ln\left(\frac{a_1 + \cdots + a_n + a_{n+1}}{n+1}\right).$$

This holds by applying Jensen's inequality for $\ln x$:

$$\alpha \ln x + (1 - \alpha)\ln y \leq \ln(\alpha x + (1 - \alpha)y)$$

with

$$\alpha = \frac{1}{n+1}, \quad x = a_{n+1} \text{ and } y = \frac{a_1 + a_2 + \cdots + a_n}{n}.$$

Hence, by the monotone convergence theorem, we see that $\lim_{n \to \infty} x_n$ exists and the limit is between 0 and 1. Now we just need to extend the above arguments to the weighted means.

Solution.

(a) Let $B_n = b_1 + b_2 + \cdots + b_n$ and $S_n = a_1 b_1 + a_2 b_2 + \cdots + a_n b_n$. Since $x_n \geq 0$ for all $n \geq 1$, by the monotone convergence theorem, it suffices to show that x_n is decreasing. To this end, we have $x_1 = 1$ and for $n \geq 1$

$$\frac{x_{n+1}}{x_n} = \frac{a_{n+1}^{b_{n+1}}(S_n/B_n)^{B_n}}{(S_{n+1}/B_{n+1})^{B_{n+1}}}.$$

It is clear that $x_{n+1}/x_n \leq 1$ if and only if

$$a_{n+1}^{b_{n+1}}(S_n/B_n)^{B_n} \leq (S_{n+1}/B_{n+1})^{B_{n+1}}.$$

Taking logarithms on both sides yields

$$b_{n+1} \ln a_{n+1} + B_n \ln(S_n/B_n) \leq B_{n+1} \ln(S_{n+1}/B_{n+1}),$$

or

$$\frac{b_{n+1}}{B_{n+1}} \ln a_{n+1} + \frac{B_n}{B_{n+1}} \ln(S_n/B_n) \leq \ln(S_{n+1}/B_{n+1}).$$

This inequality follows from Jensen's inequality

$$\alpha \ln x + (1 - \alpha) \ln y \leq \ln(\alpha x + (1 - \alpha)y)$$

with $\alpha = b_{n+1}/B_{n+1}, x = a_{n+1}$, and $y = S_n/B_n$.

(b) By the weighted AM-GM inequality, we see that $\lim_{n\to\infty} x_n \in [0,1]$. Now we show that $[0,1]$ is exactly the set of limit points of $\{x_n\}$; i.e., if $c \in [0,1]$, then there exist sequences $\{a\}$ and $\{b\}$ such that x_n converges to c. If $c = 0$, letting $a_n = n, b_n = 1$, from Stirling's formula, we have

$$x_n = \frac{n!}{\left(\frac{n+1}{2}\right)^n} \sim \sqrt{2\pi n} \left(\frac{2}{e}\right)^n \left(\frac{n}{n+1}\right)^n \to 0.$$

Next, let $a_n = 1 + t/n, b_n = 1$ with $t \geq 0$. We find that

$$x_n = \frac{\prod_{k=1}^n \left(1 + \frac{t}{k}\right)}{\left(1 + \frac{tH_n}{n}\right)^n} = \frac{\prod_{k=1}^n \left(1 + \frac{t}{k}\right)}{\left(1 + \frac{tH_n}{n}\right)^{(n/tH_n)\cdot tH_n}}$$

$$\to \frac{1}{te^{\gamma t}\Gamma(t)} = \frac{1}{e^{\gamma t}\Gamma(t+1)},$$

where we have used the gamma product formula:

$$\frac{1}{\Gamma(t)} = te^{\gamma t} \prod_{k=1}^{\infty} \left\{\left(1 + \frac{t}{k}\right)e^{-t/n}\right\}.$$

By the intermediate value theorem, the continuous function $1/e^{\gamma t}\Gamma(t + 1)$ attains every value in $(0,1]$ as $t \in [0,\infty)$. This completes the proof. $\square$

Remark. In the published solution [61], Kouba gave an elementary proof on Part **(b)** as follows: if $c \in (0,1)$, let $b_n = 1$ for all n and

$$a_1 = \frac{1 + \sqrt{1-c}}{\sqrt{c}}, \quad a_2 = \frac{1 - \sqrt{1-c}}{\sqrt{c}}, \quad a_n = \frac{1}{\sqrt{c}} \text{ for } n \geq 3.$$

Direct computation shows that $x_n = c$ for all $n \geq 2$ and so $x_n \to c$.

Here is one more problem for your practice. Let $\{a_n\}$ be an arithmetic sequence of positive numbers. Show the limit set of

$$\lim_{n\to\infty} \frac{\sqrt[n]{a_1 a_2 \cdots a_n}}{\frac{a_1 + a_2 + \cdots + a_n}{n}}$$

is either 1 or $2/e$.

1.9   Limits of mean recurrences

Problem 12057 (Proposed by P. Kórus, 125(7), 2018).

(a) Calculate the limit of the sequence defined by $a_1 = 1, a_2 = 2$, and

$$a_{2k+1} = \frac{a_{2k-1} + a_{2k}}{2} \quad \text{and} \quad a_{2k+2} = \sqrt{a_{2k}a_{2k+1}}$$

for positive integers k.

(b) Calculate the limit of the sequence defined by $b_1 = 1, b_2 = 2$, and

$$b_{2k+1} = \frac{b_{2k-1} + b_{2k}}{2} \quad \text{and} \quad b_{2k+2} = \frac{2b_{2k}b_{2k+1}}{b_{2k} + b_{2k+1}}$$

for positive integers k.

Discussion.
Normally, we expect to find formulas for a_{2k} and a_{2k+1} in terms of a_1 and a_2. However, direct calculations lead to nested radical sequences. Because they are patternless, there is no way to condense them into a closed form. The key insights needed are to use trigonometric substitution in **(a)** and to apply the homogeneous property of the means in **(b)**.

Solution.
(a) By induction, based on the definition of the sequences, it is easy to show that
$$a_{2k-1} \leq a_{2k+1} \leq a_{2k+2} \leq a_{2k}, \quad \text{for } k = 1, 2, \ldots.$$

The monotone convergence theorem, together with the recurrences, implies that the two sequences a_{2k+1} and a_{2k} converge to a common limit $M(a_1, a_2)$. We now show that $M(1, 2) = 3\sqrt{3}/\pi$. Indeed, we can determine $M(a_1, a_2)$ for any positive initial values a_1 and a_2. We distinguish three cases.
(i) If $a_1 = a_2$, the given recursions imply that $a_n = a_1$ for all $n \in \mathbb{N}$. Thus, $M(a_1, a_1) = a_1$.
(ii) If $a_1 < a_2$, we construct a right triangle with adjacent side a_1 and hypotenuse a_2. Let $\theta \in (0, \pi/2)$ be the angle such that $\cos \theta = a_1/a_2$ and let a be the opposite side. Then

$$a_1 = a \cot \theta, \quad a_2 = a \csc \theta; \tag{1.12}$$

$$\theta = \cos^{-1}(a_1/a_2), \quad a = \sqrt{a_2^2 - a_1^2}.$$

We claim

$$a_{2k-1} = \frac{a}{2^{k-1}} \cot(\theta/2^{k-1}), \quad a_{2k} = \frac{a}{2^{k-1}} \csc(\theta/2^{k-1}), \quad \text{for } k = 1, 2, \ldots. \tag{1.13}$$

The base case $k = 1$ follows from (1.12). Assume (1.13) holds for some k. By the trigonometric identities, for $0 < \alpha < \pi/2$,

$$\frac{\cot(2\alpha) + \csc(2\alpha)}{2} = \frac{1}{2}\cot\alpha, \quad \sqrt{\csc(2\alpha) \cdot \frac{\cot\alpha}{2}} = \frac{1}{2}\csc\alpha,$$

we find that

$$a_{2k+1} = \frac{a}{2^{k-1}}\frac{\cot(\theta/2^{k-1}) + \csc(\theta/2^{k-1})}{2} = \frac{a}{2^k}\cot(\theta/2^k),$$

$$a_{2k} = \frac{a}{2^{k-1}}\sqrt{\cot(\theta/2^{k-1}) \cdot \csc(\theta/2^{k-1})} = \frac{a}{2^k}\csc(\theta/2^k).$$

By induction (1.13) holds for all $k \geq 1$. As $x \to 0$, $\cot x \sim \csc x \sim x^{-1}$. Therefore

$$\lim_{k\to\infty} a_{2k-1} = \lim_{k\to\infty}\frac{a}{2^{k-1}}\cot(\theta/2^{k-1}) = \frac{a}{\theta};$$

$$\lim_{k\to\infty} a_{2k} = \lim_{k\to\infty}\frac{a}{2^{k-1}}\csc(\theta/2^{k-1}) = \frac{a}{\theta}.$$

Thus

$$M(a_1, a_2) = \frac{a}{\theta} = \frac{\sqrt{a_2^2 - a_1^2}}{\cos^{-1}(a_1/a_2)}.$$

In particular, if $a_1 = 1, a_2 = 2$, we obtain $M(1, 2) = 3\sqrt{3}/\pi$ as claimed.
(iii) If $a_1 > a_2$, using the identities

$$\frac{\coth(2\alpha) + \operatorname{csch}(2\alpha)}{2} = \frac{1}{2}\coth\alpha, \quad \sqrt{\operatorname{csch}(2\alpha) \cdot \frac{\coth\alpha}{2}} = \frac{1}{2}\operatorname{csch}\alpha,$$

similarly, we find that

$$M(a_1, a_2) = \frac{\sqrt{a_1^2 - a_2^2}}{\cosh^{-1}(a_1/a_2)}.$$

(b) Similarly, by induction we can show that

$$b_{2k-1} \leq b_{2k+1} \leq b_{2k+2} \leq b_{2k}, \quad \text{for } k = 1, 2, \ldots.$$

Thus the two sequences b_{2k+1} and b_{2k} also converge to a common limit $M(b_1, b_2)$. By the definition of the sequence, we have

$$M(\alpha b_1, \alpha b_2) = \alpha M(b_1, b_2)$$

for any positive α, b_1, b_2. Thus, without loss of generality, it suffices to consider the case where $b_1 = 1 + x, b_2 = 1$ with $x > -1$. Applying the definition of the sequence yields

$$b_3 = 2^{-1}(2 + x), \quad b_4 = 2\frac{2 + x}{2^2 + x}, \quad b_5 = 2^{-2}\frac{(2 + x)(2^3 + x)}{2^2 + x},$$

$$b_6 = 2^2\frac{(2 + x)(2^3 + x)}{(2^2 + x)(2^4 + x)}.$$

In general, by induction, we readily conclude that

$$b_{2n+1} = 2^{-n} \prod_{k=1}^{n}(2^{2k-1}+x)/\prod_{k=1}^{n-1}(2^{2k}+x), \quad n=1,2,\ldots,$$

$$b_{2n+2} = 2^{n} \prod_{k=1}^{n}\left[(2^{2k-1}+x)/(2^{2k}+x)\right], \quad n=1,2,\ldots.$$

So

$$M(1+x,1) = \lim_{n\to\infty} 2^{n} \prod_{k=1}^{n} \frac{2^{2k-1}+x}{2^{2k}+x} = \prod_{k=1}^{\infty} \frac{1+x/2^{2k-1}}{1+x/2^{2k}}.$$

In particular, letting $x = -1/2$ yields

$$M(1,2) = 2M(1-1/2,1) = 2\prod_{k=1}^{\infty}\left[\left(1-\frac{1}{4^k}\right)/\left(1-\frac{1}{2\cdot 4^k}\right)\right] = \frac{(1/4;1/4)_\infty}{(1/2;1/4)_\infty},$$

where $(a;q)_\infty$ is the q-Pochhammer symbol, so called a q-series (see [8], Chapter 10). □

Remark. The early history of using mean recursions can be traced back to the calculation of accurate values of π. Archimedes' process (250 BCE) constitutes the first rigorous algorithm for π. This process can be stated as the following harmonic-geometric mean recursion. Set $a_1 = 3\sqrt{3}, b_1 = 3\sqrt{3}/2$. Then define

$$a_{n+1} = \frac{2a_n b_n}{a_n + b_n}, \quad b_{n+1} = \sqrt{a_{n+1}b_n}, \quad \text{for } n \geq 1.$$

Geometrically, a_n and b_n denote the semi-perimeters of circumscribed and inscribed regular $(n+2)$-gons of the unit circle. For a regular 96-gon, these recursions give an intriguing estimate for π:

$$3.1408 \simeq 3\frac{10}{71} < \pi < 3\frac{1}{7} \simeq 3.1429.$$

No wonder Archimedes is viewed as the first numerical analyst.

Another famous mean recursion is the Gauss arithmetic-geometric mean (AGM) ([20], p. 5-7), which is defined by the two-term recurrence:

$$a_{n+1} = \frac{a_n + b_n}{2}, \quad b_{n+1} = \sqrt{a_n b_n}, \quad \text{for } n \geq 0,$$

Gauss observed that the common limit M satisfies

$$M(ta,tb) = tM(a,b), \quad \text{for any } t > 0, \qquad \text{(homogeneous)};$$

$$M(a,b) = M\left(\frac{a+b}{2}, \sqrt{ab}\right), \qquad \text{(invariant)},$$

and found that

$$M(a, b) = \left(\frac{2}{\pi} \int_0^{\pi/2} \frac{d\theta}{\sqrt{a^2 \cos^2 \theta + b^2 \sin^2 \theta}} \right)^{-1}.$$

This led to his discovery of the theory of elliptic functions.

Back to the proposed problem (**b**). Using the Jacobi triple product identity ([8], p. 497, [25], p. 50), we have

$$
\begin{aligned}
\sum_{k=0}^{\infty} x^{k(k+1)} &= \frac{1}{2} \sum_{k=-\infty}^{\infty} x^{k(k+1)} \\
&= \frac{1}{2} \prod_{n=1}^{\infty} \left[(1 - x^{2n})(1 + x^{2n})(1 + x^{2(n-1)}) \right] \\
&= \prod_{n=1}^{\infty} \left[(1 - x^{2n})(1 + x^{2n})^2 \right] \\
&= \prod_{n=1}^{\infty} \frac{(1 - x^{4n})^2}{1 - x^{2n}} \\
&= \prod_{n=1}^{\infty} \frac{1 - x^{4n}}{1 - x^{4n-2}}.
\end{aligned}
$$

Setting $x = 2^{-1/2}$, we find another alternative representation of $M(1, 2)$:

$$M(1, 2) = 2M(1/2, 1) = \prod_{k=1}^{\infty} \frac{1 - 1/2^{2k}}{1 - 1/2^{2k-1}} = \sum_{n=0}^{\infty} \frac{1}{2^{n(n+1)/2}}.$$

Historically, the q-series is associated with the *Ramanujan theta function* ([8], p. 501-505):

$$f(a, b) := \sum_{n=-\infty}^{\infty} a^{n(n+1)/2} b^{n(n-1)/2}.$$

For example, $M(1, 2)$ is the direct consequence (with $q = 1/2$) of the formula

$$f(q, q^3) = \sum_{n=0}^{\infty} q^{n(n+1)/2} = \frac{(q^2; q^2)_\infty}{(q; q^2)_\infty}.$$

As in (**a**), trigonometric substitution is particularly useful in solving recurrence relations. Here we present one more example.

Find the limits of the sequences defined by

$$a_{n+1} = \sqrt{a_n b_n} \quad \text{and} \quad b_{n+1} = \frac{2a_{n+1}b_n}{a_{n+1} + b_n}, \qquad \text{for } n \geq 0.$$

Along the same lines in (**a**), first, let $0 < a_0 < b_0$. Construct a right triangle with the adjacent side $\sqrt{a_0}$ and the hypotenuse $\sqrt{b_0}$. Let $\theta \in (0, \pi/2)$ be the angle such that $\cos^2 \theta = a_0/b_0$ and let

$$a = \sqrt{\frac{a_0}{b_0 - a_0}}\, b_0 = b_0 \cot \theta.$$

Then $b_0 = a \tan \theta$ and

$$
\begin{aligned}
a_1 &= \sqrt{a_0 b_0} = \sqrt{\frac{a_0}{b_0 - a_0}}\, b_0 \cdot \frac{\sqrt{b_0 - a_0}}{\sqrt{b_0}} = a \sin \theta; \\
b_1 &= \frac{2a_1 b_0}{a_1 + b_0} = \frac{2a^2 \tan \theta \sin \theta}{a \sin \theta + a \tan \theta} \\
&= 2a \frac{\sin \theta}{1 + \cos \theta} = 2a \tan(\theta/2).
\end{aligned}
$$

Based on these results and the calculations of a few next iterations, we conjecture that

$$a_n = 2^{n-1} a \sin\left(\theta/2^{n-1}\right), \quad b_n = 2^n a \tan\left(\theta/2^n\right), \quad (\text{for } n \geq 1). \quad (1.14)$$

We have already proved that (1.14) holds for $n = 1$. Assume (1.14) holds for some k. Then

$$
\begin{aligned}
a_{k+1} &= \sqrt{a_k b_k} = \sqrt{2^{2k-1} a^2 \sin\left(\theta/2^{k-1}\right) \tan\left(\theta/2^k\right)} \\
&= \sqrt{2^{2k} a^2 \sin\left(\theta/2^k\right) \cos\left(\theta/2^k\right) \tan\left(\theta/2^k\right)} \\
&= 2^k a \sin\left(\theta/2^k\right); \\
b_{k+1} &= \frac{2a_{k+1} b_k}{a_{k+1} + b_k} = \frac{2^{2k+1} a^2 \sin\left(\theta/2^k\right) \tan\left(\theta/2^k\right)}{2^k a \sin\left(\theta/2^k\right) + 2^k a \tan\left(\theta/2^k\right)} \\
&= \frac{2^{k+1} a \sin\left(\theta/2^k\right)}{1 + \cos\left(\theta/2^k\right)} = 2^{k+1} a \tan\left(\theta/2^{k+1}\right).
\end{aligned}
$$

By induction (1.14) holds for all $n \in \mathbb{N}$.

As $x \to 0$, $\sin x \sim \tan x \sim x$. We find that

$$\lim_{n \to \infty} a_n = \lim_{n \to \infty} 2^{n-1} a \sin\left(\theta/2^{n-1}\right) = a\theta;$$

$$\lim_{n \to \infty} b_n = \lim_{n \to \infty} 2^n a \tan\left(\theta/2^n\right) = a\theta.$$

Hence,

$$M(a_0, b_0) = a\theta = \sqrt{\frac{a_0}{b_0 - a_0}}\, b_0 \cdot \cos^{-1}(\sqrt{a_0/b_0}).$$

In particular, if $a_0 = 2, b_0 = 4$, we have

$$M(2, 4) = 4 \cdot \frac{\pi}{4} = \pi.$$

Thus (1.14) presents another algorithm for π. For example, we have

$$3.141591 \simeq a_{10} < \pi < b_{10} \simeq 3.141593.$$

Second, if $0 < b_0 < a_0$, we have

$$\begin{aligned} a_n &= 2^{n-1} a \sinh\left(\theta/2^{n-1}\right); \\ b_n &= 2^n a \tanh\left(\theta/2^n\right), \end{aligned}$$

where

$$\frac{a_0}{b_0} = \cosh^2 \theta, \quad a = \sqrt{\frac{a_0}{a_0 - b_0}}\, b_0.$$

Thus

$$M(a_0, b_0) = \sqrt{\frac{a_0}{a_0 - b_0}}\, b_0 \cdot \cosh^{-1}(\sqrt{a_0/b_0}).$$

We leave the details to the reader. We end this section with two additional practice problems.

1. In part **(b)**, for any positive initial values b_1 and b_2, prove that

$$M(b_1, b_2) = \frac{(\alpha; 1/4)_\infty}{(\beta; 1/4)_\infty},$$

   where $\alpha = 1/2 - b_1/(2b_2)$, $\beta = 1 - b_1/b_2$.

2. Let a be a positive constant. It is well-known that, for any initial term $x_1 > 0$, the sequence $\{x_n\}$ defined for $n \geq 1$ by

$$x_{n+1} = \frac{1}{2}\left(x_n + \frac{a}{x_n}\right)$$

   always converges to $\sqrt{a}$. Consider the sequence y_n defined for $n \geq 1$ by

$$y_{n+1} = \frac{1}{2}\left(y_n - \frac{a}{y_n}\right).$$

   For any positive integer $p > 1$, show that one can choose the initial term y_1 such that $\{y_n\}$ is periodic with the period p.

1.10 A disguised half-angle iteration

Problem 10973 (Proposed by L. D. Servi, 109(9), 2002). With $R_k(s)$ defined as below, prove that $\lim_{k\to\infty} R_k(2)/R_k(3) = 3/2$.

$$R_k(s) = \overbrace{\sqrt{2 - \sqrt{2 + \sqrt{2 + \sqrt{\cdots + \sqrt{2 + \sqrt{s}}}}}}}^{k \text{ square roots}}.$$

Discussion.

Let $f(x) = \sqrt{2+x}$. Define $f^1 = f$, $f^k = f \circ f^{k-1}$ for all $k \geq 2$. Observe that

$$f^k(x) = \overbrace{\sqrt{2 + \sqrt{2 + \sqrt{2 + \sqrt{\cdots + \sqrt{2+x}}}}}}^{k \text{ square roots}}.$$

To find a closed form for f^k, the key insight comes from the proof of the Viète formula:

$$\sqrt{\frac{1}{2}} \cdot \sqrt{\frac{1}{2} + \frac{1}{2}\sqrt{\frac{1}{2}}} \cdot \sqrt{\frac{1}{2} + \frac{1}{2}\sqrt{\frac{1}{2} + \sqrt{\frac{1}{2}}}} \cdots = \frac{2}{\pi}.$$

Setting $t = \pi/2$ in the well-known limit

$$\lim_{n \to \infty} \cos\frac{t}{2} \cos\frac{t}{2^2} \cdots \cos\frac{t}{2^n} = \frac{\sin t}{t},$$

we find that

$$\lim_{n \to \infty} \cos\frac{\pi}{2^2} \cos\frac{\pi}{2^3} \cdots \cos\frac{t}{2^{n+1}} = \frac{2}{\pi}.$$

The Viète formula now follows from

$$\cos\frac{\pi}{4} = \sqrt{\frac{1}{2}} \quad \text{and} \quad \cos\frac{\theta}{2} = \sqrt{\frac{1}{2} + \frac{1}{2}\cos\theta}.$$

Once we rewrite the last half angle formula as

$$\sqrt{2 + 2\cos\theta} = 2\cos\frac{\theta}{2},$$

similarly, we generate f^k as the required nested radicals.

Solution.

Let $f(x) = \sqrt{2+x}$ and $x = 2\cos\alpha$ for $\alpha \in (0, \pi/2)$. Since

$$f(2\cos\alpha) = \sqrt{2 + 2\cos\alpha} = 2\cos(\alpha/2),$$

it follows that

$$f^2(2\cos\alpha) = f(2\cos(\alpha/2)) = 2\cos(\alpha/2^2).$$

By induction, for all $k \geq 1$, we find that

$$f^k(x) = f^k(2\cos\alpha) = 2\cos(\alpha/2^k) = 2\cos[\cos^{-1}(x/2)/2^k].$$

Thus

$$R_k(4\cos^2\alpha) = \sqrt{2 - f^{k-2}(2\cos\alpha)} = \sqrt{2 - 2\cos(\alpha/2^{k-2})} = 2\sin\left(\alpha/2^{k-1}\right).$$

In general, for any $s, t \in (0, 4)$, we have

$$\lim_{k \to \infty} \frac{R_k(s)}{R_k(t)} = \lim_{k \to \infty} \frac{2 \sin[2^{1-k} \cos^{-1}(\sqrt{s}/2)]}{2 \sin[2^{1-k} \cos^{-1}(\sqrt{t}/2)]} = \frac{\cos^{-1}(\sqrt{s}/2)}{\cos^{-1}(\sqrt{t}/2)}.$$

In particular,

$$\lim_{k \to \infty} \frac{R_k(2)}{R_k(3)} = \frac{\cos^{-1}(\sqrt{2}/2)}{\cos^{-1}(\sqrt{3}/2)} = \frac{\pi/4}{\pi/6} = \frac{3}{2}.$$

$\square$

Remark. There is another way to find f^k by using the Chebyshev polynomials. If $x \in [-2, 2]$, the inverse function of f is $f^{-1}(x) = x^2 - 2$ for $x \in [0, 2]$. Recall that the Chebyshev polynomial of the first kind of degree 2 is

$$T_2(x) = 2x^2 - 1.$$

Then $f^{-1}(x) = 2T_2(x/2)$. Since

$$T_2^k(x) = T_{2^k}(x) \quad \text{and} \quad T_n(x) = \cos(n \cos^{-1} x),$$

it implies that

$$f^{-k}(x) = 2T_2^k(x/2) = 2T_{2^k}(x/2) = 2\cos[2^k \cos^{-1}(x/2)].$$

Once again we find that

$$f^k(x) = 2\cos[\cos^{-1}(x/2)/2^k].$$

In general, consider a class of periodic continued radicals of the form

$$a_0 \sqrt{2 + a_1 \sqrt{2 + a_2 \sqrt{2 + a_3 \sqrt{2 + \cdots}}}}, \tag{1.15}$$

where $a_{n+k} = a_k$ for some positive integer n and $a_k = \pm 1$ for $k = 0, 1, 2, \ldots, n-1$. By induction, we can show that these radicals given by (1.15) have limits two times the fixed points of the Chebyshev polynomials $T_{2^n}(x)$. Explicitly, we have

$$a_0 \sqrt{2 + a_1 \sqrt{2 + a_2 \sqrt{2 + a_3 \sqrt{2 + \cdots + a_{n-1}\sqrt{2}}}}}$$
$$= 2 \sin\left[\left(a_0 + \frac{a_0 a_1}{2} + \cdots + \frac{a_0 a_1 \cdots a_{n-1}}{2^{n-1}}\right) \frac{\pi}{4}\right].$$

Since the sequence

$$a_0 + \frac{a_0 a_1}{2} + \cdots + \frac{a_0 a_1 \cdots a_{n-1}}{2^{n-1}}$$

converges absolutely, let its limit be α. Hence, the original radical converges to $2\sin(\alpha\pi/4)$. This immediately yields a solution to the following question.

Problem 12129 (Proposed by H. Ohtsuka, 126(7), 2019). Compute

$$\sqrt{2+\sqrt{2+\sqrt{2+\cdots+\sqrt{2-\sqrt{2+\cdots}}}}},$$

where the sequence of signs consists of $n-1$ plus signs followed by a minus sign and repeats with period n.

Having found the limits of (1.15), the next natural question is to determine the limit of the radical

$$a_0\sqrt{b+a_1\sqrt{b+a_2\sqrt{b+a_3\sqrt{b+\cdots}}}}$$

for values of the variable b that make the radical (and the limit) well defined. However, a direct application of the above method fails and so far a convenient variation has been elusive. Therefore, the limit of the last radical in the general case remains an open problem although it is known at least in one case. **Putnam Problem 1953 A-6** ([47], p. 39). Show that the sequence

$$\sqrt{7},\ \sqrt{7-\sqrt{7}},\ \sqrt{7-\sqrt{7+\sqrt{7}}},\ \sqrt{7-\sqrt{7+\sqrt{7-\sqrt{7}}}},\ldots$$

converges, and evaluate the limit.

Recently, Fernández-Sánchez and Trutschnig in an elegant paper [39] show how various aforementioned results on the nested square roots can be derived easily via a topological conjugacy linking the following tent map and the logistic map. See Figure1.1.

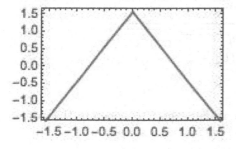 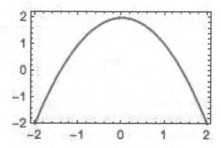

FIGURE 1.1
The tent (left) and logistic (right) maps

Moreover, applying this clever approach, they find some new and striking results for the radicals which are not periodic.

Here is another Monthly problem which may use trigonometric substitution to compute the limits.

Problem E 2835 (Proposed by M. Golomb, 87(6), 1980). Let $-1 < a_0 < 1$, and define recursively
$a_n = \sqrt{(1 + a_{n-1})/2}$, $n > 0$. Find $\lim_{n \to \infty} 4^n(1 - a_n)$, $B := \lim_{n \to \infty} a_1 a_2 \cdots a_n$, and

$$\lim_{n \to \infty} 4^n(B - a_1 a_2 \cdots a_n).$$

A much more challenging limit problem involving radicals is as follows:

Problem 11367 (Proposed by A. Cusumano, 115(5), 2008). Let $x_1 = \sqrt{1 + 2}$, $x_2 = \sqrt{1 + 2\sqrt{1 + 3}}$, and in general, let x_{n+1} be the number obtained by replacing the innermost expression $(1 + (n + 1))$ in the nested square root formula for x_n with $1 + (n + 1)\sqrt{1 + (n + 2)}$. Show that

$$\lim_{n \to \infty} \frac{x_n - x_{n-1}}{x_{n+1} - x_n} = 2.$$

As an continuation of this Monthly problem, the reader may study the conditions on x and a_n for which

$$\sqrt{1 + x\sqrt{1 + (x+1)\sqrt{1 + \cdots + (x + n - 1)\sqrt{1 + (x + n)a_n}}}}$$

converges and then determine the limit.

1.11 Nested radicals and generalized Fibonacci numbers

Problem 12063 (Proposed by H. Ohtsuka, 125(8), 2018). Let p and q be real numbers with $p > 0$ and $q > -p^2/4$. Let $U_0 = 0, U_1 = 1$, and $U_{n+2} = pU_{n+1} + qU_n$ for $n \geq 0$. Calculate

$$\lim_{n \to \infty} \sqrt{U_1^2 + \sqrt{U_2^2 + \sqrt{U_4^2 + \sqrt{\cdots + \sqrt{U_{2^{n-1}}^2}}}}}.$$

Discussion.

For each $n \in \mathbb{N}$, the challenge is determining how to generate the nested radicals systematically. To gain insight, we analyze a simpler case:

$$\sqrt{F_1^2 + \sqrt{F_2^2 + \sqrt{F_4^2 + \sqrt{\cdots + \sqrt{F_{2^{n-1}}^2}}}}},$$

where F_n is the nth Fibonacci numbers. Recall Catalan's identity

$$F_i^2 - F_{i+k}F_{i-k} = (-1)^{i-k}F_k^2.$$

Setting $i = 2^n + 2, k = 2^n$ in this identity yields

$$F_{2^n+2}^2 - F_{2^{n+1}+2} = F_{2^n}^2.$$

Thus

$$F_{2^n+2} = \sqrt{F_{2^n}^2 + F_{2^{n+1}+2}}.$$ (1.16)

Repeatedly applying (1.16) leads to

$$2 = F_{2^0+2} = \sqrt{F_{2^0}^2 + F_{2^1+2}} = \sqrt{F_1^2 + \sqrt{F_{2^1}^2 + F_{2^2+2}}}$$

$$= \cdots = \sqrt{F_1^2 + \sqrt{F_2^2 + \sqrt{F_4^2 + \sqrt{\cdots + \sqrt{F_{2^{n-2}}^2 + \sqrt{F_{2^{n-1}}^2 + F_{2^n+2}}}}}}}.$$

That is exactly what we expect. Taking $a_n = F_{2^n+2}$, (1.16) becomes $a_n = \sqrt{F_{2^n}^2 + a_{n+1}}$. This suggests we find a proper sequence a_n which satisfies the iteration

$$a_n = \sqrt{U_{2^n}^2 + a_{n+1}}.$$

Solution.
The limit is $(p + \sqrt{4 + p^2 + 4q})/2$. To see this, let

$$\alpha = \frac{p + \sqrt{p^2 + 4q}}{2}, \quad \beta = \frac{p - \sqrt{p^2 + 4q}}{2}.$$

Similar to Binet's formula for Fibonacci numbers, we have

$$U_n = \frac{1}{\sqrt{p^2 + 4q}}(\alpha^n - \beta^n), \quad \text{for } n \geq 0.$$

Let $V_n = \alpha^n + \beta^n$ for $n \geq 0$. Direct calculations give

$$U_{2n} = U_n V_n,$$ (1.17)

$$V_n^2 + (p^2 + 4q)U_n^2 = 2V_{2n}, \quad \text{and}$$ (1.18)

$$V_{2n} = V_n^2 - 2(-q)^n.$$ (1.19)

Define

$$a_n = \frac{V_{2^n} + \sqrt{4 + p^2 + 4q}\, U_{2^n}}{2}, \quad \text{for } n \geq 0.$$

Applying (1.17) and (1.18) yields

$$
\begin{aligned}
4a_n^2 &= V_{2^n}^2 + (4 + p^2 + 4q)U_{2^n}^2 + 2\sqrt{4 + p^2 + 4q}\, U_{2^n} V_{2^n}\\
&= 4U_{2^n}^2 + V_{2^n}^2 + (p^2 + 4q)U_{2^n}^2 + 2\sqrt{4 + p^2 + 4q}\, U_{2^n} V_{2^n}\\
&= 4U_{2^n}^2 + 2V_{2^{n+1}} + 2\sqrt{4 + p^2 + 4q}\, U_{2^{n+1}}\\
&= 4U_{2^n}^2 + 4a_{n+1}.
\end{aligned}
$$

Thus,

$$
a_n = \sqrt{U_{2^n}^2 + a_{n+1}}.
$$

Repeatedly using this identity gives

$$
a_0 = \sqrt{U_1^2 + a_1} = \sqrt{U_1^2 + \sqrt{U_2^2 + a_2}} = \cdots
$$

$$
= \sqrt{U_1^2 + \sqrt{U_2^2 + \sqrt{\cdots + \sqrt{U_{2^{n-1}}^2 + a_n}}}}.
$$

Since $a_n > 0$, it follows

$$
a_0 > \sqrt{U_1^2 + \sqrt{U_2^2 + \sqrt{U_4^2 + \sqrt{\cdots + \sqrt{U_{2^{n-1}}^2}}}}}. \tag{1.20}
$$

On the other hand, given any $\epsilon \in (0, a_0)$, let $t = 1 - \epsilon/a_0$. Then

$$
a_0 - \epsilon = ta_0 = t\sqrt{U_1^2 + \sqrt{U_2^2 + \sqrt{\cdots + \sqrt{U_{2^{n-1}}^2 + a_n}}}}
$$

$$
= \sqrt{t^2 U_1^2 + \sqrt{t^{2^2} U_2^2 + \sqrt{\cdots + \sqrt{t^{2^n}(U_{2^{n-1}}^2 + a_n)}}}}
$$

$$
< \sqrt{U_1^2 + \sqrt{U_2^2 + \sqrt{\cdots + \sqrt{t^{2^n}(U_{2^{n-1}}^2 + a_n)}}}}.
$$

From (1.18) and (1.20), we have

$$
\begin{aligned}
\frac{a_n}{U_{2^{n-1}}^2} &= \frac{V_{2^n} + \sqrt{4 + p^2 + 4q}\, U_{2^n}}{2U_{2^{n-1}}^2}\\
&= \frac{V_{2^{n-1}}^2 - 2(-q)^n + \sqrt{4 + p^2 + 4q}\, U_{2^{n-1}} V_{2^{n-1}}}{2U_{2^{n-1}}^2}\\
&= \frac{1}{2}\left(\frac{V_{2^{n-1}}}{U_{2^{n-1}}}\right)^2 - \frac{(-q)^n}{U_{2^{n-1}}^2} + \frac{\sqrt{4 + p^2 + 4q}}{2}\, \frac{V_{2^{n-1}}}{U_{2^{n-1}}}\\
&\to \frac{1}{2}(p^2 + 4q) + \frac{\sqrt{4 + p^2 + 4q}}{2}\sqrt{p^2 + 4q}, \quad (\text{as } n \to \infty).
\end{aligned}
$$

Let

$$L := \frac{1}{2}(p^2 + 4q) + \frac{\sqrt{4 + p^2 + 4q}}{2}\sqrt{p^2 + 4q}.$$

Then

$$\frac{U_{2^{n-1}}^2}{U_{2^{n-1}}^2 + a_n} = \frac{1}{1 + a_n/U_{2^{n-1}}^2} \to \frac{1}{1 + L} > 0 \quad (\text{as } n \to \infty).$$

Since $0 < t < 1$, there exists an $N \in \mathbb{N}$ such that

$$t^{2^n} < \frac{U_{2^{n-1}}^2}{U_{2^{n-1}}^2 + a_n}, \quad (\text{as } n > N).$$

Therefore, when $n > N$, we have

$$a_0 - \epsilon < \sqrt{U_1^2 + \sqrt{U_2^2 + \sqrt{U_4^2 + \sqrt{\cdots + \sqrt{U_{2^{n-1}}^2}}}}}. \tag{1.21}$$

Since ϵ is arbitrary, from (1.20) and (1.21), by the squeeze theorem, we find

$$\lim_{n \to \infty} \sqrt{U_1^2 + \sqrt{U_2^2 + \sqrt{U_4^2 + \sqrt{\cdots + \sqrt{U_{2^{n-1}}^2}}}}} = a_0 = \frac{p + \sqrt{4 + p^2 + 4q}}{2},$$

as claimed. □

Remark. Similarly, for any $k > 0$, we have

$$\lim_{n \to \infty} \sqrt{kU_1^2 + \sqrt{kU_2^2 + \sqrt{kU_4^2 + \sqrt{\cdots + \sqrt{kU_{2^{n-1}}^2}}}}} = \frac{p + \sqrt{4k + p^2 + 4q}}{2}.$$

Nested radicals were a favorite topic of Ramanujan. One of his famous results

$$\sqrt{1 + 2\sqrt{1 + 3\sqrt{1 + 4\sqrt{1 + \cdots}}}} = 3,$$

appeared as **Putnam Problem 1966 A-6** ([4], p. 5).

There are three more problems for additional practice.

1. (**A Convergence Test for the Nested Radicals**) Let $\{a_n\}$ be a positive sequence and

$$R_n := \sqrt{a_1 + \sqrt{a_2 + \sqrt{a_3 + \cdots + \sqrt{a_n}}}}.$$

Assume the following limit exists:

$$L = \lim_{n \to \infty} \sup \frac{\ln(\ln a_n)}{n}.$$

•If $L < 2$, then R_n converges.

•If $L > 2$, then R_n diverges.

•If $L = 2$, the test is inconclusive.

2. This problem involves the rate of convergence: Let

$$a_n = \sqrt{1 + \sqrt{2 + \sqrt{3 + \cdots + \sqrt{n}}}}$$

for $n \in \mathbb{N}$. Prove that $\lim_{n\to\infty} a_n = a$ exists and $\lim_{n\to\infty} n \cdot \sqrt[n]{a - a_n} = \sqrt{e}/2$.

3. **Problem 11967** (Proposed by H. Ohtsuka, 124(3), 2017). Let F_n be the nth Fermat number $2^{2^n} + 1$. Find

$$\lim_{n\to\infty} \sqrt{6F_1 + \sqrt{6F_2 + \sqrt{6F_3 + \sqrt{\cdots + \sqrt{6F_n}}}}}.$$

1.12   A limit involving arctangent

Problem 11592 (Proposed by M. Ivan, 118(8), 2011). Find

$$\lim_{n\to\infty} \left(-\ln n + \sum_{k=1}^{n} \arctan \frac{1}{k} \right).$$

Discussion.
Since there is no closed form for $\sum_{k=1}^{n} \arctan \frac{1}{k}$, based on the well-known limit

$$\lim_{n\to\infty} \left(\sum_{k=1}^{n} \frac{1}{k} - \ln n \right) = \gamma, \tag{1.22}$$

we can replace the arctan sum by $\sum_{k=1}^{n} \frac{1}{k}$. Another observation is

$$\int_0^1 \frac{k}{x^2 + k^2} \, dx = \arctan \frac{1}{k} = \operatorname{Im}(\ln(1 + i/k)) = \operatorname{Arg}(1 + i/k),$$

where $i = \sqrt{-1}$. This makes the problem more manageable in the complex domain. Here we give two solutions. One uses the gamma function while the other uses the digamma function.

Solution I.
Let L denote the desired limit. From (1.22) we have

$$L = \gamma - \lim_{n \to \infty} \sum_{k=1}^{\infty} \left(\frac{1}{k} - \arctan \frac{1}{k} \right).$$

Recall the product representation of the gamma function

$$\frac{1}{\Gamma(z)} = z e^{\gamma z} \prod_{k=1}^{\infty} \left(1 + \frac{z}{k} \right) e^{-z/k}.$$

Taking the logarithm of both sides, and then setting $z = i$, we obtain

$$\ln \Gamma(i) = -i\gamma - \ln i + \sum_{k=1}^{\infty} \left(\frac{i}{k} - \ln \left(1 + \frac{i}{k} \right) \right).$$

From $\text{Im}(\ln(1 + i/k)) = \text{Arg}(1 + i/k) = \arctan \frac{1}{k}$, so equating the imaginary parts yields

$$\text{Arg}(\Gamma(i)) = -\gamma - \frac{\pi}{2} + \sum_{k=1}^{\infty} \left(\frac{1}{k} - \arctan \frac{1}{k} \right),$$

which implies that

$$L = -\pi/2 - \text{Arg}(\Gamma(i)) = -\text{Arg}(\Gamma(1+i)),$$

with the principle branch cut. □

Solution II.
Recall two properties of the digamma function ψ:

P1 $\psi(z + n) = \frac{1}{z} + \frac{1}{z+1} + \cdots + \frac{1}{z+n-1} + \psi(z)$ for $n \in \mathbb{N}$ ([8], p. 13, Theorem 1.2.7).

P2 $\psi(z) = \ln z + O(1/z)$ for $|\text{arg} z| < \pi - \delta, \delta > 0$ ([8], p. 22, Corollary 1.4.5).

By partial fractions and **P1-2**, we have

$$\sum_{k=1}^{n} \frac{k}{x^2 + k^2} = \frac{1}{2} \sum_{k=1}^{n} \left(\frac{1}{k + ix} + \frac{1}{k - ix} \right)$$

$$= \frac{1}{2} (\psi(n + 1 + ix) - \psi(1 + ix) + \psi(n + 1 - ix) - \psi(1 - ix))$$

$$= -\frac{1}{2} (\psi(1 + ix) + \psi(1 - ix)) + \frac{1}{2} \ln(n^2 + x^2) + O(1/n).$$

Thus

$$\sum_{k=1}^{n} \arctan \frac{1}{k} = \int_0^1 \left(\sum_{k=1}^{n} \frac{k}{x^2 + k^2} \right) dx$$

$$= \frac{i}{2} \left(\ln \Gamma(1 + ix) - \ln \Gamma(1 - ix) \right) |_0^1 + \frac{1}{2} \ln(n^2 + 1)$$
$$+ n \arctan(1/n) - 1 + O(1/n)$$

$$= \frac{i}{2} \left(\ln \Gamma(1 + i) - \ln \Gamma(1 - i) \right) + \frac{1}{2} \ln(n^2 + 1)$$
$$+ n \arctan(1/n) - 1 + O(1/n).$$

This implies

$$\lim_{n \to \infty} \left(-\ln n + \sum_{k=1}^{n} \arctan \frac{1}{k} \right) = \frac{i}{2} \ln \frac{\Gamma(1 + i)}{\Gamma(1 + i)} = -\text{Arg}(\Gamma(1 + i)).$$

□

Remark. Beckwith noted this problem is a special case of Formula 6.1.27 in [1]:

$$\text{Arg}(\Gamma(x + iy)) = y\psi(x) + \sum_{k=0}^{\infty} \left(\frac{y}{x + k} - \arctan \left(\frac{y}{x + k} \right) \right).$$

Replacing arctan with arctanh in the proposed problem, then telescoping based on $\text{arctanh} x = \frac{1}{2} \ln \left(\frac{1+x}{1-x} \right)$, we have a surprisingly simple result

$$\lim_{n \to \infty} \left(-\ln n + \sum_{k=1}^{n} \text{arctanh} \frac{1}{k} \right) = -\frac{1}{2} \ln 2.$$

Here are five more problems for additional practice.

1. Find

$$\lim_{n \to \infty} \sqrt{n} \prod_{k=1}^{n} \frac{e^{1-1/k}}{\left(1 + \frac{1}{k} \right)^k}.$$

2. **Problem 11494** (Proposed by O. Furdui, 117(3), 2010). Let A be the *Glaisher-Kinkelin constant*, given by

$$A = \lim_{n \to \infty} n^{-n^2/2 - n/2 - 1/2} e^{n^2/4} \prod_{k=1}^{n} k^k = 1.2824\ldots.$$

Prove that

$$\prod_{n=1}^{\infty} \left(\frac{n!}{\sqrt{2\pi n}(n/e)^n} \right)^{(-1)^{n-1}} = \frac{A^3}{2^{7/12}\pi^{1/4}}.$$

3. **Problem 11612** (Proposed by P. Bracken, 118(10), 2011). Evaluate in closed from

$$\prod_{n=1}^{\infty} \left(\frac{n+z+1}{n+z} \right)^n e^{(2z-2n+1)/(2n)}.$$

4. **Problem 12029** (Proposed by H. Ohtsuka, 125(3), 2018). For $a > 0$, evaluate

$$\lim_{n\to\infty} \prod_{k=1}^{n} \left(a + \frac{k}{n} \right).$$

5. **Problem 11677** (Proposed by A. Stadler, 119(10), 2012). Evaluate

$$\prod_{n=1}^{\infty} \left(1 + 2e^{-n\pi\sqrt{3}} \cosh\left(\frac{n\pi}{\sqrt{3}} \right) \right).$$

Hint. Recall the *eta function* $\eta(z) = e^{\pi i z/12} \prod_{n=1}^{\infty} (1 - e^{2n\pi i z})$. Express this product in terms of a ratio of eta function values. Stenger [89] presents an interesting solution to this problem via the approach of "Experimental Math."

1.13   Summing to the double factorials

Problem 11821 (Proposed by F. Holland and C. Koester, 122(2), 2015). Let p be a positive integer. Prove that

$$\lim_{n\to\infty} \frac{1}{2^n n^p} \sum_{k=0}^{n} (n-2k)^{2p} \binom{n}{k} = \prod_{j=1}^{p} (2j-1).$$

Discussion.
We can proceed with this problem in two different ways. Let

$$S_p(n) = \sum_{k=0}^{n} (n-2k)^{2p} \binom{n}{k}.$$

First, to get a feel for this sequence, we compute the first few terms by *Math-*

ematica:

$$S_1(n) = \sum_{k=0}^{n} (n - 2k)^2 \binom{n}{k} = 2^n n,$$

$$S_2(n) = \sum_{k=0}^{n} (n - 2k)^4 \binom{n}{k} = 2^n (3n^2 - 2n),$$

$$S_3(n) = \sum_{k=0}^{n} (n - 2k)^6 \binom{n}{k} = 2^n (15n^3 - 30n^2 + 16n),$$

$$S_4(n) = \sum_{k=0}^{n} (n - 2k)^8 \binom{n}{k} = 2^n (105n^4 - 410n^3 + 588n^2 - 272n).$$

The emerging pattern suggests that $S_p(n) = 2^n P_p(n)$, where $P_p(n)$ is a polynomial in n with the leading coefficient $n^p \prod_{j=1}^{p} (2j - 1)$, from which the proposed limit follows immediately.

The second approach is to show that $2^{-n} S_p(n)$ is a $(2p)$th derivative of a well-known function.

Solution I.

Let

$$S_p(n) = \sum_{k=0}^{n} (n - 2k)^{2p} \binom{n}{k} \qquad \text{for } n, p > 0.$$

We claim that $S_p(n)$ satisfies the recurrence

$$S_{p+1}(n) = n^2 S_p(n) - 4n(n - 1)S_p(n - 2). \qquad (1.23)$$

Indeed, we have

$$
\begin{aligned}
4n(n - 1)S_p(n - 2) &= \sum_{k=0}^{n-2} 4n(n - 1)(n - 2 - 2k)^{2p} \binom{n - 2}{k} \\
&= \sum_{k=1}^{n-1} 4n(n-1)(n-2k)^{2p} \binom{n - 2}{k - 1} \quad \text{(replacing } k + 1 \text{ by } k) \\
&= \sum_{k=1}^{n-1} 4k(n - k)(n - 2k)^{2p} \binom{n}{k}
\end{aligned}
$$

(using binomial coefficient formula).

Thus (1.23) follows from

$$
\begin{aligned}
n^2 S_p(n) - 4n(n - 1)S_p(n - 2) &= \sum_{k=0}^{n} (n^2 - 4k(n - k))(n - 2k)^{2p} \binom{n}{k} \\
&= \sum_{k=0}^{n} (n - 2k)^{2(p+1)} \binom{n}{k} = S_{p+1}(n).
\end{aligned}
$$

Next, let $M_p(n) = 2^{-n}S_p(n)$. So $M_0(n) = S_0(n) = 1$. We show by induction on p that $M_p(n)$ is a polynomial of degree p in n with the leading coefficient $\prod_{j=1}^{p}(2j-1)$. From this the desired result follows immediately.

We use the notation that $\prod_{j=1}^{p}(2j-1) = (2p-1)!!$. Let c_p be the coefficient of n^{p-1} in $M_p(n)$. From (1.23), the inductive computation for $p \geq 0$ is

$$
\begin{aligned}
M_{p+1}(n) &= n^2 M_p(n) - n(n-1)M_p(n-2) \\
&= (2p-1)!!n^{p+2} + c_p n^{p+1} - (n^2-n)[(2p-1)!!(n-2)^p \\
&\quad + c_p(n-2)^{p-1}] + O(n^p) \\
&= c_p n^{p+1} - c_p n^{p+1} + n(2p-1)!!(n-2)^p \\
&\quad + n^2(2p)(2p-1)!!n^{p-1} + O(n^p) \\
&= (2p+1)(2p-1)!!n^{p+1} + O(n^p).
\end{aligned}
$$

$\square$

Solution II.

Let $M_p(n) = 2^{-n}S_p(n)$. Applying the binomial theorem, we have

$$
M_p(n) = \frac{1}{2^n} \sum_{m=0}^{2p} \binom{2p}{m} n^{2p-m} \cdot \sum_{k=0}^{n} \binom{n}{k}(-2k)^m.
$$

Let $[x^n]f(x)$ denote the coefficient of x^n in the power series of $f(x)$. Then

$$
\left[\frac{x^{2p-m}}{(2p-m)!}\right]e^{nx} = n^{2p-m} \quad \text{and} \quad \left[\frac{x^m}{m!}\right](1+e^{-2x})^n = \sum_{k=0}^{n}\binom{n}{k}(-2k)^m.
$$

By the Leibniz rule for derivatives, we deduce that

$$
\begin{aligned}
M_p(n) &= \frac{1}{2^n} \sum_{m=0}^{2p} \binom{2p}{m}\left[\frac{x^{2p-m}}{(2p-m)!}\right]e^{nx} \cdot \left[\frac{x^m}{m!}\right](1+e^{-2x})^n \\
&= \frac{1}{2^n}\left[\frac{x^{2p}}{(2p)!}\right]e^{nx}(1+e^{-2x})^n = \left[\frac{x^{2p}}{(2p)!}\right]\left(\frac{e^x+e^{-x}}{2}\right)^n \\
&= D^{2p}(\cosh^n(x))(0),
\end{aligned}
$$

where D^{2p} indicates the $(2p)$th derivative. We now use the following Faà di Bruno's formula [55] to compute the derivative:

$$
D^{2p}(f \circ g)(x) = \sum \frac{(2p)!}{k_1! \cdots k_{2p}!}(D^k f)(g(x)) \cdot \prod_{m=1}^{2p}\left(\frac{D^m g(x)}{m!}\right)^{k_m},
$$

where $k = k_1 + k_2 + \cdots + k_{2p}$ and the sum is over all partitions of $2p$ that satisfy $k_1 + 2k_2 + \ldots + (2p)k_{2p} = 2p$. Let $f(x) = x^n$, $g(x) = \cosh x$. Note that

$$
D^k(x^n)(\cosh(0)) = n(n-1)\cdots(n-k+1) \quad \text{and}
$$

$$
D^m(\cosh(x))(0) = \begin{cases} 1, & \text{if } m \text{ is even}, \\ 0, & \text{if } m \text{ is odd}. \end{cases}
$$

This implies that $k_1 = k_3 = \cdots = k_{2p-1} = 0$. Thus $k = \sum_{m=1}^{p} k_{2m}$ and the constraint becomes $\sum_{m=1}^{p} mk_{2m} = p$. Furthermore,

$$D^{2p}(\cosh^n(x))(0) = \sum \frac{(2p)!}{\prod_{m=1}^{p} (k_{2m}!)[(2m)!]^{k_{2m}}} \prod_{m=0}^{k-1} (n-m).$$

Note that the right hand side is a p-th degree polynomial in n. The degree is attained only at $k_2 = p, k_4 = \cdots = k_{2p} = 0$. Thus we finally obtain

$$\lim_{n\to\infty} \frac{M_p(n)}{n^p} = \lim_{n\to\infty} \frac{D^{2p}(\cosh^n(x))(0)}{n^p} = \frac{(2p)!!}{p!\,(2!)^p} = (2p-1)!!.$$

$\square$

Remark. It is possible to find an explicit representation of $S_p(n)$ in terms of the Stirling numbers of the second kind $S(i, j)$ ([1], p. 822). To see this, recall the generating function of $S(i, j)$:

$$x^i = \sum_{j=0}^{i} S(i, j)(x)_j,$$

where $(x)_j = x(x-1)\cdots(x-j+1)$, called the *falling factorial*. This implies that

$$\sum_{k=0}^{n} \binom{n}{k} k^i = \sum_{j=1}^{i} 2^{n-j} S(i, j)(n)_j.$$

Hence, we have

$$S_p(n) = \sum_{k=0}^{n} \left(\sum_{i=0}^{2p} \binom{2p}{i} (-2k)^i n^{2p-i} \right) \binom{n}{k}$$

$$= \sum_{i=0}^{2p} (-2)^i \binom{2p}{i} n^{2p-i} \left(\sum_{k=0}^{n} \binom{n}{k} k^i \right)$$

$$= 2^n \sum_{i=0}^{2p} (-1)^i \binom{2p}{i} n^{2p-i} \left(\sum_{j=1}^{i} 2^{i-j} S(i, j)(n)_j \right)$$

Moreover,

$$M_p(n) = 2^{-n} S_p(n) = \sum_{i=0}^{2p} (-1)^i \binom{2p}{i} \left(\sum_{j=1}^{i} 2^{i-j} S(i, j) n^{2p-i}(n)_j \right).$$

Faà di Bruno's formula generalizes the chain rule to higher derivatives. Let $|A|$ be the cardinality of the set A. The formula has a "combinatorial" form

$$D^n(f \circ g)(x) = \sum_{\pi \in \Pi} D^{|\pi|} f(g(x)) \cdots \prod_{B \in \pi} D^{|B|}(x),$$

where π runs through the set Π of all partitions of the set $\{1, 2, \ldots, n\}$, $B \in \pi$ means the variable B runs through the list of all of the blocks of the partition π. There is a nice explanation via an explicit example in https://en.wikipedia.org/wiki/Fa_di_Bruno\%27s_formula, which also contains a memorizable scheme for $1 \le n \le 4$.

1.14   A fractional part sum with Euler's constant

Problem 11206 (Proposed by M. Ivan and A. Lupas, 113(2), 2006). Find

$$\lim_{n \to \infty} \frac{1}{n} \sum_{k=1}^{n} \left\{ \frac{n}{k} \right\}^2,$$

where $\{x\} = x - \lfloor x \rfloor$ denotes the fractional part of x.

Discussion.
Note that

$$\frac{1}{n} \sum_{k=1}^{n} \left\{ \frac{n}{k} \right\}^2 = \sum_{k=1}^{n} \left\{ \frac{1}{k/n} \right\}^2 \cdot \frac{1}{n}$$

is a Riemann sum for $\int_0^1 \{1/x\}^2 \, dx$. Since $\{1/x\}^2$ is bounded and continuous on $(0, 1)$ except at $\{1/n : n \in \mathbb{N}\}$, it is Riemann integrable. In this way, we are led to evaluate the integral $\int_0^1 \{1/x\}^2 \, dx$.

Solution.
We show that

$$\lim_{n \to \infty} \frac{1}{n} \sum_{k=1}^{n} \left\{ \frac{n}{k} \right\}^2 = \int_0^1 \left\{ \frac{1}{x} \right\}^2 \, dx = \ln(2\pi) - \gamma - 1.$$

For $x \in \left(\frac{1}{n+1}, \frac{1}{n} \right)$,

$$\left\{ \frac{1}{x} \right\} = \frac{1}{x} - \left\lfloor \frac{1}{x} \right\rfloor = \frac{1}{x} - n.$$

Then

$$\int_0^1 \left\{ \frac{1}{x} \right\}^2 \, dx = \sum_{n=1}^{\infty} \int_{1/(n+1)}^{1/n} \left(\frac{1}{x} - n \right)^2 \, dx$$

$$= \sum_{n=1}^{\infty} \left(1 - 2n \ln \frac{n+1}{n} + \frac{n}{n+1} \right).$$

Let the N-partial sum be S_N. Rewrite $1 + \frac{n}{n+1} = 2 - \frac{1}{n+1}$ and note that

$$\sum_{n=1}^{N} n \ln \frac{n+1}{n} = \sum_{n=1}^{N} (n \ln(n+1) - (n-1) \ln n - \ln n)$$

$$= N \ln(N+1) - \sum_{n=1}^{N} \ln n = N \ln(N+1) - \ln(N!).$$

We obtain that

$$S_N = 2N - 2N \ln(N+1) + 2 \ln(N!) - H_{N+1} + 1,$$

where $H_n = \sum_{k=1}^{n} 1/k$ is the kth harmonic number. By Stirling's formula, we have

$$\ln(N!) = \left(N + \frac{1}{2}\right) \ln N - N + \frac{1}{2} \ln(2\pi) + O(1/N).$$

This, in addition to the fact that $H_N = \ln N + \gamma + O(1/N)$, where γ is Euler's constant, implies that

$$\begin{aligned} S_N &= 2N \ln\left(\frac{N}{N+1}\right) + (\ln N - H_N) + \ln(2\pi) + 1 + O(1/N) \\ &= \ln(2\pi) - \gamma - 1 + O(1/N). \end{aligned}$$

Letting $N \to \infty$ yields the claimed limit. $\qquad\qquad\qquad\qquad\qquad\square$

Remark. One well-known fractional part integral is

$$\int_0^1 \left\{\frac{1}{x}\right\} dx = 1 - \gamma.$$

One may wonder: Are there other similar formulas? Is it possible to extend these equalities from one dimension to multiple dimensions? Furdui's Book [43] provides a beautiful collection of these types of problems and solutions. Their solutions cover a host of mathematical topics including integrals, infinite series, exotic constants, and special functions.

It is interesting to see that we can explicitly represent $\int_0^1 \{1/x\}^k \, dx$ in terms of the Riemann zeta function for any positive integer k. To this end, by the substitution $u = 1/x$, we have

$$\begin{aligned} \int_0^1 \left\{\frac{1}{x}\right\}^k dx &= \int_1^\infty \frac{1}{u^2} \{u\}^k \, du \\ &= \sum_{n=1}^{\infty} \int_n^{n+1} \frac{(u-n)^k}{u^2} \, du \quad (\text{let } y = u - n) \\ &= \sum_{n=1}^{\infty} \int_0^1 \frac{y^k}{(n+y)^2} \, dy \\ &= \int_0^1 y^k \left(\sum_{n=1}^{\infty} \frac{1}{(y+n)^2}\right) dy. \end{aligned}$$

In view of

$$\frac{1}{(y+n)^2} = \int_0^\infty te^{-(y+n)t}\, dt,$$

we have

$$\sum_{n=1}^\infty \frac{1}{(y+n)^2} = \int_0^\infty \left(\sum_{n=1}^\infty te^{-(y+n)t} \right) dt = \int_0^\infty \frac{te^{-yt}}{e^t - 1}\, dt,$$

and so

$$\int_0^1 \left\{ \frac{1}{x} \right\}^k dx = \int_0^\infty \frac{t}{e^t - 1} \left(\int_0^1 y^k e^{-yt}\, dy \right) dt.$$

Since

$$\int_0^1 y^k e^{-yt}\, dy = k! e^{-t} \sum_{i=1}^\infty \frac{t^{i-1}}{(k+i)!},$$

we finally obtain

$$\int_0^1 \left\{ \frac{1}{x} \right\}^k dx = k! \sum_{i=1}^\infty \frac{i!}{(k+i)!} (\zeta(i+1) - 1)$$

as expected.

Here are three more Monthly problems for additional practice:

1. **Problem 11637** (Proposed by O. Furdui, 119(4), 2012). Let $m \geq 1$ be a nonnegative integer. Prove that

$$\int_0^1 \left\{ \frac{1}{x} \right\}^m x^m\, dx = 1 - \frac{1}{m+1} \sum_{k=1}^m \zeta(k+1)$$

   where ζ is the Riemann zeta function.

2. **Problem 12031** (Proposed by O. Furdui, 125(3), 2018).

   (a) Prove

$$\int_0^1 \int_0^1 \left\{ \frac{x}{1-xy} \right\} dx\,dy = 1 - \gamma,$$

   where $\{a\}$ denotes the fractional part of a, and γ is Euler's constant.

   (b) Let k be a nonnegative integer. Prove

$$\int_0^1 \int_0^1 \left\{ \frac{x}{1-xy} \right\}^k dx\,dy = \int_0^1 \left\{ \frac{1}{x} \right\}^k dx.$$

3. **Problem 12181** (Proposed by G. Apostolopoulos, 127(5), 2020). Prove

$$\sum_{k=2}^\infty \frac{1}{k} \int_0^1 \left\{ \frac{1}{\sqrt[k]{x}} \right\} dx = \gamma,$$

   where γ is Euler's constant.

1.15   A Putnam/Monthly limit problem

Problem 11837 (Proposed by I. Pinelis, 124(1), 2017). Let $a_0 = 1$, and for $n \geq 0$ let $a_{n+1} = a_n + e^{-a_n}$. Let $b_n = a_n - \ln n$. For $n \geq 0$, show that $0 < b_{n+1} < b_n$; also show that $\lim_{n\to\infty} b_n = 0$.

Discussion.
This problem is a modification of **Putnam Problem 2012 B-4**. The original problem asked whether b_n has a finite limit as $n \to \infty$. Two published solutions indeed show that b_n converges to 0 as $n \to \infty$. Please refer [57] and the solution by Kedlaya and Ng in http://kskedlaya.org/putnam-archive/2012s.pdf Here we give another proof based on the estimates of two Riemann sums.

Solution.
Note that a_n is strictly increasing by its definition. Partition $[a_0, a_n]$ with subintervals $[a_{k-1}, a_k], k = 1, 2, \ldots, n$. Since e^x is increasing, from $a_k - a_{k-1} = e^{-a_{k-1}}$, the Riemann sum with the left-endpoint rule gives

$$e^{a_n} - e = \int_{a_0}^{a_n} e^x \, dx = \sum_{k=1}^n \int_{a_{k-1}}^{a_k} e^x \, dx$$

$$> \sum_{k=1}^n e^{a_{k-1}} (a_k - a_{k-1}) = \sum_{k=1}^n 1 = n. \qquad (1.24)$$

It follows that $a_n > \ln(e + n) > \ln(n + 1) > \ln n$ for all $n \in \mathbb{N}$. Since

$$0 < a_{n+1} - a_n = e^{-a_n} < \frac{1}{n+1} < \int_n^{n+1} \frac{dx}{x} = \ln(n + 1) - \ln n,$$

this shows that $0 < b_{n+1} < b_n$. Thus b_n is decreasing, and so b_n has a finite limit as $n \to \infty$. We now show that $b_n \to 0$ as $n \to \infty$. It suffices to show that e^{a_n}/n converges to 1 as $n \to \infty$. By (1.24), we have

$$\frac{e^{a_n}}{n} > 1 + \frac{e}{n}. \qquad (1.25)$$

On the other hand, the Riemann sum with the right-end point yields

$$e^{a_n} - e = \int_{a_0}^{a_n} e^x \, dx = \sum_{k=1}^n \int_{a_{k-1}}^{a_k} e^x \, dx$$

$$< \sum_{k=1}^n e^{a_k} (a_k - a_{k-1}) = \sum_{k=1}^n e^{a_k - a_{k-1}} = \sum_{k=1}^n e^{e^{-a_{k-1}}}.$$

Notice that $e^{a_{k-1}} > (k - 1) + e > k$ from (1.24). Thus, $e^{-a_{k-1}} < 1/k$ and

$$e^{a_n} - e \leq \sum_{k=1}^n e^{1/k} \leq e + \int_1^n e^{1/x} \, dx.$$

Hence,

$$\frac{e^{a_n}}{n} \leq \frac{2e}{n} + \frac{1}{n} \int_1^n e^{1/x} \, dx. \tag{1.26}$$

Notice that, for example, using the Stolz-Cesàro theorem,

$$\lim_{n \to \infty} \frac{1}{n} \int_1^n e^{1/x} \, dx = 1.$$

Now, $e^{a_n}/n \to 1$ as $n \to \infty$ follows from (1.25), (1.26), and the squeeze theorem. $\square$

Remark. One reviewer of this book offered an alternative solution based on the following inequality

$$0 < a_n - \ln n < \ln \frac{(\sqrt{n} + e)^2}{n}.$$

The proof is a nice application of the mean value theorem. As an exercise, the interested reader may derive this inequality.

In general, this problem is an example of the principle that one can often predict the asymptotic behavior of a recursive sequence by studying solutions of a sufficiently similar-looking differential equation. In this proposed problem, as Kedlaya and Ng suggested in `http://kskedlaya.org/putnam-archive/` `2012s.pdf`: We can start with the equation $a_{n+1} - a_n = e^{-a_n}$, then replace a_n with function $y(x)$ and replace the difference $a_{n+1} - a_n$ with the derivative $y'(x)$ to obtain the differential equation $y' = e^{-y}$, which has the solution $y = \ln x$.

2

Infinite Series

We select 12 infinite series problems that have appeared in the Monthly and try to present solutions in a cohesive and engaging way. Through various solutions and proofs, we illustrate how problem-solving evolve over time — from the specific to the general, from the simplified scenario to the theoretical framework, from the concrete to the abstract. You may find some problems ultimately reach results related to current research. Along the way, we hope you can learn how to solve hard problems and the motivation behind them, and expand the breadth and depth of your mathematical knowledge.

2.1   Wilf wants us thinking rationally

Problem 11068 (Proposed by H. Wilf, 111(3), 2004). For a rational number x that equals a/b in lowest terms, let $f(x) = ab$.

(a) Show that

$$\sum_{x \in \mathbb{Q}^+} \frac{1}{f^2(x)} = \frac{5}{2},$$

where the sum extends over all positive rationals.

(b) More generally, exhibit an infinite sequence of distinct rational exponents s such that $\sum_{x \in \mathbb{Q}^+} f^{-s}(x)$ is rational.

Discussion.
Note that every positive rational number $x = a/b$ can be viewed as a ordered pair (a, b). Let $d = \gcd(a, b)$. Then there exists a unique ordered pair (p, q) such that $\gcd(p, q) = 1, a = dp, b = dq$. By the definition, we have

$$ab = d^2(pq) \quad \text{and} \quad f(a/b) = pq.$$

It is easy to verify that $(a, b) \mapsto (p, q, d)$ is bijective. Collecting together those a, b in $\sum_{a,b=1}^{\infty} \frac{1}{(ab)^2}$ which have the same $\gcd(a, b)$ yields

$$\sum_{a,b=1}^{\infty} \frac{1}{(ab)^2} = \left(\sum_{x \in \mathbb{Q}^+} \frac{1}{f^2(x)} \right) \cdot \left(\sum_{d=1}^{\infty} \frac{1}{d^4} \right).$$

Thus we see the rationality of the required series depends on the ratio of the Riemann zeta function values.

Solution.

(a) Based on the above discussion, for any $s > 1$, we have

$$\left(\sum_{a=1}^{\infty} \frac{1}{a^s}\right) \cdot \left(\sum_{b=1}^{\infty} \frac{1}{b^s}\right) = \sum_{a,b=1}^{\infty} \frac{1}{(ab)^s}$$

$$= \sum_{d=1}^{\infty} \left(\sum_{p,q=1,\, \gcd(p,q)=1}^{\infty} \frac{1}{d^{2s}(pq)^s}\right)$$

$$= \sum_{d=1}^{\infty} \frac{1}{d^{2s}} \cdot \left(\sum_{p,q=1,\, \gcd(p,q)=1}^{\infty} \frac{1}{(pq)^s}\right)$$

$$= \sum_{d=1}^{\infty} \frac{1}{d^{2s}} \cdot \sum_{x \in \mathbb{Q}^+} \frac{1}{f^s(x)}.$$

In terms of the Riemann zeta function $\zeta(s) = \sum_{k=1}^{\infty} 1/k^s$, it follows that

$$\sum_{x \in \mathbb{Q}^+} \frac{1}{f^s(x)} = \frac{\zeta^2(s)}{\zeta(2s)}. \tag{2.1}$$

In particular, since $\zeta(2) = \pi^2/6$ and $\zeta(4) = \pi^4/90$, From (2.1) we find that

$$\sum_{x \in \mathbb{Q}^+} \frac{1}{f^2(x)} = \frac{(\pi^2/6)^2}{(\pi^4/90)} = \frac{5}{2}.$$

Here are a few more exponents such that the corresponding series have rational values:

$$\sum_{x \in \mathbb{Q}^+} \frac{1}{f^4(x)} = \frac{7}{6}, \quad \sum_{x \in \mathbb{Q}^+} \frac{1}{f^6(x)} = \frac{715}{691}, \quad \sum_{x \in \mathbb{Q}^+} \frac{1}{f^8(x)} = \frac{7297}{7234}.$$

(b) In view of the above results for exponents $s = 4, 6, 8$, it is natural to think of even exponents. Recall Euler's famous result on the Riemann zeta function at even positive integers:

$$\zeta(2n) = (-1)^{n-1} \frac{2^{2n-1} B_{2n} \pi^{2n}}{(2n)!}, \tag{2.2}$$

where B_k is the kth Bernoulli number. Let $s = 2n$. By (2.1) and (2.2), we find that

$$\sum_{x \in \mathbb{Q}^+} \frac{1}{f^{2n}(x)} = \frac{\zeta^2(2n)}{\zeta(4n)} = \frac{1}{2}\binom{4n}{2n}\frac{B_{2n}^2}{|B_{4n}|},$$

which is rational because B_k is rational for all $k \in \mathbb{N}$. ☐

Remark. Since there is no closed form similar to (2.2) for $\zeta(2n+1)$, it is unknown whether there exists any other rational number $s > 1$ such that $\zeta^2(s)/\zeta(2s)$ is rational.

The question whether $\zeta(2n+1)$ is irrational has been asked since the time of Euler. For almost 200 years there had been no progress until 1978 when Apéry [9] proved that $\zeta(3)$ is irrational. Despite considerable effort we still know very little about the irrationality of $\zeta(2n+1)$ for $n \geq 2$. Apéry, who used the series for $\zeta(3)$ below in his proof of irrationality of $\zeta(3)$, suggested representing $\zeta(n)$ via the series involving the central binomial coefficients. For example,

$$\zeta(2) = 3 \sum_{n=1}^{\infty} \frac{1}{n^2 \binom{2n}{n}},$$

$$\zeta(3) = \frac{5}{2} \sum_{n=1}^{\infty} \frac{(-1)^{n-1}}{n^3 \binom{2n}{n}},$$

$$\zeta(4) = \frac{36}{17} \sum_{n=1}^{\infty} \frac{1}{n^4 \binom{2n}{n}}.$$

But this kind of Apéry-like analogous formula fails at

$$\zeta(5) = 2 \cdot \sum_{n=1}^{\infty} \frac{(-1)^{n-1}}{n^5 \binom{2n}{n}} - \frac{5}{2} \sum_{n=1}^{\infty} \frac{(-1)^{n-1}}{n^3 \binom{2n}{n}} \sum_{k=1}^{n-1} \frac{1}{k^2}.$$

Thus it would seem we can't find $\zeta(2n+1)$ in terms of any numbers whose names we have already known. They are actually "new" numbers!

Another path to follow is to replace the function $f(x)$. For example, let $f(x) = ab(a+b)$ if $\gcd(a, b) = 1$. Then

$$\sum_{a,b=1}^{\infty} \frac{1}{ab(a+b)} = \sum_{d=1}^{\infty} \left(\sum_{p,q=1,\, \gcd(p,q)=1}^{\infty} \frac{1}{d^3 pq(p+q)} \right)$$

$$= \sum_{d=1}^{\infty} \frac{1}{d^3} \cdot \left(\sum_{p,q=1,\, \gcd(p,q)=1}^{\infty} \frac{1}{pq(p+q)} \right)$$

$$= \zeta(3) \cdot \sum_{x \in \mathbb{Q}^+} \frac{1}{f(x)}.$$

On the other hand, we have

$$\sum_{a,b=1}^{\infty} \frac{1}{ab(a+b)} = \sum_{a,b=1}^{\infty} \frac{1}{ab} \int_0^1 x^{a+b-1}\, dx$$

$$= \int_0^1 \left(\sum_{a=1}^{\infty} \frac{x^a}{a}\right)\left(\sum_{b=1}^{\infty} \frac{x^b}{b}\right) \frac{dx}{x}$$

$$= \int_0^1 \frac{\ln^2(1-x)}{x}\, dx = \int_0^{\infty} \frac{u^2 e^{-u}}{1-e^{-u}}\, du \quad (\text{using } x = 1 - e^{-u})$$

$$= \int_0^{\infty} u^2 \left(\sum_{k=1}^{\infty} e^{-ku}\right) du = \sum_{k=1}^{\infty} \frac{2}{k^3} = 2\zeta(3).$$

Here the interchange of integration and summation is justified by the positivity of the summands. Thus we obtain

$$\sum_{x \in \mathbb{Q}^+} \frac{1}{f(x)} = 2,$$

a rational number!

It is interesting to see that $f(x) = ab$ and $f(x) = ab(a+b)$ both are just special cases of the following Witten zeta function [94]

$$\mathcal{W}(r,s,t) := \sum_{a,b=1}^{\infty} \frac{1}{a^r b^s (a+b)^t},$$

which is a function associated to a root system that encodes the degrees of the irreducible representations of the corresponding Lie group. Witten's work has launched extensive research to obtain exact values of $\mathcal{W}(r,s,t)$ in terms of the Riemann zeta function. Some typical evaluations include

$$\mathcal{W}(2n, 2n, 2n) = \frac{4}{3} \sum_{k=0}^{n} \binom{4n-2k-1}{2n-1} \zeta(2k)\zeta(6n-2k), \quad (2.3)$$

$$\mathcal{W}(2n+1, 2n+1, 2n+1) = -4 \sum_{k=0}^{n} \binom{4n-2k+1}{2n} \zeta(2k)\zeta(6n-2k+3).$$

$$(2.4)$$

Let $f(x) = a^r b^s (a+b)^t$ if $\gcd(a,b) = 1$. Then

$$\mathcal{W}(r,s,t) = \zeta(r+s+t) \cdot \sum_{x \in \mathbb{Q}^+} \frac{1}{f(x)}.$$

From (2.2) and (2.3) we see that $\sum_{x \in \mathbb{Q}^+} \frac{1}{f(x)}$ is rational when $r = s = t = 2n$. But, note that

$$\mathcal{W}(1,1,3) = \int_0^1 \frac{\ln^2 x \ln^2(1-x)}{x} = -2\zeta(2)\zeta(3) + 4\zeta(5),$$

we don't know whether

$$\sum_{p,q=1,\,\gcd(p,q)=1}^{\infty} \frac{1}{pq(p+q)^3} = 4 - 2\frac{\zeta(2)\zeta(3)}{\zeta(5)}$$

is rational.

Here we skim the terms in Riemann zeta function to produce a rational sum. By Riemann's theorem on conditionally convergent series, we can rearrange a conditional convergent series that sum to any rational number (indeed, any number!) we wish. For example, with the alternating harmonic series, we have

$$1 - \frac{1}{2} - \frac{1}{4} - \frac{1}{6} - \frac{1}{8} + \frac{1}{3} - \frac{1}{10} - \frac{1}{12} - \frac{1}{14} - \frac{1}{16} + \cdots = 0.$$

In general, as early as 1953, in the following Monthly problem, Klamkin asked to characterize the rational sum by rearrangement. We leave it to the reader for additional practice.

Problem 4552 (Proposed by M. S. Klamkin, 60(7), 1953). What derangement of terms of

$$\sum_{n=1}^{\infty} \frac{(-1)^{n+1}}{n}$$

will produce a sum which is rational?

2.2   Old wine in a new bottle

Problem 4305 (Proposed by H. F. Sandham, 55(7), 1948). Prove that

$$1 + \left(\frac{1+1/2}{2}\right)^2 + \left(\frac{1+1/2+1/3}{3}\right)^2 + \left(\frac{1+1/2+1/3+1/4}{4}\right)^2 + \cdots = \frac{17\pi^4}{360}.$$

Discussion.
This problem was initially published as the Monthly Advanced Problem in 1948. Let H_n be the nth harmonic number. The stated identity can be rewritten in the form nowadays they are called *Euler sums*:

$$S := \sum_{n=1}^{\infty} \frac{H_n^2}{n^2} = \frac{17\pi^4}{360} = \frac{17}{4}\zeta(4). \qquad (2.5)$$

This identity apparently remained unnoticed until 1993 when Au-Yeung, an undergraduate at the University of Waterloo, numerically rediscovered (2.5). Shortly thereafter it was rigorously proven true by Borwein and Borwein in [16]. This empirical result launched a fruitful search for Euler sums through a profusion of methods: Combinatorial, analytic, and algebraic. Here we give four proofs to (2.5). The first algebraic proof is a modification of the original 1950 published solution by Kneser [58]. He was the solo solver beside the Proposer. Applying the Riemann zeta function and the approach we used in the previous section makes the original proof more compact. The second solution is based on logarithmic integrals and Abel's summation formula. The third solution uses Parseval's identity in Fourier analysis. In contrast to Borweins' original complex-variable proof in [16], here we present a real-variable proof. Bearing in mind Hadamard's dictum ([52], p. 123) "The shortest and best way between two truths of the real domain often passes through the imaginary one." We give the final proof by using complex residue theory.

Before proceeding the proofs, we give a brief historical account of Euler sums. In response to a letter from Goldbach in 1742, for integers $m \geq 1$ and $n \geq 2$, Euler [37] studied sums of the form

$$S(m,n) := \sum_{k=1}^{\infty} \frac{1}{(k+1)^n} \left(1 + \frac{1}{2} + \cdots + \frac{1}{k} \right)^m = \sum_{k=1}^{\infty} \frac{H_k^m}{(k+1)^n}$$

and established the following beautiful formula

$$S(1,n) = \frac{n}{2}\zeta(n+1) - \frac{1}{2}\sum_{k=1}^{n-2} \zeta(k+1)\zeta(n-k),$$

or equivalently,

$$\sum_{k=1}^{\infty} \frac{H_k}{k^n} = \frac{n+2}{2}\zeta(n+1) - \frac{1}{2}\sum_{k=1}^{n-2} \zeta(k+1)\zeta(n-k). \qquad (2.6)$$

However, over the centuries progress on evaluating $S(m,n)$ for $m \geq 2$ (often called *nonlinear Euler sums*) has been minimal. For example, using the method of partial fractions, Nielsen (1906) found the formula for $S(m,n)$ when $m = n$ [71]; Georghiou and Philippou (1983) established [45]

$$\sum_{k=1}^{\infty} \frac{H_k^{(2)}}{k^{2n+1}} = \zeta(2)\zeta(2n+1) - \frac{(n+2)(2n+1)}{2}\zeta(2n+3)$$

$$+ 2\sum_{j=2}^{n+1} (j-1)\zeta(2j-1)\zeta(2n+4-2j), \qquad n \geq 1,$$

where $H_k^{(2)} = 1 + \frac{1}{2^2} + \cdots + \frac{1}{k^2}$.

Until 1993, especially after the publication of the very insightful solution to this Monthly problem [16], Borweins and their co-researchers have generated a revival of interest in Euler sums. Since then, a large class of Euler sums and their variations have been explicitly evaluated in terms of the Riemann zeta function values. The derivations make a very pleasant journey through diverse topics in classical analysis include generating functions, special integrals, and special functions. For example, Borwein and Bradley [21] collected 32 intriguing proofs of $S(1,2) = \zeta(3)$. For reasonably detailed accounts of the recent developments of Euler sums, please refer the Chapter 3 in Borwein-Bailey-Grigensohn's book [19] .

Solution I — The initial proof by Kneser.
First, by partial fractions, telescoping yields

$$\sum_{i=1}^{\infty} \frac{1}{i(i+n)} = \sum_{i=1}^{\infty} \frac{1}{n}\left(\frac{1}{i} - \frac{1}{i+n}\right) = \frac{H_n}{n}.$$

Then

$$
\begin{aligned}
S &= \sum_{n=1}^{\infty}\left(\sum_{i=1}^{\infty} \frac{1}{i(i+n)}\right)^2 = \sum_{n=1}^{\infty}\left(\sum_{i,j=1}^{\infty} \frac{1}{ij(i+n)(j+n)}\right) \\
&= \sum_{n=1}^{\infty}\left(\sum_{i} \frac{1}{i^2(i+n)^2} + 2\sum_{1\leq i<j} \frac{1}{ij(i+n)(j+n)}\right) \quad (\text{let } j = i+k) \\
&= \sum_{1\leq i<j} \frac{1}{i^2 j^2} + 2\sum_{n=1}^{\infty}\left(\sum_{i,k=1} \frac{1}{i(i+k)(i+k+n)(i+n)}\right) \\
&= \frac{1}{2}\left[\left(\sum_{i=1}^{\infty} \frac{1}{i^2}\right)^2 - \sum_{i=1}^{\infty} \frac{1}{i^4}\right] + 2\sum_{i,k,n=1} \frac{(i+k)(i+n) - i(i+k+n)}{ikn(i+k)(i+k+n)(i+n)} \\
&= \frac{1}{2}(\zeta^2(2) - \zeta(4)) + 2\sum_{i,k,n=1} \frac{1}{ikn(i+k+n)} - 2\sum_{i,k,n=1} \frac{1}{kn(i+k)(i+n)} \\
&= \frac{1}{2}(\zeta^2(2) - \zeta(4)) + 2\sum_{i,k,n=1} \frac{1}{ikn(i+k+n)} - 2S.
\end{aligned}
$$

It follows that, using $\zeta^2(2) = \frac{5}{2}\zeta(4)$,

$$S = \frac{1}{4}\zeta(4) + \frac{2}{3}\sum_{i,k,n=1} \frac{1}{ikn(i+k+n)}. \tag{2.7}$$

As we did in the remark of the previous section, we calculate the series in

(2.7) as follows:

$$\sum_{i,k,n=1} \frac{1}{ikn(i+k+n)} = \sum_{i,k,n=1} \int_0^1 \frac{x^{i+k+n-1}}{ikn} \, dx$$

$$= \int_0^1 \left(\sum_{n=1}^\infty \frac{x}{n} \right)^3 \frac{dx}{x} = -\int_0^1 \frac{\ln^3(1-x)}{x} \, dx$$

$$= -\int_0^1 \frac{\ln^3 u}{1-u} \, du = -\sum_{n=1}^\infty \int_0^1 u^{n-1} \ln^3 u \, du$$

$$= \sum_{n=1}^\infty \frac{6}{n^4} = 6\zeta(4).$$

Here the interchange the integration and summation is justified by the positivity of the summands. This, together with (2.7), concludes

$$S = \frac{1}{4}\zeta(4) + 4\zeta(4) = \frac{17}{4}\zeta(4) = \frac{17\pi^4}{360}.$$

□

Solution II — By logarithmic integrals and Abel's summation formula.

We first establish three auxiliary identities. Let $n \in \mathbb{N}$. Then

(a) $\int_0^1 x^{n-1} \ln(1-x) \, dx = -\frac{H_n}{n}$.

(b) $\int_0^1 x^{n-1} \ln^2(1-x) \, dx = \frac{2}{n} \sum_{k=1}^n \frac{H_k}{k} = \frac{H_n^2 + H_n^{(2)}}{n}$, where $H_n^{(2)} = \sum_{k=1}^n 1/k^2$.

(c) $\sum_{n=1}^\infty \frac{H_n^{(2)}}{n^2} = \frac{7}{4}\zeta(4)$.

For **(a)**, we compute

$$\int_0^1 x^{n-1} \ln(1-x) \, dx = \int_0^1 x^{n-1} \left(\int_0^x -\frac{dt}{1-t} \right) dx$$

$$= -\int_0^1 \frac{1}{1-t} \left(\int_t^1 x^{n-1} \, dx \right) dt$$

$$= -\frac{1}{n} \int_0^1 \frac{1-t^n}{1-t} \, dt$$

$$= -\frac{1}{n} \int_0^1 (1 + t + \cdots + t^{n-1}) \, dt = -\frac{H_n}{n}.$$

This proves the claimed identity. Similarly, we have

$$
\begin{aligned}
\int_0^1 x^{n-1} \ln^2(1-x)\, dx &= \int_0^1 x^{n-1} \left(\int_0^x -2\frac{\ln(1-t)\, dt}{1-t} \right) dx \\
&= -2\int_0^1 \frac{\ln(1-t)}{1-t} \left(\int_t^1 x^{n-1}\, dx \right) dt \\
&= -\frac{2}{n} \int_0^1 \ln(1-t)\frac{1-t^n}{1-t}\, dt \\
&= \frac{-2}{n} \int_0^1 \ln(1-t)(1+t+t^2+\cdots+t^{n-1})\, dt \\
&= \frac{2}{n} \sum_{k=1}^n \frac{H_k}{k},
\end{aligned}
$$

where (**a**) is used in the last equation above. Next, we have

$$
H_n^2 = \sum_{i=1}^n \sum_{j=1}^n \frac{1}{i\,j} = 2\sum_{i=1}^n \sum_{j=1}^i \frac{1}{i\,j} - H_n^{(2)} = 2\sum_{i=1}^n \frac{H_i}{i} - H_n^{(2)},
$$

from which (**b**) follows. To prove (**c**), recall the limit version of Abel's summation formula

$$
\sum_{n=1}^\infty a_n b_n = \lim_{n\to\infty}(A_n b_{n+1}) + \sum_{n=1}^\infty A_n(b_n - b_{n+1}),
$$

where $A_n = \sum_{k=1}^n a_k$. Letting $a_n = 1/n^2, b_n = H_n^{(2)}$ yields

$$
\begin{aligned}
T &:= \sum_{n=1}^\infty \frac{1}{n^2} H_n^{(2)} \\
&= \lim_{n\to\infty} H_n^{(2)} H_{n+1}^{(2)} - \sum_{n=1}^\infty \frac{1}{(n+1)^2} H_n^{(2)} \\
&= \zeta^2(2) - \sum_{n=1}^\infty \frac{1}{(n+1)^2} H_{n+1}^{(2)} + \sum_{n=1}^\infty \frac{1}{(n+1)^4} \\
&= \zeta^2(2) - T + \zeta(4)
\end{aligned}
$$

from which (**c**) follows because $\zeta^2(2) = \frac{5}{2}\zeta(4)$.

Now we are ready to prove (2.5). By (**b**) we have

$$
\int_0^1 \frac{1}{n} x^{n-1} \ln^2(1-x)\, dx = \frac{1}{n^2} H_n^2 + \frac{1}{n^2} H_n^{(2)}
$$

and

$$
\sum_{n=1}^\infty \int_0^1 \frac{1}{n} x^{n-1} \ln^2(1-x)\, dx = \sum_{n=1}^\infty \frac{1}{n^2} H_n^2 + \sum_{n=1}^\infty \frac{1}{n^2} H_n^{(2)}. \qquad (2.8)
$$

Note that

$$\sum_{n=1}^{\infty} \int_0^1 \frac{1}{n} x^{n-1} \ln^2(1-x)\, dx = \int_0^1 \sum_{n=1}^{\infty} \frac{1}{n} x^{n-1} \ln^2(1-x)\, dx$$

$$= -\int_0^1 \frac{\ln^3(1-x)}{x}\, dx = -\int_0^1 \frac{\ln^3 u}{1-u}\, du = 6\zeta(4).$$

Using (2.8) and (c), we finally find that

$$\sum_{n=1}^{\infty} \frac{H_n^2}{n^2} = 6\zeta(4) - \frac{7}{4}\zeta(4) = \frac{17}{4}\zeta(4) = \frac{17\pi^4}{360}.$$

$\square$

Solution III — By Fourier analysis.
Based on Parseval's identity, the desired Fourier series is

$$\sum_{n=1}^{\infty} \frac{H_n}{n} \sin(nx).$$

Observe that this series is the imaginary part of the power series $\sum_{n=1}^{\infty} \frac{H_n}{n} z^n$ with $z = e^{ix}$. Recall the generating function of H_n:

$$\sum_{n=1}^{\infty} H_n z^n = -\frac{\ln(1-z)}{1-z}.$$

This implies that

$$\sum_{n=1}^{\infty} \frac{H_n}{n} z^n = -\int_0^z \frac{\ln(1-t)}{t(1-t)}\, dt = \frac{1}{2} \ln^2(1-z) + \sum_{n=1}^{\infty} \frac{z^n}{n^2}$$

or

$$\sum_{n=1}^{\infty} \left(\frac{H_n}{n} - \frac{1}{n^2} \right) z^n = \frac{1}{2} \ln^2(1-z), \tag{2.9}$$

where the series converges for $|z| < 1$. Let $z = re^{i\theta}$ with $0 < r < 1, 0 < \theta \le \pi$. Then the imaginary part of the left-hand side of (2.9) becomes

$$\sum_{n=1}^{\infty} \left(\frac{H_n}{n} - \frac{1}{n^2} \right) r^n \sin n\theta$$

which, by Abel's limit theorem, tends to

$$\sum_{n=1}^{\infty} \left(\frac{H_n}{n} - \frac{1}{n^2} \right) \sin n\theta$$

as $r \to 1^-$. On the other hand, we have

$$\lim_{r \to 1^-} \operatorname{Im} \left(\frac{1}{2} \ln^2(1 - re^{i\theta}) \right) = \frac{1}{2}(\theta - \pi) \ln(2\sin(\theta/2)).$$

Thus, we obtain a superficially similar Fourier series

$$\frac{1}{2}(\theta - \pi) \ln(2\sin(\theta/2)) = \sum_{n=1}^{\infty} \left(\frac{H_n}{n} - \frac{1}{n^2} \right) \sin n\theta.$$

Applying Parseval's identity yields

$$\frac{1}{2\pi} \int_0^\pi (\theta - \pi)^2 \ln^2(2\sin(\theta/2)) \, d\theta = \sum_{n=1}^{\infty} \left(\frac{H_n}{n} - \frac{1}{n^2} \right)^2. \qquad (2.10)$$

Note that

$$\sum_{n=1}^{\infty} \left(\frac{H_n}{n} - \frac{1}{n^2} \right)^2 = \sum_{n=1}^{\infty} \frac{H_n^2}{n^2} - 2 \sum_{n=1}^{\infty} \frac{H_n}{n^3} + \zeta(4).$$

Letting $n = 3$ in (2.6) yields

$$\sum_{n=1}^{\infty} \frac{H_n}{n^3} = \frac{5}{4} \zeta(4).$$

We now see that (2.5) holds if

$$\frac{1}{2\pi} \int_0^\pi (\theta - \pi)^2 \ln^2(2\sin(\theta/2)) \, d\theta = \frac{11}{4} \zeta(4).$$

Equivalently, replacing θ by $\pi - \theta$ leads to

$$I := \frac{1}{2\pi} \int_0^\pi \theta^2 \ln^2(2\cos(\theta/2)) \, d\theta = \frac{11}{4} \zeta(4). \qquad (2.11)$$

In contrast to Borweins' original proof [16], here we give a real-variable proof of (2.11). Recall that

$$\ln(2\cos(\theta/2)) = \sum_{n=1}^{\infty} \frac{(-1)^{n+1}}{n} \cos n\theta.$$

We have

$$\ln^2(2\cos(\theta/2)) = \sum_{m,n=1}^{\infty} \frac{(-1)^{m+n}}{mn} \cos m\theta \cos n\theta.$$

Thus

$$I = \frac{1}{2\pi} \sum_{m,n=1}^{\infty} \frac{(-1)^{m+n}}{mn} \int_0^{\pi} \theta^2 \cos m\theta \cos n\theta \, d\theta$$

$$= \frac{1}{2\pi} \left(\sum_{n=1}^{\infty} \frac{1}{n^2} \int_0^{\pi} \theta^2 \cos^2 n\theta \, d\theta \right.$$

$$\left. + \sum_{m,n=1, m\neq n}^{\infty} \frac{(-1)^{m+n}}{mn} \int_0^{\pi} \theta^2 \cos m\theta \cos n\theta \, d\theta \right).$$

We compute

$$\sum_{n=1}^{\infty} \frac{1}{n^2} \int_0^{\pi} \theta^2 \cos^2 n\theta \, d\theta = \sum_{n=1}^{\infty} \frac{1}{n^2} \left(\frac{\pi^3}{6} + \frac{\pi}{4n^2} \right)$$

$$= \frac{\pi^3}{6} \zeta(2) + \frac{\pi}{4} \zeta(4) = \frac{11}{360} \pi^5.$$

Since

$$\int_0^{\pi} \theta^2 \cos m\theta \cos n\theta \, d\theta = \pi \left(\frac{(-1)^{m+n}}{(m+n)^2} + \frac{(-1)^{m-n}}{(m-n)^2} \right),$$

we have

$$\sum_{m,n=1, m\neq n}^{\infty} \frac{1}{mn(m+n)^2} = \sum_{m,n=1}^{\infty} \frac{1}{mn(m+n)^2} - \sum_{m=n=1}^{\infty} \frac{1}{mn(m+n)^2}$$

$$= \frac{1}{2} \sum_{m,n=1}^{\infty} \frac{(m+n)^2 - m^2 - n^2}{m^2 n^2 (m+n)^2} - \frac{1}{4} \zeta(4)$$

$$= \frac{1}{2} \left(\sum_{m,n=1}^{\infty} \frac{1}{m^2 n^2} - \sum_{m,n=1}^{\infty} \frac{1}{n^2(m+n)^2} \right.$$

$$\left. - \sum_{m,n=1}^{\infty} \frac{1}{m^2(m+n)^2} \right) - \frac{1}{4} \zeta(4)$$

$$= \frac{1}{2} \left(\sum_{m,n=1}^{\infty} \frac{1}{m^2 n^2} - \sum_{m>n} \frac{1}{n^2 m^2} - \sum_{m<n} \frac{1}{n^2 m^2} \right)$$

$$- \frac{1}{4} \zeta(4)$$

$$= \frac{1}{2} \sum_{m=n=1}^{\infty} \frac{1}{m^2 n^2} - \frac{1}{4} \zeta(4)$$

$$= \frac{1}{4} \zeta(4) = \frac{1}{360} \pi^4, \quad \text{and}$$

$$\sum_{m,n=1,m\neq n}^{\infty} \frac{1}{mn(m-n)^2} = \sum_{n=1}^{\infty}\sum_{m=1}^{n-1} \frac{1}{mn(m-n)^2} + \sum_{n=1}^{\infty}\sum_{m=n+1}^{\infty} \frac{1}{mn(m-n)^2}$$

$$= 2\sum_{n=1}^{\infty}\sum_{m=n+1}^{\infty} \frac{1}{mn(m-n)^2} \quad \text{(by symmetry)}$$

$$= 2\sum_{n=1}^{\infty}\sum_{k=1}^{\infty} \frac{1}{nk^2(n+k)^2} \quad \text{(let } k = m-n\text{)}$$

$$= 2\sum_{k=1}^{\infty} \frac{1}{k^2}\sum_{n=1}^{\infty} \frac{1}{n(n+k)^2}$$

$$= 2\sum_{k=1}^{\infty} \frac{1}{k^2}\sum_{n=1}^{\infty} \frac{1}{k}\left(\frac{1}{n} - \frac{1}{n+k}\right)$$

$$= 2\sum_{k=1}^{\infty} \frac{1}{k^3} H_k = \frac{\pi^4}{36}.$$

Combining the above results, in summary, we conclude

$$I = \frac{1}{2\pi}\left(\frac{11}{360}\pi^5 + \frac{1}{360}\pi^5 + \frac{1}{36}\pi^5\right) = \frac{11}{360}\pi^4 = \frac{11}{4}\zeta(4)$$

as desired. $\square$

Solution IV — By contour integral.

This approach is based on Flajolet and Salvy's paper [41]. Recall the digamma function $\psi(x)$ defined by

$$\psi(x) = \frac{d}{dx}\ln\Gamma(x) = -\gamma - \frac{1}{x} + \sum_{n=1}^{\infty}\left(\frac{1}{n} - \frac{1}{n+x}\right).$$

It satisfies the complement formula

$$\psi(x) - \psi(-x) = -\frac{1}{x} - \pi\cot\pi x.$$

Furthermore, it has an expansion at $x = 0$

$$\psi(-x) + \gamma = \frac{1}{x} - \zeta(2)x - \zeta(3)x^2 - \zeta(4)x^3 - \zeta(5)x^4 - \zeta(6)x^5 + O(x^6)$$

and at $x = n \in \mathbb{N}$

$$\psi(-x) + \gamma = \frac{1}{x-n} + H_n + \sum_{k=1}^{\infty}\left((-1)^k H_n^{(k+1)} - \zeta(k+1)\right)(x-n)^k.$$

This implies that

$$(\psi(-z) + \gamma)^3 = \frac{1}{(z-n)^3} + \frac{3H_n}{(z-n)^2} + \frac{3H_n^2}{z-n} - \frac{3H_n^{(2)}}{z-n} - \frac{\pi^2}{2(z-n)} + \cdots.$$

Let $f(z) = (\psi(-z) + \gamma)^3/z^2$. We have

$$\text{Res}(f(z))_{z=n} = \lim_{z \to n} \text{Res}\left(\frac{1}{(z-n)^3 z^2} + \frac{3H_n}{(z-n)^2 z^2} + \frac{3H_n^2}{(z-n)z^2}\right.$$
$$\left. - \frac{3H_n^{(2)}}{(z-n)z^2} - \frac{\pi^2}{2(z-n)z^2}\right)$$

$$= \frac{3}{n^4} - \frac{6H_n}{n^3} + \frac{3H_n^2}{n^2} - \frac{3H_n^{(2)}}{n^2} - \frac{\pi^2}{2}\frac{1}{n^2}$$

and

$$\text{Res}(f(z))_{z=0} = \lim_{z \to 0} \text{Res}\left(\frac{1}{z^5} - \frac{3\zeta(2)}{z^3} - \frac{3\zeta(3)}{z^2} + \frac{\pi^4}{20z}\right) = \frac{\pi^4}{20}.$$

Applying the residue theorem to $f(z)$, where O is a circle of arbitrarily large radius centered at the origin, we have

$$\frac{1}{2\pi i} \oint_O f(z)\, dz = 0.$$

This deduces

$$\sum_{n=1}^{\infty} \left(\frac{3}{n^4} - \frac{6H_n}{n^3} + \frac{3H_n^2}{n^2} - \frac{3H_n^{(2)}}{n^2} - \frac{\pi^2}{2}\frac{1}{n^2}\right) + \frac{\pi^4}{20} = 0.$$

Using $\sum_{n=1}^{\infty} H_n/n^3 = \frac{5}{4}\zeta(4)$ and $\sum_{n=1}^{\infty} H_n^{(2)}/n^2 = \frac{7}{4}\zeta(4)$, again, we find that

$$\sum_{n=1}^{\infty} \frac{H_n^2}{n^2} = \frac{17}{4}\zeta(4).$$

$\square$

Remark. The study of modifications and generalizations of Euler sums has also attracted the attention in the last two decades. We single out two topics based on author's interests. First, motivated by the established identities for the classical Euler sums and Ramanujan's constant $G(1)$ (see Problem 4 in this practice problem set), replacing H_n by

$$h_n = 1 + \frac{1}{3} + \cdots + \frac{1}{2n-1} \tag{2.12}$$

leads to a class of variant Euler sums

$$S(m,n) = \sum_{k=1}^{\infty} \frac{h_k^m}{k^n}.$$

Many closed formulas including

$$\sum_{k=1}^{\infty} \frac{h_k}{k^2} = \frac{7}{4}\zeta(3) \quad \text{and} \quad \sum_{k=1}^{\infty} \frac{h_k^2}{k^2} = \frac{45}{16}\zeta(4)$$

and the alternating versions

$$\sum_{k=1}^{\infty} (-1)^{k+1} \frac{h_k}{k} = \frac{3}{8} \zeta(2) \quad \text{and} \quad \sum_{k=1}^{\infty} (-1)^{k+1} \frac{h_k^2}{k} = \frac{7}{16} \zeta(3)$$

can be found in [28]. It seems that it is still an open question to find a closed form of $S(m, n)$ for all $m, n \geq 3$ in terms of the Riemann zeta function. We will list a few partial results in the independent study.

Second, Conway and Guy (1996) introduced the *hyperharmonic numbers* as follows [33]:

$$\mathcal{H}_n^{(1)} = H_n, \quad \mathcal{H}_n^{(k)} = \sum_{m=1}^{n} \mathcal{H}_m^{(k-1)} \quad \text{for } k > 1. \tag{2.13}$$

These numbers can be expressed by the binomial coefficients and the regular harmonic numbers:

$$\mathcal{H}_n^{(k)} = \binom{n+k-1}{k-1} (H_{n+k-1} - H_{k-1}).$$

The generating function is given by

$$\sum_{n=0}^{\infty} \mathcal{H}_n^{(k)} x^n = -\frac{\ln(1-x)}{(1-x)^k}.$$

Similar to Euler sums, it is interesting to determine the series in the form of

$$\mathcal{H}(k, m) := \sum_{n=1}^{\infty} \frac{\mathcal{H}_n^{(k)}}{n^m}.$$

For $k \geq 2$, since

$$\mathcal{H}_n^{(k)} \sim \frac{1}{(k-1)!} (n^{k-1} \ln n) \quad \text{as } n \to \infty,$$

it follows that $\mathcal{H}(k, m)$ converges if $m > k$.

We now conclude this section with some problems for additional practice and independent study.

1. **Fourier series.** Prove that

   (a) $\sum_{n=1}^{\infty} \frac{\sin n}{n} = \sum_{n=1}^{\infty} \left(\frac{\sin n}{n}\right)^2 = \frac{\pi-1}{2}$.

   (b) $\sum_{n=1}^{\infty} \frac{\sin^2 n}{n^4} = \frac{(\pi-1)^2}{6}$.

2. **Problem 4946** (Proposed by M. S. Klamkin, 68(1), 1961). Let $S_n = 1 + 1/2 + \cdots + 1/n$. Sum $\sum_{n=1}^{\infty} S_n/n!$.

3. **Problem 11885** (Proposed by C. I. Vălean, 123(1), 2016). Prove that

$$\sum_{p=1}^{\infty}\sum_{n=1}^{\infty}\sum_{m=1}^{\infty} \frac{1}{(m+n)^4 + ((m+n)(m+p))^2} = \frac{3}{2}\zeta(3) - \frac{5}{4}\zeta(4).$$

4. **Ramanujan's constant.** Ramanujan [12] claimed that, without a proof,

$$G(1) := \frac{1}{8}\sum_{k=1}^{\infty}\frac{h_k}{k^3} = \frac{\pi}{4}\sum_{k=0}^{\infty}\frac{(-1)^k}{(4k+1)^3} - \frac{\pi}{3\sqrt{3}}\sum_{k=0}^{\infty}\frac{1}{(2k+1)^3},$$

where $h_k = \sum_{i=1}^{k} 1/(2i-1)$. Can you prove or disprove it?

5. **Evaluate log-sine and log-cosine integrals in closed form.** Find the values of (m, n, x) to calculate the following integrals in closed form.

$$\mathrm{Lc}(m, n, x) = \int_0^x \theta^m \ln^n(2\cos(\theta/2))\, d\theta,$$

$$\mathrm{Ls}(m, n, x) = \int_0^x \theta^m \ln^n(2\sin(\theta/2))\, d\theta.$$

Some known results for your tests:

$$\int_0^{\pi/2} \theta \ln(2\sin(\theta/2))\, d\theta = \frac{35}{32}\zeta(3) - \frac{1}{2}\pi G,$$

$$\int_0^{\pi/3} \theta \ln^2(2\sin(\theta/2))\, d\theta = \frac{17}{6480}\pi,$$

where G denotes the Catalan constant defined by $G = \sum_{k=0}^{\infty}\frac{(-1)^k}{(2k+1)^2}$.

6. **Independent study** on $\mathcal{S}(m, n) = \sum_{k=1}^{\infty}\frac{h_k^m}{k^n}$, where $h_k = \sum_{i=1}^{k} 1/(2i-1)$.

(a) Show that

$$\sum_{k=1}^{\infty}(-1)^{k+1}\frac{h_k}{k^2} = \pi G - \frac{7}{4}\zeta(3),$$

where G is the Catalan constant.

(b) When n is even, show that

$$\mathcal{S}(1, n) = \frac{2^{n+1} - 1}{4}\zeta(n+1) - \frac{1}{2}\sum_{j=1}^{n/2-1}(2^{n-2j+1}-1)\zeta(2j)\zeta(n-2j+1).$$

In particular,

$$\sum_{k=1}^{\infty} \frac{h_k}{k^4} = \frac{31}{4}\zeta(5) - \frac{7}{2}\zeta(2)\zeta(3).$$

(c) For $n \geq 2$, let

$$J(n) := \frac{(-1)^{n-1}}{2^n(n-1)!} \int_0^1 \frac{\ln^{n-1}(1-x)}{x\sqrt{1-x}}\,dx.$$

Show that

$$J(n) = \frac{1}{2}\left(\zeta(n) + \eta(n)\right) = \sum_{k=1}^{\infty} \frac{1}{(2k-1)^n},$$

where η is the Dirichlet eta function, an alternating zeta function. i.e.,

$$\eta(x) := \sum_{k=1}^{\infty} \frac{(-1)^{k+1}}{k^x} = (1 - 2^{1-x})\zeta(x).$$

(d) Show that, for $n \geq 1$,

$$\sum_{k=1}^{\infty} \frac{h_k}{(2k-1)^{2n+1}} + \sum_{k=1}^{\infty} \frac{h_k}{(2k)^{2n+1}}$$

$$= \frac{1}{2}J(2n+2) + J(2n+1)\ln 2 - \sum_{i=1}^{n-1} \frac{2^{2i+1}-1}{2^{2n+2}}\zeta(2i+1)\zeta(2n-2i+1).$$

(e) Let $G(1)$ be Ramanujan's constant defined in Problem 4. Show that

$$G(1) = -\frac{1}{8}\int_0^1 \frac{\ln x}{x}\ln^2\left(\frac{1+x}{1-x}\right)\,dx$$

$$= \frac{3}{16}\zeta(4) - \frac{1}{4}\sum_{k=1}^{\infty}(-1)^{n+1}\frac{H_k^{(2)}}{(n+1)^2}.$$

(f) Show that

$$\sum_{k=1}^{\infty}(-1)^{n+1}\frac{H_k^{(2)}}{(n+1)^2} = \frac{65}{16}\zeta(4) - \frac{7}{2}\zeta(3)\ln 2 + \zeta(2)\ln^2 2$$

$$- \frac{1}{6}\ln^4 2 - 4\operatorname{Li}_4(1/2),$$

where $\operatorname{Li}_n(x) = \sum_{k=1}^{\infty} x^n/k^n$ denotes the polylogarithm function.

7. **Independent study** on hyperharmonic numbers.
   Let $\mathcal{H}_n^{(k)}$ be the hyperharmonic numbers defined by (2.13).

   (a) Recall the *Hurwitz zeta function* defined by

   $$\zeta(m, n) = \sum_{i=0}^{\infty} \frac{1}{(n+i)^m}.$$

   For $m \geq k + 1$, show that

   $$\sum_{n=1}^{\infty} \frac{\mathcal{H}_n^{(k)}}{n^m} = \sum_{n=1}^{\infty} \mathcal{H}_n^{(k-1)} \zeta(m, n).$$

   (b) $\sum_{n=1}^{\infty} \frac{\mathcal{H}_n^{(k)}}{n} x^n = \sum_{n=1}^{\infty} \frac{H_n}{n} x^n + \sum_{j=1}^{k-1} \left(\frac{1}{j} \sum_{n=0}^{\infty} \mathcal{H}_n^{(k)} x^n - \frac{1}{j^2} \sum_{n=0}^{\infty} \binom{n+j-1}{n} x^n \right).$

   (c) $\mathcal{H}(k, m) = \mathcal{H}(1, m) + \sum_{i=1}^{k-1} \left(\frac{1}{i} \mathcal{H}(k, m-1) - \frac{1}{i^2} \sum_{n=0}^{\infty} \binom{n+i-1}{n} \frac{1}{n^{m-1}} \right).$

   (d) $\mathcal{H}(2, 3) = -\zeta(2) + 2\zeta(3) + \frac{5}{4}\zeta(4); \quad \mathcal{H}(3, 4) = -\frac{3}{4}\zeta(2) - \left(\frac{1}{4} + \zeta(2) \right) \zeta(3) + \frac{15}{8}\zeta(4) + 3\zeta(5).$

2.3   Another Euler sum

Problem 11810 (Proposed by O. Furdui, 122(1), 2015). Let $H_n = \sum_{k=1}^{n} 1/k$, and let ζ be the Riemann zeta function. Find

$$\sum_{n=1}^{\infty} H_n \left(\zeta(3) - \sum_{k=1}^{n} \frac{1}{k^3} \right).$$

Discussion.
Let $b_n = \zeta(3) - \sum_{k=1}^{n} 1/k^3$ be the nth tail (remainder) in the series expansion of $\zeta(3)$. Applying Abel's summation formula with $a_n = H_n$ yields

$$\sum_{n=1}^{n} H_n b_n = \left(\sum_{k=1}^{n} H_k \right) b_n - \sum_{k=1}^{n-1} \left(\sum_{i=1}^{k} H_i \right) (b_{k+1} - b_k).$$

The identity $\sum_{i=1}^{k} H_i = (k+1)H_k - k$ makes the limit process manageable.
   An alternative approach is based on the fact that

$$\int_0^1 x^{k-1} \ln^2 x \, dx = \frac{2}{k^3},$$

from which we have

$$b_n = \frac{1}{2} \int_0^1 \frac{x^n \ln^2 x}{1-x} \, dx.$$

This transforms the problem into the computation of a definite integral.

Solution I.

We claim the answer is $2\zeta(3) - \zeta(2)$. To prove this, we begin with

$$\sum_{k=1}^n H_k = \sum_{k=1}^n \sum_{i=1}^k \frac{1}{i} = \sum_{i=1}^n \sum_{k=i}^n \frac{1}{i} = \sum_{i=1}^n \frac{n+1-i}{i}$$

$$= (n+1)H_n - n = (n+1)(H_{n+1} - 1). \tag{2.14}$$

Let $a_n = H_n, b_n = \zeta(3) - \sum_{k=1}^n 1/k^3$ and $A_n = \sum_{k=1}^n H_k$. Applying Abel's summation formula and (2.14), we find that

$$\sum_{n=1}^n H_n b_n = A_n b_n - \sum_{k=1}^{n-1} A_k(b_{k+1} - b_k) = A_n b_n + \sum_{k=1}^{n-1} \frac{H_{k+1} - 1}{(k+1)^2}. \tag{2.15}$$

Using (2.14) and $H_n = O(\ln n)$, we have $A_n = (n+1)(H_{n+1} - 1) = O(n \ln n)$. Note that $b_n = \sum_{k=n+1}^\infty 1/k^3$. We have

$$\frac{1}{2(n+1)^2} = \int_{n+1}^\infty \frac{1}{x^3} \, dx \le b_n \le \int_n^\infty \frac{1}{x^3} \, dx = \frac{1}{2n^2},$$

and so $b_n = O(1/n^2)$. Letting $n \to \infty$ in (2.15) we obtain

$$\sum_{n=1}^\infty H_n b_n = \sum_{k=1}^\infty \frac{H_{k+1} - 1}{(k+1)^2} = \sum_{k=1}^\infty \frac{H_k}{k^2} - \zeta(2) = 2\zeta(3) - \zeta(2)$$

as claimed, where we have used the well-known identity of Euler (see (2.6)):

$$\sum_{k=1}^\infty \frac{H_k}{k^2} = 2\zeta(3). \tag{2.16}$$

For completeness, we give an elementary proof of (2.16). Since

$$\sum_{k=1}^\infty \frac{H_k}{k^2} = \sum_{k=1}^\infty \frac{H_{k-1} + 1/k}{k^2} = \sum_{k=2}^\infty \sum_{i=1}^{k-1} \frac{1}{ik^2} + \zeta(3),$$

it suffices to show that

$$A := \sum_{k=2}^\infty \sum_{i=1}^{k-1} \frac{1}{ik^2} = \zeta(3).$$

In fact, we have

$$A = \sum_{i,j=1}^{\infty} \frac{1}{i(i+j)^2} \quad (\text{let } k = i+j)$$

$$= \sum_{i,j=1}^{\infty} \frac{1}{i^2} \left(\frac{1}{j} - \frac{1}{i+j} - \frac{i}{(i+j)^2} \right).$$

Notice that, for every $i \in \mathbb{N}$, telescoping yields

$$\sum_{j=1}^{\infty} \left(\frac{1}{j} - \frac{1}{i+j} \right) = \sum_{j=1}^{i} \frac{1}{j}.$$

Thus

$$A = \sum_{i=1}^{\infty} \left(\frac{1}{i^2} \sum_{j=1}^{i} \frac{1}{j} - \sum_{j=1}^{\infty} \frac{1}{i(i+j)^2} \right)$$

$$= \sum_{i=1}^{\infty} \sum_{j=1}^{i} \frac{1}{ji^2} - \sum_{k=2}^{\infty} \sum_{i=1}^{k-1} \frac{1}{ik^2}$$

$$= \sum_{i=1}^{\infty} \frac{1}{i^3} = \zeta(3).$$

$\square$

Solution II.
Since

$$\int_0^1 x^{k-1} \ln^2 x \, dx = \frac{2}{k^3}, \quad (\text{for } k = 1, 2, \ldots),$$

we have

$$\zeta(3) - \sum_{k=1}^{n} \frac{1}{k^3} = \sum_{k=n+1}^{\infty} \frac{1}{k^3} = \frac{1}{2} \sum_{k=n+1}^{\infty} \int_0^1 x^{k-1} \ln^2 x \, dx = \frac{1}{2} \int_0^1 \frac{x^n \ln^2 x}{1-x} \, dx.$$

Thus

$$\sum_{n=1}^{\infty} H_n \left(\zeta(3) - \sum_{k=1}^{n} \frac{1}{k^3} \right) = \frac{1}{2} \sum_{n=1}^{\infty} H_n \int_0^1 \frac{x^n \ln^2 x}{1-x} \, dx$$

$$= \frac{1}{2} \int_0^1 \left(\sum_{n=1}^{\infty} H_n x^n \right) \frac{\ln^2 x}{1-x} \, dx$$

$$= -\frac{1}{2} \int_0^1 \frac{\ln(1-x) \ln^2 x}{(1-x)^2} \, dx.$$

Here interchanging the order of the summation and integration is asserted by the positivity of the general terms, and $-\frac{\ln(1-x)}{1-x} = \sum_{n=1}^{\infty} H_n x^n$ is used in the last equation. Moreover, using integration by parts, we have

$$\int_0^1 \frac{\ln(1-x)\ln^2 x}{(1-x)^2}\, dx = \int_0^1 \frac{\ln^2 x}{(1-x)^2}\, dx - 2\int_0^1 \frac{\ln x \ln(1-x)}{x(1-x)}\, dx. \quad (2.17)$$

Recall that $1/(1-x)^2 = \sum_{n=1}^{\infty} n x^{n-1}$. We find the first integral on the right-hand side of (2.17)

$$\int_0^1 \frac{\ln^2 x}{(1-x)^2}\, dx = \sum_{n=1}^{\infty} n \int_0^1 x^{n-1} \ln^2 x\, dx = \sum_{n=1}^{\infty} \frac{2}{n^2} = 2\zeta(2).$$

To compute the second integral on the right-hand side of (2.17), in view of the facts:

$$\frac{1}{x(1-x)} = \frac{1}{x} + \frac{1}{1-x}, \quad \int_0^1 \frac{\ln x \ln(1-x)}{x}\, dx = \int_0^1 \frac{\ln x \ln(1-x)}{1-x}\, dx,$$

and

$$\ln(1-x) = -\sum_{n=1}^{\infty} \frac{x^n}{n},$$

we have

$$\int_0^1 \frac{\ln x \ln(1-x)}{x(1-x)}\, dx = 2\int_0^1 \frac{\ln x \ln(1-x)}{x}\, dx$$

$$= -2\sum_{n=1}^{\infty} \frac{1}{n} \int_0^1 x^{n-1} \ln x\, dx = 2\zeta(3).$$

In summary, we finally find that

$$\sum_{n=1}^{\infty} H_n \left(\zeta(3) - \sum_{k=1}^{n} \frac{1}{k^3} \right) = -\frac{1}{2}\left(2\zeta(2) - 4\zeta(3)\right) = 2\zeta(3) - \zeta(2).$$

$\square$

Remark. It is interesting to see that the identity (2.16) and its generalizations have appeared in the Monthly multiple times. For example,
Advanced Problem 4431 (Proposed by M. S. Klamkin, 58(2), 1951). If $S_n = \sum_{k=1}^{n} 1/k$, prove:

$$\sum_{n=1}^{\infty} \frac{S_n}{n^2} = 2\sum_{n=1}^{\infty} \frac{1}{n^3} = 2\sum_{n=1}^{\infty} \frac{S_n}{(n+1)^2}.$$

Advanced Problem 4564 (Proposed by M. S. Klamkin, 62(2), 1955). If $S_n = \sum_{k=1}^{n} 1/k$, prove:

$$\sum_{n=1}^{\infty} \frac{S_n}{n^3} = \frac{\pi^4}{72}.$$

Problem 10635 (Proposed by N. R. Farnum, 105(1), 1998). Show that the value of $\zeta(3) = \sum_{k=1}^{\infty} k^{-3}$, also called Apéry's constant, can be expressed as $\zeta(3) = \sum_{n=1}^{\infty} r_n/n$, where $r_n = (\pi^2/6) - \sum_{k=1}^{n} k^{-2}$ is the nth remainder of the series expansion of $\zeta(2)$.

Problem 10754 (Proposed by P. Bracken, 106(8), 1999). Let $\zeta(s) = \sum_{k=1}^{\infty} k^{-s}$, and let $\rho(s,n) = \sum_{k=n+1}^{\infty} k^{-s}$. Show that for positive integers $s \geq 2$,

$$\sum_{k=1}^{\infty} \frac{\rho(s,k)}{k} = \frac{s}{2} \zeta(s+1) - \frac{1}{2} \sum_{k=1}^{s-2} \zeta(s-k)\zeta(k+1). \qquad (2.18)$$

You may notice that (2.18) is actually a rearranged form of Euler's formula (2.6) because

$$\sum_{k=1}^{\infty} \frac{\rho(s,k)}{k} = \sum_{k=1}^{\infty} \sum_{i=k}^{\infty} \frac{1}{k(i+1)^s} = \sum_{i=1}^{\infty} \frac{1}{(i+1)^s} \sum_{k=1}^{i} \frac{1}{k} = S(1,s).$$

Recall (2.16): $2\zeta(3) = \sum_{k=1}^{\infty} H_k/k^2$. Motivated by (2.18), we define the nth tail of $2\zeta(3)$ as

$$t_n = 2\zeta(3) - \sum_{k=1}^{n} \frac{H_k}{k^2} = \sum_{k=n+1}^{\infty} \frac{H_k}{k^2}.$$

We challenge the reader to prove

$$\sum_{k=1}^{n} \frac{t_k}{k} = \sum_{k=1}^{\infty} \frac{1}{k} \left(2\zeta(3) - \frac{H_1}{1^2} - \frac{H_2}{2^2} - \cdots - \frac{H_k}{k^2} \right) = 3\zeta(4) = \frac{\pi^4}{30}.$$

Applying the approach in Solution I, we can easily prove the following identity:

Problem 12102 (Proposed by O. Furdui and A. Sîntămărian, 126(3), 2019). Prove

$$\sum_{n=1}^{\infty} H_n^2 \left(\zeta(2) - \sum_{k=1}^{n} \frac{1}{k^2} - \frac{1}{n} \right) = 2 - \zeta(2) - 2\zeta(3),$$

where H_n is the nth harmonic number.

Moreover, along the same lines in Solution II, we can generalize **Problem 10635** above as follows: For $k \in \mathbb{N}$,

$$\sum_{n_1=1}^{\infty} \sum_{n_2=1}^{\infty} \cdots \sum_{n_k=1}^{\infty} \frac{1}{n_1 n_2 \cdots n_k} \sum_{p=n_1+n_2+\cdots+n_k+1}^{\infty} \frac{1}{p^2} = k!\zeta(k+2).$$

We now end this section with another extension of the problem for your exercise: Let

$$T(k) := \sum_{n=1}^{\infty} H_n \left(\zeta(k) - \sum_{j=1}^{n} \frac{1}{j^k} \right).$$

For $k \geq 3$, show that

$$T(k) = \frac{k+1}{2}\zeta(k) - \zeta(k-1) - \frac{1}{2}\sum_{j=1}^{k-3}\zeta(k-1-j)\zeta(j+1),$$

where the last sum is empty if $k = 3$.

2.4 Reciprocal Catalan number sums

Problem 11765 (Proposed by D. Beckwith, 121(3), 2014). Let C_n be the nth Catalan number, given by $C_n = \frac{1}{n+1}\binom{2n}{n}$. Show that

(a) $\sum_{n=0}^{\infty}\frac{2^n}{C_n} = 5 + \frac{3}{2}\pi$,

(b) $\sum_{n=0}^{\infty}\frac{3^n}{C_n} = 22 + 8\sqrt{3}\,\pi$.

Discussion.
Since the *Mathematica* command

`Sum[2^n/CatalanNumber[n], {n, 0, Infinity}] // Expand`

gives the exact stated answer $5 + 3\pi/2$, it is more ambitious to find a closed form of the generating function for the reciprocal of the Catalan numbers:

$$\sum_{n=0}^{\infty}\frac{1}{C_n}x^n.$$

Here we demonstrate two approaches. The first is based on an integral representation which is established by the Euler beta function. The second is based on manipulation of the well-known power series of $\arcsin^2(x/2)$, which Lehmer [65] extensively used to find a large class of *interesting* series; interesting in the sense that the series has a closed form in terms of well-known constants.

Solution I.
Recall that the Euler beta function is defined by

$$B(\alpha, \beta) = \int_0^1 t^{\alpha-1}(1-t)^{\beta-1}\, dt.$$

Using

$$B(\alpha, \beta) = \frac{\Gamma(\alpha)\Gamma(\beta)}{\Gamma(\alpha+\beta)},$$

we have

$$\frac{1}{C_n} = \frac{(n+1)(2n)!}{n!n!} = n(n+1)B(n, n+1) = n(n+1)\int_0^1 t^{n-1}(1-t)^n\, dt.$$

Thus

$$\sum_{n=0}^{\infty} \frac{1}{C_n} x^n = 1 + \sum_{n=1}^{\infty} n(n+1)x^n \int_0^1 t^{n-1}(1-t)^n \, dt$$

$$= 1 + \int_0^1 \sum_{n=1}^{\infty} n(n+1)x^n t^{n-1}(1-t)^n \, dt$$

$$= 1 + \int_0^1 \frac{2x(1-t)}{(1-xt(1-t))^3} \, dt,$$

where the interchange of summation and integration is justified by the positivity of the general terms. In the last equation we have used the identity

$$\sum_{n=1}^{\infty} n(n+1)z^n = \frac{2z}{(1-z)^3}.$$

Calculating the integral by partial fractions, we find that

$$\sum_{n=0}^{\infty} \frac{1}{C_n} x^n = 2 \frac{(8+x)\sqrt{4-x} + 12\sqrt{x}\arcsin(\sqrt{x}/2)}{(4-x)^{5/2}}. \qquad (2.19)$$

By the ratio test, the series in (2.19) converges for $|x| < 4$. Thus, setting $x = 2$ and 3 in (2.19), respectively yields the desired formulas for (a) and (b). $\quad\square$

Solution II.

Begin with the power series [65]

$$f(x) := \arcsin^2(x/2) = \frac{1}{2} \sum_{n=1}^{\infty} \frac{1}{n^2 \binom{2n}{n}} x^{2n},$$

which converges for $|x| < 2$. Let $D_x = \frac{d}{dx}$. Direct calculation yields

$$D_x(x^3 D_x(x D_x f)) = D_x \left\{ x^3 D_x \left[x D_x \left(\frac{1}{2} \sum_{n=1}^{\infty} \frac{1}{n^2 \binom{2n}{n}} x^{2n} \right) \right] \right\}$$

$$= D_x \left[x^3 D_x \left(\sum_{n=1}^{\infty} \frac{1}{n \binom{2n}{n}} x^{2n} \right) \right]$$

$$= 2D_x \left(\sum_{n=1}^{\infty} \frac{1}{\binom{2n}{n}} x^{2n+2} \right)$$

$$= 4x \sum_{n=1}^{\infty} \frac{n+1}{\binom{2n}{n}} x^{2n}$$

$$= 4x \sum_{n=1}^{\infty} \frac{1}{C_n} x^{2n}.$$

Thus

$$\sum_{n=0}^{\infty} \frac{1}{C_n} x^{2n} = 1 + \frac{1}{4x} D_x(x^3 D_x(x D_x))[\arcsin^2(x/2)].$$

This leads to

$$\sum_{n=0}^{\infty} \frac{1}{C_n} x^{2n} = 2 \frac{(8+x^2)\sqrt{4-x^2} + 12x \arcsin(x/2)}{(4-x^2)^{5/2}}.$$

Replacing x^2 by x yields (2.19) immediately. Thus, setting $x = \sqrt{2}$ and $\sqrt{3}$ yields the desired formulas for **(a)** and **(b)** again. □

Remark. The Catalan numbers are probably one of the most ubiquitous sequence of numbers in mathematics. There are many counting problems in combinatorics whose solution is given by the Catalan numbers. The monograph by Stanley [86] compiles much fascinating information on the Catalan numbers which includes 66 different combinatorial interpretations of the Catalan numbers.

Since $\arcsin t = \arctan(t/\sqrt{1-t^2})$, we have an alternative formula from (2.19)

$$\sum_{n=0}^{\infty} \frac{1}{C_n} x^n = 2 \frac{(8+x)\sqrt{4-x} + 12\sqrt{x} \arctan(\sqrt{x/(4-x)})}{(4-x)^{5/2}}.$$

Lehmer [65] used

$$\sum_{n=1}^{\infty} \frac{(2x)^{2n}}{n\binom{2n}{n}} = \sum_{n=1}^{\infty} \frac{(2x)^{2n}}{n(n+1)C_n} = \frac{2x \arcsin x}{\sqrt{1-x^2}}$$

to produce many interesting series including

$$\sum_{n=1}^{\infty} \frac{2^n}{\binom{2n}{n}} = \frac{\pi}{2} + 1, \quad \sum_{n=1}^{\infty} \frac{n\, 2^n}{\binom{2n}{n}} = \pi + 3,$$

$$\sum_{n=1}^{\infty} \frac{3^n}{\binom{2n}{n}} = \frac{4\pi\sqrt{3}}{3} + 3, \quad \sum_{n=1}^{\infty} \frac{n\, 3^n}{\binom{2n}{n}} = \frac{20\pi\sqrt{3}}{3} + 18.$$

The formulas **(a)** and **(b)** now also follow easily from these results above. Similarly, we can find additional interesting series based on (2.19). For example, we have the alternating version for **(a)** and **(b)**:

$$\sum_{n=0}^{\infty} \frac{(-1)^n}{C_n} = \frac{14}{25} - \frac{24\sqrt{5}}{125} \ln\left(\frac{1+\sqrt{5}}{2}\right),$$

$$\sum_{n=0}^{\infty} \frac{(-2)^n}{C_n} = \frac{1}{3} - \frac{\sqrt{3}}{9} \ln(2+\sqrt{3}),$$

$$\sum_{n=0}^{\infty} \frac{(-3)^n}{C_n} = \frac{10}{49} - \frac{36\sqrt{21}}{1029} \ln\left(\frac{5+\sqrt{21}}{2}\right).$$

Finally, we end this section with some problems for additional practice.

1. Show that

$$C_n = \frac{1}{\pi} \int_0^2 x^{2n} \sqrt{4 - x^2} \, dx = \frac{2}{\pi} \int_0^\infty \frac{x^2}{(x^2 + 1/4)^{n+2}} \, dx.$$

2. Let $y(x)$ be the generating function of the reciprocal of the Catalan numbers. Show that $y(x)$ satisfies

$$x(x - 4)y' + 2(x + 1)y = 2.$$

Then solve this differential equation to obtain (2.19).

3. Prove that the exponential generating function of the reciprocal Catalan numbers is given by

$$\sum_{n=0}^\infty \frac{1}{C_n} \frac{x^n}{n!} = 1 + \frac{x}{4} + \frac{1}{8} e^{z/4} \sqrt{\pi x}(x + 6) \mathrm{erf}\left(\frac{\sqrt{x}}{2}\right),$$

where erf is the *error function* defined by

$$\mathrm{erf}(x) = \frac{2}{\sqrt{\pi}} \int_0^x e^{-t^2} \, dt.$$

4. **Problem 11509** (Proposed by W. Stanford, 117(6), 2010). Let m be a positive integer. Prove that

$$\sum_{k=m}^{m^2-m+1} \frac{\binom{m^2-m+1}{k-m}}{k\binom{m^2}{k}} = \frac{1}{m\binom{2m-1}{m}}.$$

Hint: Use $\int_0^1 x^{m-1}(1 - x)^{m-1} \, dx = \int_0^1 x^{m+(m-1)^2-1}(1 - x)^{m-1} x^{-(m-1)^2} \, dx$.

5. **Problem 11716** (Proposed by O. Knill, 120(6), 2013). Let $\alpha = (\sqrt{5} - 1)/2$. Let p_n and q_n be the numerator and denominator of the nth continued fraction convergent to α. Thus, $p_n = F_{n-1}$ and $q_n = F_n$, where F_n is nth Fibonacci number defined by the recurrence $F_0 = 0, F_1 = 1$, and $F_{n+1} = F_n + F_{n-1}$ for $n > 1$. Show that

$$\sqrt{5}\left(\alpha - \frac{p_n}{q_n}\right) = \sum_{k=0}^\infty \frac{(-1)^{(n+1)(k+1)} C_k}{q_n^{2k+2} 5^k},$$

where C_k denotes the kth Catalan number, given by $C_k = \frac{1}{k+1}\binom{2k}{k}$.

2.5   Catalan generating function

Problem 11832 (Proposed by Donald Knuth, 122(4), 2015). Let $C(z) = \sum_{n=0}^{\infty} \binom{2n}{n} \frac{z^n}{n+1}$ (thus $C(z)$ is the generating function of the Catalan numbers). Prove that

$$\ln^2(C(z)) = \sum_{n=1}^{\infty} \binom{2n}{n} (H_{2n-1} - H_n) \frac{z^n}{n}.$$

Here, $H_k = \sum_{j=1}^{k} 1/j$; that is, H_k is the kth harmonic number.

Discussion.

From the well-known generating function of the central binomial coefficients

$$\sum_{n=0}^{\infty} \binom{2n}{n} z^n = \frac{1}{\sqrt{1-4z}},$$

integrating both sides yields the generating function of the Catalan numbers

$$C(z) = \frac{2}{1 + \sqrt{1-4z}} = \frac{1 - \sqrt{1-4z}}{2z}.$$

Logarithmic differentiation implies that

$$\frac{C'(z)}{C(z)} = \frac{1}{2z} \left(\frac{1}{\sqrt{1-4z}} - 1 \right),$$

which enables us to find a power series expansion of $\ln C(z)$. After some manipulations, the problem remains to show that

$$\frac{1}{2} \sum_{k=1}^{n-1} \binom{2k}{k} \binom{2n-2k}{n-k} \frac{1}{k} = \binom{2n}{n} (H_{2n-1} - H_n).$$

For this purpose, here we present two proofs — one by the method of generating function, another by the Wilf-Zeilberger algorithm [74] with telescoping sums.

Solution I.

Note that

$$\frac{C'(z)}{C(z)} = \frac{1}{2z} \left(\frac{1}{\sqrt{1-4z}} - 1 \right) = \frac{1}{2} \sum_{n=1}^{\infty} \binom{2n}{n} z^{n-1}.$$

Integrating both sides yields

$$\ln C(z) = \frac{1}{2} \sum_{n=1}^{\infty} \binom{2n}{n} \frac{z^n}{n}. \qquad (2.20)$$

By the Cauchy product, we find that

$$\ln^2 C(z) = \frac{1}{4} \sum_{n=1}^{\infty} \left(\sum_{k=1}^{n-1} \binom{2k}{k} \binom{2(n-k)}{n-k} \frac{1}{k(n-k)} \right) z^n.$$

Since

$$\frac{1}{k(n-k)} = \frac{1}{n} \left(\frac{1}{k} + \frac{1}{n-k} \right),$$

by symmetry, we have

$$\ln^2 C(z) = \frac{1}{2} \sum_{n=1}^{\infty} \left(\sum_{k=1}^{n-1} \frac{1}{k} \binom{2k}{k} \binom{2(n-k)}{n-k} \right) \frac{z^n}{n}.$$

Comparing this with the proposed identity, we need to show

$$\frac{1}{2} \sum_{k=1}^{n-1} \frac{1}{k} \binom{2k}{k} \binom{2(n-k)}{n-k} = \binom{2n}{n} (H_{2n-1} - H_n).$$

Adding $\frac{1}{2n} \binom{2n}{n}$ both sides yields

$$\frac{1}{2} \sum_{k=1}^{n} \frac{1}{k} \binom{2k}{k} \binom{2(n-k)}{n-k} = \binom{2n}{n} (H_{2n} - H_n). \tag{2.21}$$

To prove (2.21), we invoke Formula 7.43 in the Table 351 ([50], p. 351) that

$$\frac{1}{(1-z)^{m+1}} \ln \frac{1}{1-z} = \sum_{n=0}^{\infty} \binom{m+n}{n} (H_{m+n} - H_m) z^n.$$

Let $m = n$. From

$$\frac{1}{(1-z)^{n+1}} = \sum_{k=0}^{\infty} \binom{n+k}{n} z^k,$$

matching the coefficients of z^n on both sides yields

$$\sum_{k=1}^{n} \frac{1}{k} \binom{2n-k}{n} = \binom{2n}{n} (H_{2n} - H_n).$$

Therefore it suffices to show that

$$\sum_{k=1}^{n} \frac{1}{k} \binom{2n-k}{n} = \frac{1}{2} \sum_{k=1}^{n} \frac{1}{k} \binom{2k}{k} \binom{2(n-k)}{n-k}. \tag{2.22}$$

We now prove (2.22) by using the method of generating functions. We compute

$$\sum_{n=1}^{\infty}\left(\sum_{k=1}^{n}\frac{1}{k}\binom{2n-k}{n}\right)z^n = \sum_{k=1}^{\infty}\frac{1}{k}\left(\sum_{n=k}^{\infty}\binom{2n-k}{n}z^n\right)$$

$$= \sum_{k=1}^{\infty}\frac{1}{k}\left(\sum_{m=0}^{\infty}\binom{2m+k}{m+k}z^{m+k}\right)$$

(let $n = m + k$)

$$= \sum_{k=1}^{\infty}\frac{z^k}{k}\left(\sum_{m=0}^{\infty}\binom{2m+k}{m}z^m\right)$$

$$= \sum_{k=1}^{\infty}\frac{z^k}{k}\left(\frac{1}{\sqrt{1-4z}}\left(\frac{1-\sqrt{1-4z}}{2z}\right)^k\right)$$

$$= -\frac{1}{\sqrt{1-4z}}\ln\left(\frac{1+\sqrt{1-4z}}{2}\right)$$

$$= \frac{1}{\sqrt{1-4z}}\ln C(z),$$

where we have used Formula 5.72 ([50], p. 203) that

$$\sum_{m=0}^{\infty}\binom{2m+k}{m}z^m = \frac{1}{\sqrt{1-4z}}\left(\frac{1-\sqrt{1-4z}}{2z}\right)^k.$$

By (2.20), in the Cauchy product of $\frac{1}{\sqrt{1-4z}}\ln C(z)$, we extract the coefficient of z^n as

$$\frac{1}{2}\sum_{k=1}^{n}\frac{1}{k}\binom{2k}{k}\binom{2(n-k)}{n-k}$$

and prove (2.22) as desired. □

Solution II.
We prove

$$\frac{1}{2}\sum_{k=1}^{n-1}\binom{2k}{k}\binom{2(n-k)}{n-k}\frac{1}{k} = \binom{2n}{n}(H_{2n-1}-H_n)$$

by using the Wilf-Zeilberger algorithm. First, we rewrite the above identity as

$$\sum_{k=1}^{n-1}F(n,k) = 2(H_{2n-1}-H_n), \tag{2.23}$$

where

$$F(n,k) = \frac{1}{k}\binom{2k}{k}\binom{2(n-k)}{n-k}\binom{2n}{n}^{-1},$$

which is hypergeometric in both arguments. Invoking the Wilf-Zeilberger algorithm — EKHAD (available at `http://sites.math.rutgers.edu/ ~zeilberg/tokhniot/EKHAD`), we find that

$$F(n+1,k) - F(n,k) = G(n,k+1) - G(n,k),$$

where

$$G(n,k) = -\frac{k^2(2n-2k+1)F(n,k)}{(n+1)(2n+1)(n+1-k)}.$$

Thus

$$
\begin{aligned}
\sum_{k=1}^{n-1} F(n,k) &= \sum_{k=1}^{n-1} F(n-1,k) + \sum_{k=1}^{n-1}(G(n-1,k+1) - G(n-1,k)) \\
&= 2(H_{2n-3} - H_{n-1}) + F(n-1,n-1) + G(n-1,n) \\
&\quad - G(n-1,1) \\
&= 2(H_{2n-3} - H_{n-1}) + \frac{1}{n-1} + 0 + \frac{(2n-3)G(n-1,1)}{n(2n-1)(n-1)} \\
&= 2(H_{2n-3} - H_{n-1}) + \frac{1}{n-1} + \frac{1}{n(2n-1)} \\
&= 2(H_{2n-1} - H_n).
\end{aligned}
$$

$\square$

Remark. As Solution II indicates, by using the Wilf-Zeilberger algorithm, we can now follow a mechanical procedure to evaluate sums of binomial coefficients and discover the answer quite systematically. We refer the interested reader to [70] or [74] for details.

In fact, these procedures enable us to find many similar formulas that arise frequently in practice. Here we illustrate how to combine Abel's summation formula with the Wilf-Zeilberger algorithm to derive the following identity on harmonic numbers:

$$\sum_{k=0}^{n} (-1)^{n-k} \binom{n}{k}\binom{n+k}{k} H_k^{(2)} = 2\sum_{k=1}^{n} \frac{(-1)^{k-1}}{k^2}.$$

Let

$$S(n) := \sum_{k=0}^{n} (-1)^{n-k} \binom{n}{k}\binom{n+k}{k} H_k^{(2)}, \quad F(n,k) = (-1)^{n-k}\binom{n}{k}\binom{n+k}{k}.$$

The Wilf-Zeilberger algorithm gives

$$F(n+1,k) - F(n,k) = G(n,k+1) - G(n,k),$$

where

$$G(n,k) = \frac{2(-1)^{n-k}k^2\binom{n}{k}\binom{n+k}{k}}{(n-k+1)(n+1)}.$$

Thus
$$S(n+1) - S(n) = \sum_k \left(G(n, k+1) - G(n, k)\right) H_k^{(2)}.$$

Applying Abel's summation formula on the right-hand side, we have

$$
\begin{aligned}
S(n+1) - S(n) &= \sum_k \frac{-G(n, k+1)}{(k+1)^2} \\
&= \sum_{k \geq 0} \left(T(k+1) - T(k)\right) \\
&= 2\frac{(-1)^n}{(n+1)^2},
\end{aligned}
$$

where
$$T(k) = \frac{2(-1)^{n-k-1}(k+1)^2 \binom{n}{k+1}\binom{n+k+1}{k+1}}{(n-k)(n+1)^3}.$$

Since $S(0) = 0$, we find that

$$S(n) = S(0) + 2\sum_{k=1}^n \frac{(-1)^{k-1}}{k^2} = 2\sum_{k=1}^n \frac{(-1)^{k-1}}{k^2}.$$

Another powerful approach to study sums involving harmonic numbers is to express them in terms of differentiation of binomial coefficients. Let

$$H_0(x) = 0 \quad \text{and} \quad H_n(x) = \sum_{k=1}^n \frac{1}{x+k} \quad \text{for } n = 1, 2, \ldots.$$

For a differentiable function $f(x)$, denote

$$D_x f(x) = \frac{d}{dx} f(x) \quad \text{and} \quad D_0 f(x) = \frac{d}{dx} f(x)|_{x=0}.$$

If $m \leq n$, applying logarithmic differentiation yields

$$D_x \binom{x+n}{m} = \binom{x+n}{m} \sum_{i=1}^m \frac{1}{1+x+n-i} = \binom{x+n}{m} \left(H_n(x) - H_{n-m}(x)\right).$$

In particular, we have

$$D_0 \binom{x+n}{m} = \binom{n}{m} \left(H_n - H_{n-m}\right).$$

Chu and Donno [32] applied these derivative operators to hypergeometric series, and derived many striking summation identities involving harmonic numbers like

$$\sum_{k=0}^n \binom{n}{k}^2 H_k = \binom{2n}{n}(2H_n - H_{2n}).$$

Indeed, the featured solution to **Problem 11832** due to James Smith also employed this technique to the identity

$$\sum_{k=0}^{n} \frac{x}{x + 2k} \binom{x + 2k}{k} \binom{2(n - k)}{n - k} = \binom{x + 2n}{n}$$

to obtain (2.21).

Here we compile a list of problems for your practice in using generating functions:

1. **Problem 11343** (Proposed by D. Backwith, 115(2), 2008). Show that when n is a positive integer,

   $$\sum_{k \geq 0} \binom{n}{k} \binom{2k}{k} = \sum_{k \geq 0} \binom{n}{2k} \binom{2k}{k} 3^{n - 2k}.$$

   You may generalize this identity to

   $$\sum_{k \geq 0} \binom{n}{k} \binom{2k}{k + m} = \sum_{k \geq 0} \binom{n}{2k + m} \binom{2k + m}{k} 3^{n - (2k + m)}.$$

2. **Problem 11356** (Proposed by M. Poghosyan, 115(4), 2008). Prove that for any positive integer n,

   $$\sum_{k=0}^{n} \frac{\binom{n}{k}^2}{(2k + 1)\binom{2n}{2k}} = \frac{2^{4n}(n!)^4}{(2n)!(2n + 1)!}.$$

3. **Problem 11406** (Proposed by A. A. Dzhumadil'daeva, 116(1), 2009). Let $n!!$ denote the product of all positive integers not greater than n and congruent to n mod 2, and $0!! = (-1)!! = 1$. For positive integer n, find

   $$\sum_{i=0}^{n} \binom{n}{i} (2i - 1)!![(2(n - i) - 1]!!$$

   in closed form.

4. **Problem 11757** (Proposed by I. Gessel, 121(2), 2014). Let $[x^a y^b]$ be the coefficient of $x^a y^b$ in the Taylor series expansion of f. Show that

   $$[x^n y^n] \frac{1}{(1 - 3x)(1 - y - 3x + 3x^2)} = 9^n.$$

5. **Problem 11862** (Proposed by D. A. Cox and U. Thieu, 122(8), 2015). For positive integers n and k, evaluate

   $$\sum_{i=0}^{k} (-1)^i \binom{k}{i} \binom{kn - in}{k + 1}.$$

6. **Problem 11897** (Proposed by P. P. Dályay, 123(3), 2016). Prove for $n \geq 0$,

$$\sum_{k+j=n, k, j \geq 0} \frac{\binom{2k}{k}\binom{2j+2}{j+1}}{k+1} = 2\binom{2n+2}{n}.$$

A list of problems for your practice in using the Wilf-Zeilberger algorithm:

1. **Problem 11274** (Proposed by D. Knuth, 114(2), 2007). Prove that for nonnegative integers m and n,

$$\sum_{k=0}^{n} 2^k \binom{2m-k}{m+n} = 4^m - \sum_{j=1}^{n} \binom{2m+1}{m+j}.$$

2. **Problem 11940** (Proposed by H. Ohtsuka, 123(9), 2016). Let $T_n = n(n+1)/2$ and $C(n,k) = (n-2k)\binom{n}{k}$. For $n \geq 1$, prove

$$\sum_{k=0}^{n-1} C(T_k, k)C(T_{k+1}, k) = \frac{n^3 - 2n^2 + 4n}{n+2}\binom{T_n}{n}\binom{T_{n+1}}{n}.$$

3. **Problem 12049** (Proposed by Z. K. Silagadze, 125(6), 2018). For all nonnegative integers m and n with $m \leq n$, prove

$$\sum_{k=m}^{n} \frac{(-1)^{k+m}}{2k+1}\binom{n+k}{n-k}\binom{2k}{k-m} = \frac{1}{2n+1}.$$

For $n, p \in \mathbb{N}, n \geq p$, prove the following identities

4. $\displaystyle\sum_{k=0}^{n}(-1)^{n-k}\binom{n}{k}\binom{n+k}{k}\binom{k}{p}H_k = (-1)^n\binom{n+p}{p}\binom{n}{p}(2H_n - H_p).$

5. $\displaystyle\sum_{k=0}^{n}\binom{n}{k}^2\binom{k}{p}H_k = \binom{2n-p}{n}\binom{n}{p}(2H_n - H_{2n-p}).$

6. $\displaystyle\sum_{k=0}^{n}(-1)^k\binom{2n}{k}^2 H_k = \frac{(-1)^n}{2}\binom{2n}{n}\binom{n}{p}(H_n + H_{2n}).$

A list of problems for your practice in using derivative operators:

1. For all $n \in \mathbb{N}$, prove that

$$\sum_{k=0}^{n}\binom{n}{k}^2\binom{2n+k}{k}(H_{2n+k} - H_k) = 2\binom{2n}{n}^2(H_{2n} - H_n).$$

2. Let $(x)_n = x(x+1)\cdots(x+n-1)$ for $n \in \mathbb{N}$. Prove that

$$\sum_{k=0}^{m}(-1)^k\binom{m}{k}\binom{x+k}{k}\frac{H_{n+k}-H_k(x)}{\binom{n+k}{k}} = \frac{(n-x)_m}{(n+1)_m}H_{m+n}.$$

3. For all $n \in \mathbb{N}$, prove that

$$\sum_{k=1}^{\infty}\frac{n!}{(k)_{n+1}}(H_{k+n}-H_n) = \sum_{k=0}^{\infty}\sum_{m=0}^{k}\binom{k}{m}\frac{(-1)^m}{(m+n+1)^2} = \frac{1}{n^2}.$$

4. **Problem 11026** (Proposed by J. Sondow, 110(7), 2003). Let H_n denote the nth harmonic number. Let $H_0 = 0$. Prove that for positive integers n and k with $k \leq n$,

$$\sum_{i=0}^{k-1}\sum_{j=k}^{n}(-1)^{i+j-1}\binom{n}{i}\binom{n}{j} = \sum_{i=0}^{k-1}\binom{n}{i}^2(H_{n-i}-H_i).$$

5. **Problem 11164** (Proposed by D. Barrero, 112(6), 2005). Show that if n is a positive integer then

$$\sum_{k=1}^{n}(-1)^{k+1}\binom{n}{k}\sum_{1\leq i\leq j\leq k}\frac{1}{ij} = \frac{1}{n^2}. \qquad (2.24)$$

Comments. Tauraso (https://www.mat.uniroma2.it/~tauraso/AMM/AMM11164.pdf) generalized (2.24) to

$$\sum_{k=1}^{n}(-1)^{k+1}\binom{n}{k}\sum_{1\leq i_1\leq i_2\leq\cdots\leq i_l\leq k}\prod_{j=1}^{l}\frac{1}{i_j} = \frac{1}{n^l}.$$

Subsequently, introducing an extra variable x, the above identity can be extended further:

$$\sum_{k=0}^{n}(-1)^k\binom{n}{k}\binom{x+k}{k}^{-1}\sum_{1\leq i_1\leq i_2\leq\cdots\leq i_l\leq k}\prod_{j=1}^{l}\frac{1}{x+i_j} = \frac{x}{(x+n)^{l+1}}.$$

The interested reader is encouraged to prove this identity and pursue new results in this direction.

2.6   A series with log and harmonic numbers

Problem 11499 (Proposed by O. Kouba, 117(2), 2010). Let H_n be the nth harmonic number, given by $H_n = \sum_{k=1}^{n} 1/k$. Let

$$S_k = \sum_{n=1}^{\infty} (-1)^{n-1}(\ln k - (H_{kn} - H_n)).$$

Prove that for $k \geq 2$,

$$S_k = \frac{k-1}{2k} \ln 2 + \frac{1}{2} \ln k - \frac{\pi}{2k^2} \sum_{l=1}^{[k/2]} (k+1-2l) \cot\left(\frac{(2l-1)\pi}{2k}\right).$$

Discusion.

We attempt to evaluate this series by integration. How can we obtain it? The key step is to represent $\ln k - (H_{kn} - H_n)$ in terms of an integral. Let $P_k(x) = 1 + x + \cdots + x^{k-1}$. Then

$$\int_0^1 \frac{P_k'(x)}{P_k(x)} \, dx = \ln P_k(x)|_0^1 = \ln k.$$

Combining this result with

$$\int_0^1 \frac{1 - x^m}{1 - x} \, dx = \int_0^1 (1 + x + \cdots + x^{m-1}) \, dx = H_m,$$

we arrive at the desired integral formula (2.25) below.

Solution.

Let $P_k(x) = 1 + x + \cdots + x^{k-1}$. First, we prove that

$$\ln k - (H_{kn} - H_n) = \int_0^1 \frac{P_k'(x)}{P_k(x)} x^{kn} \, dx. \tag{2.25}$$

Since $(1-x)P_k(x) = 1 - x^k$, differentiating gives $(1-x)P_k'(x) = P_k(x) - kx^{k-1}$. Thus

$$
\begin{aligned}
\frac{P_k'(x)}{P_k(x)} (1 - x^{kn}) &= (1 - x)P_k'(x) \frac{1 - x^{kn}}{1 - x^k} \\
&= (P_k(x) - kx^{k-1}) \frac{1 - x^{kn}}{1 - x^k} \\
&= (1 + x + \cdots + x^{k-1} - kx^{k-1}) \\
&\quad (1 + x^k + x^{2k} + \cdots + x^{(n-1)k}) \\
&= \sum_{i=1}^{kn} x^{i-1} - k \sum_{i=1}^{n} x^{ki-1}.
\end{aligned}
$$

Integrating this expression yields (2.25) as desired. Next, we have

$$S_k = \sum_{n=1}^{\infty} (-1)^{n-1} \int_0^1 \frac{P_k'(x)}{P_k(x)} x^{kn} \, dx$$

$$= \int_0^1 \frac{P_k'(x)}{P_k(x)} \sum_{n=1}^{\infty} (-1)^{n-1} x^{kn} \, dx$$

$$= \int_0^1 \frac{P_k'(x)}{P_k(x)} \frac{x^k}{1+x^k} \, dx.$$

Since the series above converges absolutely for $0 \leq x < 1$, the dominated convergence theorem justifies to exchange the order of summation and integration. Appealing to $2x^k = 1 + x^k - (1-x)P_k(x)$ and

$$\int_0^1 \frac{x^{k-1} \, dx}{1+x^k} = \frac{1}{k} \int_0^1 \frac{dt}{1+t} = \frac{1}{k} \ln 2,$$

we find that

$$S_k = \frac{1}{2} \int_0^1 \frac{(1+x^k)P_k'(x) - (1-x)P_k'(x)P_k(x)}{P_k(x)(1+x^k)} \, dx$$

$$= \frac{1}{2} \left(\int_0^1 \frac{P_k'(x)}{P_k(x)} \, dx - \int_0^1 \frac{(1-x)P_k'(x)}{1+x^k} \, dx \right)$$

$$= \frac{1}{2} \left(\int_0^1 \frac{P_k'(x)}{P_k(x)} \, dx - \int_0^1 \frac{P_k(x) - kx^{k-1}}{1+x^k} \, dx \right)$$

$$= \frac{1}{2} \left(\int_0^1 \frac{P_k'(x)}{P_k(x)} \, dx + k \int_0^1 \frac{x^{k-1} \, dx}{1+x^k} - \int_0^1 \frac{P_k(x)}{1+x^k} \, dx \right)$$

$$= \frac{1}{2} \left(\ln k + \ln 2 - T_k \right),$$

where

$$T_k := \int_0^1 \frac{P_k(x)}{1+x^k} \, dx.$$

We now compute T_k by partial fractions. Let

$$\frac{1 + x + \cdots + x^{k-1}}{1+x^k} = \sum \frac{A_j}{x - \xi_j},$$

where the sum is over the kth roots of -1. Then, by L'Hôpital's rule, for $1 \leq j \leq k$,

$$A_j = \lim_{x \to \xi_i} \frac{(1-x^k)(x-\xi_j)}{(1-x)(1+x^k)} = \frac{2}{k} \frac{\xi_j}{\xi_j - 1}.$$

Except for $\xi = -1$ when k is odd, the summands in the partial fraction decomposition are in conjugate pairs. If $\xi = e^{i\theta}$, then

$$\frac{\xi}{\xi - 1} \frac{1}{x - \xi} + \frac{\bar{\xi}}{\bar{\xi} - 1} \frac{1}{x - \bar{\xi}} = \frac{1+x}{(x - \cos\theta)^2 + \sin^2\theta}.$$

Moreover, we have

$$\int_0^1 \frac{1+x}{(x-\cos\theta)^2 + \sin^2\theta}\,dx$$

$$= \left[\frac{1}{2}\ln(x^2 - 2x\cos\theta + 1) + \frac{1+\cos\theta}{\sin\theta}\tan^{-1}\left(\frac{x-\cos\theta}{\sin\theta}\right)\right]\Bigg|_0^1$$

$$= \frac{1}{2}\ln(2-2\cos\theta) + \cot\frac{\theta}{2}\left(\tan^{-1}\left(\frac{1-\cos\theta}{\sin\theta}\right) + \tan^{-1}\left(\frac{\cos\theta}{\sin\theta}\right)\right)$$

$$= \ln\left(2\sin\frac{\theta}{2}\right) + \cot\frac{\theta}{2}\left(\frac{\pi-\theta}{2}\right).$$

Let $\theta = (2j-1)\pi/k$. We get

$$T_k = \frac{2}{k}\sum_{j=1}^{[k/2]}\left(\ln\left(2\sin\frac{(2j-1)\pi}{2k}\right) + \frac{(k-(2j-1))\pi}{2k}\cot\frac{(2j-1)\pi}{2k}\right)$$

$$= \frac{2}{k}\sum_{j=1}^{[k/2]}\ln\left(2\sin\frac{(2j-1)\pi}{2k}\right) + \frac{\pi}{k^2}\sum_{j=1}^{[k/2]}(k-(2j-1))\cot\frac{(2j-1)\pi}{2k}.$$

If k is odd, then $j = (k+1)/2$ in the first sum corresponds to $\xi = -1$. In view of

$$\frac{2}{k}\int_0^1 \frac{1/2}{1+x}\,dx = \frac{1}{k}\ln\left(2\sin\frac{\pi}{2}\right),$$

and the well-known fact that $\prod_{j=1}^k \sin\frac{(2j-1)\pi}{2k} = 1/2^{k-1}$, the first sum becomes

$$\frac{1}{k}\sum_{j=1}^k \ln\left(2\sin\frac{(2j-1)\pi}{2k}\right) = \frac{1}{k}\ln 2.$$

In summary, we find that

$$T_k = \frac{1}{k}\ln 2 + \frac{\pi}{k^2}\sum_{j=1}^{[k/2]}(k+1-2j)\cot\frac{(2j-1)\pi}{2k}.$$

Substituting T_k into S_k yields the proposed identity. $\qquad\square$

Remark. There is another representation for T_k:

$$T_k = \frac{1}{k}\ln 2 + \frac{\pi}{2k}\sum_{l=1}^{k-1}\csc\left(\frac{l\pi}{k}\right). \tag{2.26}$$

To this end, notice that

$$\int_0^1 \frac{P_k(x) - x^{k-1}}{1+x^k}\,dx = \int_0^1 \frac{1 + x + \cdots + x^{k-2}}{1+x^k}\,dx$$

$$= \frac{1}{2}\sum_{l=1}^{k-1}\int_0^1 \frac{x^{l-1} + x^{k-l-1}}{1+x^k}\,dx.$$

The substitution $x = 1/t$ yields

$$\int_0^1 \frac{x^{k-l-1}}{1 + x^k}\, dx = \int_1^\infty \frac{t^{l-1}}{1 + t^k}\, dt.$$

Thus, we have

$$\int_0^1 \frac{x^{l-1} + x^{k-l-1}}{1 + x^k}\, dx = \int_0^\infty \frac{x^{l-1}}{1 + x^k}\, dx = \frac{\pi}{k} \csc\left(\frac{l\pi}{k}\right)$$

and so

$$\int_0^1 \frac{P_k(x) - x^{k-1}}{1 + x^k}\, dx = \frac{\pi}{2k} \sum_{l=1}^{k-1} \csc\left(\frac{l\pi}{k}\right),$$

from which (2.26) follows. These two different ways to calculate T_k indeed present a solution to the following
Problem 11873 (Proposed by E. Ionascu, 122(10), 2015). Show that for $n \in \mathbb{N}$ with $n \geq 2$,

$$\sum_{j=1}^n \left(1 - \frac{2j-1}{n}\right) \cot \frac{(2j-1)\pi}{2n} = \sum_{j=1}^{n-1} \csc\left(\frac{j\pi}{n}\right).$$

A nice generalization is

$$\sum_{j=1}^n \frac{\sin((n-2j+1)y)}{\sin ny} \cot \frac{(2j-1)\pi}{2n} = \sum_{j=1}^{n-1} \csc\left(y + \frac{j\pi}{n}\right),$$

for any $y \neq m\pi/n$ for some $m \in \mathbb{N}$. Can you provide a proof?
  Let ψ be the digamma function. Applying

$$\psi(x + n) = \frac{1}{x} + \frac{1}{x+1} + \cdots + \frac{1}{x - n - 1} + \psi(x),$$

we also find that

$$S_k = -\frac{1}{2k} \sum_{j=1}^{k-1} \psi\left(\frac{1}{2} + \frac{j}{2k}\right) - \frac{k-1}{2k}\gamma,$$

where γ is Euler's constant. The proposed identity can be derived by formulas $\psi(x) - \psi(1 - x) = -\pi \cot(\pi x)$ and

$$\psi(p/q) = -\gamma - \frac{\pi}{2} \cot \frac{p\pi}{q} - \ln q + 2 \sum_{n=1}^{[q/2]} \cos \frac{2np\pi}{q} \ln\left(2 \sin \frac{n\pi}{q}\right)$$

where $0 < p < q$.

One similar problem appeared in the SIAM Problems and Solutions Online (no longer active, for archival purposes only now): For $k \geq 2$, find a closed form for

$$\mathcal{S}_k = \sum_{n=1}^{\infty} \frac{\ln k - (H_{kn} - H_n)}{n}.$$

By using the *Multisection formula* and polylogarithm functions, Rousseau obtained

$$\mathcal{S}_k = \frac{(k-1)(k+2)}{4k} \zeta(2) - \frac{1}{2} \ln^2 k - \frac{1}{2} \sum_{j=1}^{k-1} \ln^2 \left(2 \sin \frac{j\pi}{k} \right).$$

See `https://archive.siam.org/journals/problems/downloadfiles/06-007s.pdf`

Indeed, by (2.25), we obtain

$$\mathcal{S}_k = -\int_0^1 \frac{P_k'(x)}{P_k(x)} \ln(1 - x^k) \, dx = -\frac{1}{2} \ln^2 k - \int_0^1 \frac{P_k'(x)}{P_k(x)} \ln(1 - x) \, dx.$$

Now, it remains to show that

$$\int_0^1 \frac{P_k'(x)}{P_k(x)} \ln(1 - x) \, dx = \frac{(k-1)(k+2)}{4k} \zeta(2) - \frac{1}{2} \sum_{j=1}^{k-1} \ln^2 \left(2 \sin \frac{j\pi}{k} \right).$$

I leave the proof to the reader. As additional practice, the reader may try

1. Determine a closed form for

$$\mathcal{T}_k = \sum_{n=1}^{\infty} (-1)^{n-1} \frac{H_{kn}}{n} \quad \text{for } k \geq 1.$$

2. **Problem 12189** (Proposed by H. Katsuura, 127(6), 2020). Evaluate

$$\int_0^1 \frac{(k+1)x^k - \sum_{m=0}^k x^{mk}}{x^{k(k+1)} - 1} \, dx,$$

where k is a positive integer.
Hint: Let $P_k(x) = \sum_{i=0}^k x^i$. Prove that

$$\int_0^1 \frac{(k+1)x^k - P_k(x)}{x^{k+1} - 1} \, dx = \int_0^1 \frac{P_k'(x)}{P_k(x)} \, dx = \ln(k+1).$$

2.7 A nonlinear harmonic sum

Problem 12060 (Proposed by O. Furdui and A. Sîntămărian, 125(7), 2018). Let $\zeta(3)$ be Apéry's constant $\sum_{n=1}^{\infty} 1/n^3$, and let H_n be the nth harmonic number $1 + 1/2 + \cdots + 1/n$. Prove

$$\sum_{n=2}^{\infty} \frac{H_n H_{n+1}}{n^3 - n} = \frac{5}{2} - \frac{\pi^2}{24} - \zeta(3).$$

Discussion.

In contrast to the series in Sections 2.3 and 2.5, this proposed problem involves products of two harmonic numbers. We now try to reformulate the nonlinear harmonic sum into linear form by manipulation of the summations.

Solution.

Let the proposed series be S. Since

$$\frac{1}{n^3 - n} = \frac{1}{2} \left(\frac{1}{n(n-1)} - \frac{1}{n(n+1)} \right),$$

we have

$$
\begin{aligned}
S &= \frac{1}{2} \left(\sum_{n=2}^{\infty} \frac{H_n H_{n+1}}{n(n-1)} - \sum_{n=2}^{\infty} \frac{H_n H_{n+1}}{n(n+1)} \right) \\
&= \frac{1}{2} \left(\sum_{n=1}^{\infty} \frac{H_{n+1} H_{n+2}}{n(n+1)} - \sum_{n=2}^{\infty} \frac{H_n H_{n+1}}{n(n+1)} \right) \\
&\qquad \text{(shifting the index in the first sum)} \\
&= \frac{1}{2} \sum_{n=1}^{\infty} \frac{H_{n+1}(H_{n+2} - H_n)}{n(n+1)} + \frac{3}{8} \\
&= \frac{1}{2} \sum_{n=1}^{\infty} \frac{H_{n+1}}{n(n+1)^2} + \frac{1}{2} \sum_{n=1}^{\infty} \frac{H_{n+1}}{n(n+1)(n+2)} + \frac{3}{8}. \quad (2.27)
\end{aligned}
$$

To proceed forward, we need the following two auxiliary results:

(a) $\displaystyle \sum_{n=1}^{\infty} \frac{H_{n+1}}{n(n+1)} = 2.$

(b) $\displaystyle \sum_{n=1}^{\infty} \frac{H_{n+1}}{(n+1)^2} = 2\zeta(3) - 1.$

In view of

$$\frac{H_{n+1}}{n(n+1)} = \frac{H_n + 1/(n+1)}{n} - \frac{H_{n+1}}{n+1},$$

(a) follows from

$$\sum_{n=1}^{\infty} \frac{H_{n+1}}{n(n+1)} = \sum_{n=1}^{\infty} \left(\frac{H_n}{n} - \frac{H_{n+1}}{n+1} \right) + \sum_{n=1}^{\infty} \frac{1}{n(n+1)} = 2.$$

(b) is the direct consequence of the well-known Euler formula (2.16)

$$\sum_{n=1}^{\infty} \frac{H_n}{n^2} = 2\zeta(3).$$

Now we are ready to prove the claimed result. Using partial fractions

$$\frac{1}{n(n+1)^2} = \frac{1}{n(n+1)} - \frac{1}{(n+1)^2},$$

then applying (a) and (b), we find that

$$\sum_{n=1}^{\infty} \frac{H_{n+1}}{n(n+1)^2} = \sum_{n=1}^{\infty} \frac{H_{n+1}}{n(n+1)} - \sum_{n=1}^{\infty} \frac{H_{n+1}}{(n+1)^2} = 3 - 2\zeta(3). \qquad (2.28)$$

Similarly, using partial fractions

$$\frac{1}{n(n+1)(n+2)} = \frac{1}{2}\left(\frac{1}{n(n+1)} - \frac{1}{(n+1)(n+2)}\right),$$

we have

$$
\begin{aligned}
\sum_{n=1}^{\infty} \frac{H_{n+1}}{n(n+1)(n+2)} &= \frac{1}{2}\sum_{n=1}^{\infty}\left(\frac{H_{n+1}}{n(n+1)} - \frac{H_{n+1}}{(n+1)(n+2)}\right) \\
&= \frac{1}{2}\left(\sum_{n=1}^{\infty}\frac{H_{n+1} - H_n}{n(n+1)} + \frac{1}{2}\right) \\
&= \frac{1}{4} + \frac{1}{2}\sum_{n=1}^{\infty}\frac{1}{n(n+1)^2} \\
&= \frac{1}{4} + \frac{1}{2}\sum_{n=1}^{\infty}\left(\frac{1}{n(n+1)} - \frac{1}{(n+1)^2}\right) \\
&= \frac{1}{4} + \frac{1}{2}\left(2 - \frac{\pi^2}{6}\right) = \frac{5}{4} - \frac{\pi^2}{12}.
\end{aligned}
$$

Plugging this result and (2.28) into (2.27) yields

$$S = \frac{3}{2} - \zeta(3) + \frac{5}{8} - \frac{\pi^2}{24} + \frac{3}{8} = \frac{5}{2} - \frac{\pi^2}{24} - \zeta(3)$$

as desired. □

Remark. We may attack this problem directly by using

$$\int_0^1 x^{n-1}\ln(1-x)\,dx = -\frac{H_n}{n} \qquad (2.29)$$

and by transforming the series into an integral. We single out one calculation as follows:

$$I := \sum_{n=2}^{\infty} \frac{H_n H_{n+1}}{n(n+1)} = \zeta(2) + 2\zeta(3).$$

To see this, by (2.29), we compute

$$
\begin{aligned}
I &= \sum_{n=1}^{\infty} \int_0^1 y^{n-1} \ln(1-y)\, dy \int_0^1 x^n \ln(1-x)\, dx \\
&= \int_0^1 \int_0^1 x \left(\sum_{n=1}^{\infty} (xy)^{n-1} \right) \ln(1-x) \ln(1-y)\, dx dy \\
&= \int_0^1 \int_0^1 \frac{x \ln(1-x) \ln(1-y)}{1-xy}\, dx dy \\
&= \int_0^1 x \ln(1-x) \left(\int_0^1 \frac{\ln(1-y)}{1-xy}\, dy \right) dx.
\end{aligned}
$$

For the inner integration, the substitution $1 - xy = u$ yields

$$
\begin{aligned}
\int_0^1 \frac{\ln(1-y)}{1-xy}\, dy &= \frac{1}{x} \int_{1-x}^1 \left(\frac{\ln(1/x)}{u} + \frac{\ln(x-1+u)}{u} \right) du \\
&= \frac{1}{x} \ln x \ln(1-x) + \frac{1}{x} \int_{1-x}^1 \frac{\ln(x-1+u)}{u}\, du.
\end{aligned}
$$

Another substitution $(1-x)u \to u$ gives

$$\int_0^1 \frac{\ln(1-y)}{1-xy}\, dy = \frac{1}{x} \left(-\frac{1}{2} \ln^2(1-x) - \mathrm{Li}_2(x) \right),$$

where $\mathrm{Li}_2(x)$ is the dilogarithm function. Thus, we have

$$I = -\frac{1}{2} \int_0^1 \ln^3(1-x)\, dx - \int_0^1 \ln(1-x) \mathrm{Li}_2(x)\, dx = \zeta(2) + 2\zeta(3).$$

The following six problems will provide additional practice:

1. Prove

$$\sum_{n=2}^{\infty} \frac{H_n^2}{n^3 - n} = \frac{7}{4} + \frac{\pi^2}{12} - \frac{3}{2}\zeta(3).$$

2. **Problem 11302** (Proposed by H. Alzer, 114(6), 2007). Find

$$\sum_{k=2}^{\infty} \frac{(2k+1)H_k^2}{(k-1)k(k+1)(k+2)}.$$

3. **Problem 11633** (Proposed by A. Sofa, 119(3), 2012). For real a, let $H_n(a) = \sum_{j=1}^n 1/j^a$. Show that for integers a, b and n with $a \geq 1, b \geq 0$, and $n \geq 1$,

$$\sum_{k=1}^n \frac{H_k(1)^2 + H_k(2)}{(k+b)^a} + 2 \sum_{k=1}^n \frac{H_k(1)H_{k+b-1}(a)}{k}$$
$$= H_{n+b}(a)(H_n(1)^2 + H_n(2)).$$

4. **Problem 11802** (Proposed by I. Mezö, 121(8), 2014). Let $H_{n,2} = \sum_{k=1}^n k^{-2}$ and let $D_n = n! \sum_{k=0}^n (-1)^k/k!$ (i.e., the *derangement number* of n). Prove

$$\sum_{n=1}^\infty \frac{(-1)^n H_{n,2}}{n!} = \frac{\pi^2}{6e} - \sum_{n=0}^\infty \frac{D_n}{n!(n+1)^2}.$$

5. **Problem 11921** (Proposed by C. I. Vălean, 123(6), 2016). Prove

$$\ln^2 2 \sum_{k=1}^\infty \frac{H_k}{(k+1)2^{k+1}} + \ln 2 \sum_{k=1}^\infty \frac{H_k}{(k+1)^2 2^k} + \sum_{k=1}^\infty \frac{H_k}{(k+1)^3 2^k}$$
$$= \frac{1}{4}\left(\zeta(4) + \ln^4 2\right).$$

Here $H_k = \sum_{j=1}^k 1/j$ and ζ denotes the Riemann zeta function.

6. **Nonlinear Sums.** Let $H_n^{(2)} = 1 + 1/2^2 + \cdots + 1/n^2$. Show that

$$\ln^2 2 \sum_{k=1}^\infty \frac{H_k^{(2)}}{(k+1)2^{k+1}} + \sum_{k=1}^\infty \frac{H_k^{(2)}}{(k+1)^2 2^{k+1}}$$
$$= \frac{1}{16}\zeta(4) + \frac{1}{4}\zeta(2)\ln^2 2 - \frac{1}{8}\ln^4 2.$$

2.8   A series involving Riemann zeta values

Problem 11400 (Proposed by P. Bracken, 115(10), 2008). Let ζ be the Riemann zeta function. Evaluate $\sum_{n=1}^\infty \zeta(2n)/n(n+1)$ in closed form.

Discussion.
One way to proceed is to use the power series of $\sum_{n=1}^\infty x^n/n(n+1)$ and to convert the problem into an infinite product. Another way to proceed is to start with the generating function of $\zeta(2n)/n$ and to transform the series into some well-known integrals. Based on the discussion, we now show the required closed form equals $\ln(2\pi) - 1/2$ in two distinct ways.

Solution I.
Denote the proposed series by S. Note that

$$S = \sum_{n=1}^{\infty} \frac{1}{n(n+1)} \sum_{k=1}^{\infty} \frac{1}{k^{2n}} = \sum_{k=1}^{\infty} \sum_{n=1}^{\infty} \frac{1}{n(n+1)k^{2n}},$$

where interchanging the order of summation is justified by the positivity of summands. Recall the power series

$$\sum_{n=1}^{\infty} \frac{x^n}{n(n+1)}$$

$$= \frac{x + \ln(1-x) - x\ln(1-x)}{x}.$$

We obtain

$$\begin{aligned}
S &= 1 + \sum_{k=2}^{\infty} \left(1 + (k^2 - 1) \ln\left(1 - \frac{1}{k^2} \right) \right) \\
&= 1 + \ln\left(\prod_{k=2}^{\infty} e \left(\frac{k^2 - 1}{k^2} \right)^{k^2 - 1} \right).
\end{aligned}$$

To compute the infinite product above, let

$$P_N = \prod_{k=2}^{N} e \left(\frac{k^2 - 1}{k^2} \right)^{k^2 - 1}.$$

We have

$$\begin{aligned}
P_N &= \left(\frac{1 \cdot 3}{2^2} \right)^{2^2 - 1} \left(\frac{2 \cdot 4}{3^2} \right)^{3^2 - 1} \cdots \left(\frac{(N-2) \cdot N}{(N-1)^2} \right)^{(N-1)^2 - 1} \\
&\times \left(\frac{(N-1) \cdot (N+1)}{N^2} \right)^{N^2 - 1}.
\end{aligned}$$

Observe that, for each $2 \le k \le N - 1$, k appears in the numerator with exponents

$$[(k-1)^2 - 1] + [(k+1)^2 - 1] = 2k^2$$

and in the denominator with exponent $2(k^2 - 1)$. After cancelation, only a factor of k^2 remains in the numerator. The factor N appears in the numerator with exponent $N^2 - 2N$ and in the denominator with exponent $2(N^2 - 1)$. After cancelation, N remains in the denominator with exponent $N^2 + 2N - 2$. The factor $(N+1)$ occurs only in the numerator with exponent $N^2 - 1$. Therefore,

$$P_N = \frac{e^{N-1}(N-1)!^2(N+1)^{N^2-1}}{N^{N^2+2N-2}} = \frac{e^{N-1}(N!)^2(N+1)^{N^2-1}}{N^{N^2+2N}}.$$

Rewrite

$$P_N = \frac{2\pi}{e^2} \cdot \left(\frac{(N+1)^{N^2-1}}{e^{N-1}N^{N^2-1}} \right) \cdot \left(\frac{e^N N!}{N^N \sqrt{2\pi N}} \right)^2.$$

Using Stirling's formula, we have

$$\lim_{N \to \infty} \frac{e^N N!}{N^N \sqrt{2\pi N}} = 1;$$

also

$$\lim_{N \to \infty} \frac{(N+1)^{N^2-1}}{e^{N-1}N^{N^2-1}} = \exp\left(\lim_{N \to \infty} \left((N^2-1)\ln(1+1/N) - N + 1 \right) \right) = e^{1/2}.$$

Hence

$$\prod_{k=2}^{\infty} e\left(\frac{k^2-1}{k^2} \right)^{k^2-1} = \lim_{N \to \infty} P_N = \frac{2\pi}{e^2} \cdot e^{1/2} = \frac{2\pi}{e^{3/2}},$$

and so

$$S = 1 + \ln\left(\frac{2\pi}{e^{3/2}} \right) = \ln(2\pi) - \frac{1}{2}.$$

$\square$

Solution II.
We begin by determining the generating function of $\{\zeta(2n)/n\}$. For $x \in [0,1)$, let

$$G(x) := \sum_{n=1}^{\infty} \frac{\zeta(2n)}{n} x^n.$$

We have

$$
\begin{aligned}
G(x) &= \sum_{n=1}^{\infty} \frac{1}{n} \left(\sum_{m=1}^{\infty} \left(\frac{\sqrt{x}}{m} \right)^{2n} \right) = \sum_{m=1}^{\infty} \left(\sum_{n=1}^{\infty} \frac{1}{n} \left(\frac{\sqrt{x}}{m} \right)^{2n} \right) \\
&= \sum_{m=1}^{\infty} -\ln\left[1 - \left(\frac{\sqrt{x}}{m} \right)^2 \right] = -\ln\left(\prod_{m=1}^{\infty} \left[1 - \left(\frac{\sqrt{x}}{m} \right)^2 \right] \right) \\
&= -\ln\left(\frac{\sin(\pi\sqrt{x})}{\pi\sqrt{x}} \right).
\end{aligned}
$$

Here interchanging the order of summation is justified by the absolutely convergence. In the last equation, we have used Euler's sine infinite product formula:

$$\frac{\sin(\pi z)}{\pi z} = \prod_{n=1}^{\infty} \left(1 - \frac{z^2}{n^2} \right).$$

Since the series converges uniformly, term by term integration yields

$$S = \int_0^1 G(x)\,dx = -\int_0^1 \ln\left(\frac{\sin(\pi\sqrt{x})}{\pi\sqrt{x}}\right)dx$$

$$= -\int_0^1 \ln(\sin(\pi\sqrt{x}))\,dx + \int_0^1 \ln(\pi\sqrt{x})\,dx \quad (\text{let } t = \pi\sqrt{x})$$

$$= -\frac{2}{\pi^2}\int_0^\pi t\ln(\sin t)\,dt + \ln\pi - \frac{1}{2}.$$

By symmetry, we have

$$\int_0^\pi t\ln(\sin t)\,dt = \int_0^\pi (\pi-t)\ln(\sin(\pi-t))\,dt = \pi\int_0^\pi \ln(\sin t)\,dt - \int_0^\pi t\ln(\sin t)\,dt.$$

From the well-known integral $\int_0^\pi \ln(\sin t)\,dt = -\pi\ln 2$, we have

$$\int_0^\pi t\ln(\sin t)\,dt = \frac{\pi}{2}\int_0^\pi \ln(\sin t)\,dt = -\frac{\pi}{2}\pi\ln 2 = -\frac{\pi^2}{2}\ln 2.$$

Hence,

$$S = -\frac{2}{\pi^2}\cdot\left(-\frac{\pi^2}{2}\ln 2\right) + \ln\pi - \frac{1}{2} = \ln(2\pi) - \frac{1}{2}.$$

$\square$

Remark. The generating function $G(x)$ in Solution II also can be derived from

$$\sum_{n=0}^\infty \zeta(2n)x^n = -\frac{1}{2}\pi\sqrt{x}\cot(\pi\sqrt{x}),$$

from which, together with the power series of cotangent, Euler found the exact value of $\zeta(2n)$.

Rewrite $G(x)$ as

$$\sum_{n=1}^\infty \frac{\zeta(2n)}{n}x^{2n} = -\ln\left(\frac{\sin(\pi x)}{\pi x}\right).$$

From here it follows that

$$\sum_{n=1}^\infty \frac{\zeta(2n)}{n2^{2n}} = \ln\frac{\pi}{2} \quad\text{and}\quad \sum_{n=1}^\infty \frac{\zeta(2n)}{n(2n+1)} = \ln(2\pi) - 1.$$

Many series involving the Riemann zeta function values have been developed by various ways in recent years [85]. Here we collect a few identities for your verification.

1. $(1 - 2^{1-s})\zeta(s) = \sum_{n=1}^\infty \frac{(s)_n}{n!}\frac{\zeta(s+n)}{2^{s+n}}$, where $(s)_n = s(s+1)\cdots(s+n-1)$.

2. $(1 - 2^{1-s})\zeta(s) = 1 - \sum_{n=1}^{\infty} (-1)^{n-1} \frac{(s)_n}{n!} \frac{\zeta(s+n)}{2^{s+n}}$.

3. $\psi(x) = -\gamma + \sum_{n=2}^{\infty} (-1)^n \zeta(n) x^{n-1}$, where ψ is the digamma function and γ denotes Euler's constant.

4. $\sum_{n=1}^{\infty} \frac{\zeta(2n)}{(n+1)(n+2)} = \frac{1}{2}$.

5. $\sum_{n=1}^{\infty} \frac{\zeta(2n+1)}{(2n+1)2^{2n}} = \ln 2 - \gamma$.

6. $\sum_{n=1}^{\infty} \frac{\zeta(2n)-1}{n} = \ln 2$, $\sum_{n=1}^{\infty} \frac{\zeta(2n)-1}{n+1} = \frac{3}{2} - \ln \pi$.

7. $\sum_{n=1}^{\infty} \frac{\zeta(2n+1)-1}{2n+3} = \frac{13}{12} - \frac{1}{3}\gamma - \frac{1}{2}\ln 2 - 2\ln A$, where A denotes the Glaisher-Kinkelin constant.

8. $\sum_{n=2}^{\infty} \frac{\alpha n+\beta}{n(n+1)}(\zeta(n) - 1) = \beta - \frac{1}{2}(\alpha + \beta)\gamma + \frac{1}{2}(\alpha - \beta)(3 - \ln(2\pi))$, where α and β are constants.

9. **Problem 11333** (Proposed by P. F. Refolio, 114(10), 2007). Show that

$$\prod_{n=2}^{\infty} \left(\left(\frac{n^2 - 1}{n^2} \right)^{2(n^2-1)} \left(\frac{n+1}{n-1} \right)^n \right) = \pi.$$

10. **Problem 11793** (Proposed by I. Mező, 121(7), 2014). Prove that

$$\sum_{n=1}^{\infty} \frac{\ln(n+1)}{n^2} = -\zeta'(2) + \sum_{n=3}^{\infty} (-1)^{n+1} \frac{\zeta(n)}{n-2},$$

where ζ denotes the Riemann zeta function and ζ' denotes its derivative.

2.9 Abel theorem continued

Problem 11755 (Proposed by P. P. Dályay, 121(2), 2014). Compute

$$\sum_{n=1}^{\infty} \frac{(-1)^n}{2n - 1} \sum_{k=n+1}^{\infty} \frac{(-1)^{k-1}}{2k - 1}.$$

Discussion.
There are many ways to attack this problem. Here we single out two. Denote the double series by S.

1. By symmetry, we have

$$S = -\frac{1}{2} \left(\sum_{n=1}^{\infty} \frac{(-1)^{n-1}}{2n - 1} \sum_{k=1}^{\infty} \frac{(-1)^{k-1}}{2k - 1} - \sum_{n=1}^{\infty} \left(\frac{(-1)^{n-1}}{2n - 1} \right)^2 \right).$$

2. Using $1/(2k-1) = \int_0^1 x^{2(k-1)}\,dx$ yields

$$\sum_{k=n+1}^{\infty} \frac{(-1)^{k-1}}{2k-1} = \int_0^1 \frac{(-1)^n x^{2n}}{1+x^2}\,dx.$$

Consequently we convert the problem into the calculation of an integral:

$$S = \frac{1}{2} \int_0^1 \frac{x\ln(1+x) - x\ln(1-x)}{1+x^2}\,dx.$$

We now present two solutions based on the above observations.

Solution I.
Since $\sum_{n=1}^{\infty} \frac{(-1)^{n-1}}{2n-1}$ only converges conditionally, to justify the rearrangement of terms of the double series by Fubini's theorem, we introduce $a_n(x) = \frac{(-1)^{n-1} x^{2n-1}}{2n-1}$. For $x \in (0,1)$, define

$$f(x) := \sum_{n=1}^{\infty} a_n(x) \sum_{k=n+1}^{\infty} a_k(x).$$

The absolute convergence of $\sum_{n=1}^{\infty} a_n(x)$ now allows interchanging the order of summation. Thus

$$f(x) = \sum_{k=1}^{\infty} a_k(x) \sum_{n=1}^{k-1} a_k(x).$$

By symmetry, we find that

$$\begin{aligned}
f(x) &= \frac{1}{2}\left(\sum_{n=1}^{\infty} a_n(x)\sum_{k=1}^{\infty} a_k(x) - \sum_{k=1}^{\infty} a_k^2(x)\right)\\
&= \frac{1}{2}\left(\arctan^2(x) - \sum_{k=1}^{\infty} a_k^2(x)\right).
\end{aligned}$$

By Abel's limit theorem, we obtain

$$f(1) = \lim_{x\to 1^-} f(x) = \frac{1}{2}\left(\left(\frac{\pi}{4}\right)^2 - \sum_{k=1}^{\infty}\frac{1}{(2k-1)^2}\right) = \frac{1}{2}\left(\frac{\pi^2}{16} - \frac{\pi^2}{8}\right) = -\frac{\pi^2}{32},$$

which implies that

$$S = -f(1) = \frac{\pi^2}{32}.$$

$\square$

Solution II.
We compute

$$
\sum_{k=n+1}^{\infty} \frac{(-1)^{k-1}}{2k-1} = \sum_{k=n+1}^{\infty} (-1)^{k-1} \int_0^1 x^{2(k-1)} dx
$$

$$
= \int_0^1 \left(\sum_{k=n+1}^{\infty} (-1)^{k-1} x^{2(k-1)} \right) dx
$$

$$
= \int_0^1 \frac{(-1)^n x^{2n}}{1+x^2} dx.
$$

Then

$$
S = \sum_{n=1}^{\infty} \frac{(-1)^n}{2n-1} \int_0^1 \frac{(-1)^n x^{2n}}{1+x^2} dx
$$

$$
= \int_0^1 \frac{1}{1+x^2} \left(\sum_{n=1}^{\infty} \frac{x^{2n}}{2n-1} \right) dx
$$

$$
= \frac{1}{2} \int_0^1 \frac{x(\ln(1+x) - \ln(1-x))}{1+x^2} dx,
$$

where we have used

$$
\sum_{n=1}^{\infty} \frac{1}{2n-1} x^{2n} = \frac{1}{2} x(\ln(1+x) - \ln(1-x)).
$$

Next, the substitution $x = \frac{1-t}{1+t}$ yields

$$
S = -\frac{1}{2} \int_0^1 \frac{(1-t)\ln t}{(1+t)(1+t^2)} dt.
$$

By partial fractions

$$
\frac{1-t}{(1+t)(1+t^2)} = \frac{1}{1+t} - \frac{t}{1+t^2},
$$

using the geometric series and integration by parts, we obtain

$$
\int_0^1 \frac{\ln t}{1+t} dt = \sum_{n=0}^{\infty} (-1)^n \int_0^1 t^n \ln t \, dt
$$

$$
= \sum_{n=0}^{\infty} (-1)^{n+1} \frac{1}{(n+1)^2} = -\frac{\pi^2}{12};
$$

$$
\int_0^1 \frac{t \ln t}{1+t^2} dt = \sum_{n=0}^{\infty} (-1)^n \int_0^1 t^{2n+1} \ln t \, dt
$$

$$
= \sum_{n=0}^{\infty} (-1)^{n+1} \frac{1}{4(n+1)^2} = -\frac{\pi^2}{48}.
$$

In summary, we find that

$$S = -\frac{1}{2}\left(-\frac{\pi^2}{12} + \frac{\pi^2}{48}\right) = \frac{\pi^2}{32}.$$

$\square$

Remark. We can avoid the absolute convergence in Solution I by considering the finite sum:

$$S_N := \sum_{n=1}^{N} a_n b_n,$$

where

$$a_n = \frac{(-1)^n}{2n-1} \quad \text{and} \quad b_n = \sum_{k=n+1}^{\infty} \frac{(-1)^{k-1}}{2k-1}.$$

By Abel's summation formula, we have

$$S_N = \frac{1}{2}\left(\left(\sum_{n=1}^{N} a_n\right) b_N + \sum_{n=1}^{\infty} a_n \sum_{n=1}^{N} a_n - \sum_{n=1}^{N} a_n^2\right).$$

Letting $N \to \infty$ yields that $S = \pi^2/32$ immediately.

Recall the dilogarithm function defined by $\mathrm{Li}_2(x) = \sum_{k=1}^{\infty} x^k/k^2$. In Solution I, we explicitly obtain

$$\sum_{k=1}^{\infty} a_k^2(x) = \frac{1}{2}\left(\mathrm{Li}_2(x^2) - \frac{1}{4}\mathrm{Li}_2(x^4)\right).$$

Following the idea Euler first used to show that $\zeta(2) = \pi^2/6$, we present another charming solution to this proposed problem. By Euler's infinite product of sine, we have

$$\cos x - \sin x = \prod_{n=1}^{\infty}\left(1 + \frac{(-1)^n 4x}{(2n-1)\pi}\right).$$

Replacing x by $\pi x/4$ yields

$$\cos\left(\frac{\pi x}{4}\right) - \sin\left(\frac{\pi x}{4}\right) = \prod_{n=1}^{\infty}\left(1 + \frac{(-1)^n x}{2n-1}\right). \tag{2.30}$$

Now we compute the coefficient of x^2 on both sides of (2.30). Clearly,

$$[x^2]\left[\cos\left(\frac{\pi x}{4}\right) - \sin\left(\frac{\pi x}{4}\right)\right]$$
$$= [x^2]\left[1 - \frac{1}{2}\left(\frac{\pi x}{4}\right)^2 + \cdots + \left(\frac{\pi x}{4}\right) - \frac{1}{3!}\left(\frac{\pi x}{4}\right)^3 + \cdots\right] = -\frac{\pi^2}{32}.$$

On the other hand, we have

$$[x^2] \prod_{n=1}^{\infty} \left(1 + \frac{(-1)^n x}{2n-1}\right) = \sum_{n=1}^{\infty} \frac{(-1)^n}{2n-1} \sum_{k=n+1}^{\infty} \frac{(-1)^k}{2k-1}.$$

Thus, we conclude the desired answer by equating the coefficient of x^2 in (2.30).

A unified treatment of the summation of certain iterated series of the form $\sum_{n=1}^{\infty} \sum_{m=1}^{\infty} a_{n+m}$ is studied in [44]. Under certain conditions, the double iterated series can be represented as the difference of two single series.

We end this section with four more Monthly problems for additional practice.

1. **Problem 11519** (Proposed by O. Furdui, 117(7), 2010). Find

$$\sum_{n=1}^{\infty} \sum_{m=1}^{\infty} (-1)^{n+m} \frac{H_{n+m}}{n+m},$$

   where H_n denotes the nth harmonic number.

2. **Problem 11682** (Proposed by O. Furdui, 119(10), 2012). Compute

$$\sum_{n=0}^{\infty} (-1)^n \left(\sum_{k=1}^{\infty} \frac{(-1)^{k-1}}{n+k}\right)^2.$$

3. **Problem 12134** (Proposed by P. Bracken, 126(8), 2019). Evaluate the series

$$\sum_{n=1}^{\infty} \left(n \left(\sum_{k=n}^{\infty} \frac{1}{k^2}\right) - 1 - \frac{1}{2n}\right).$$

4. **Problem 12194** (Proposed by M. Tetiva, 127(6), 2020). Let $\gamma_n = -\ln n + \sum_{k=1}^{n} 1/k$, and let γ be Euler's constant $\lim_{n\to\infty} \gamma_n$. Evaluate

$$\sum_{n=1}^{\infty} \left(H_n - \ln n - \gamma - \frac{1}{2n}\right).$$

2.10 A convergence test

Problem 11829 (Proposed by P. Bracken, 122(3), 2015). Let a_n be a monotone decreasing sequence of real numbers that converges to 0. Prove that $\sum_{n=1}^{\infty} a_n/n < \infty$ if and only if $a_n = O(1/\ln n)$ and $\sum_{n=1}^{\infty} (a_n - a_{n+1}) \ln n < \infty$.

Discussion.
Using Abel's summation formula, we have

$$\sum_{k=1}^{n} \frac{a_n}{n} = \sum_{k=1}^{n-1} H_k(a_k - a_{k+1}) + H_n a_n,$$

where $H_k = \sum_{i=1}^{k} 1/i$ is the kth harmonic number. Since $H_n \sim \ln n$ for sufficiently large n, by the comparison test, we see that $\sum_{n=1}^{\infty} (a_n - a_{n+1}) \ln n < \infty$ if and only if $\sum_{n=1}^{\infty} (a_n - a_{n+1}) H_n < \infty$. Our solution below is based on this observation.

Solution.
Let

$$S_n = \sum_{k=1}^{n} \frac{a_k}{k}, \quad T_n = \sum_{k=1}^{n} (a_k - a_{k+1}) \ln k.$$

By the assumption of a_n, we see both S_n and T_n are monotone increasing. Let their limits be S and T, respectively.

" $\Rightarrow$ " Assume $S < \infty$. Since a_n is decreasing, we have, for all $n \in \mathbb{N}$

$$a_n H_n = a_n \sum_{k=1}^{n} \frac{1}{n} \leq \sum_{k=1}^{n} \frac{a_k}{k} = S_n \leq S.$$

This implies that $a_n = O(1/H_n) = O(1/\ln n)$ because $H_n \sim \ln n$ for sufficiently large n. Moreover,

$$
\begin{aligned}
T_n &= \sum_{k=1}^{n} a_k \ln k - \sum_{k=1}^{n} a_{k+1} \ln k \\
&= \sum_{k=2}^{n} a_k (\ln k - \ln(k-1)) - a_{n+1} \ln n \\
&\leq \sum_{k=2}^{n} \frac{a_k}{k-1} \leq \sum_{k=2}^{n} \frac{a_{k-1}}{k-1} \\
&= S_{n-1} \leq S.
\end{aligned}
$$

This shows that $T < \infty$.

" $\Leftarrow$ " Assume that $a_n = O(1/\ln n)$ and $T < \infty$. First, we show that $a_n \ln n$ is uniformly bounded. Observer that, for $m > n \geq 1$,

$$(a_n - a_{m+1}) \ln n = \sum_{k=n}^{m} (a_k - a_{k+1}) \ln n \leq \sum_{k=n}^{m} (a_k - a_{k+1}) \ln k \leq T.$$

Since $\lim_{n \to \infty} a_n = 0$, letting $m \to \infty$ yields

$$a_n \ln n \leq T \quad \text{for all } n \geq 1.$$

Moreover, we have

$$
\begin{aligned}
S_n - a_1 &= \sum_{k=2}^{n} \frac{a_k}{k} \leq \sum_{k=2}^{n} a_k (\ln k - \ln(k-1)) \\
&= \sum_{k=2}^{n} a_k \ln k - \sum_{k=2}^{n-1} a_{k+1} \ln k \\
&= \sum_{k=2}^{n-1} (a_k - a_{k+1}) \ln k + a_n \ln n \\
&\leq T + T = 2T.
\end{aligned}
$$

This proves that $S < \infty$ as well. $\qquad\square$

Remark. A similar problem appears in the Monthly:
Problem 11865 (Proposed by G. H. Chung, 122(9), 2015). Let a_n be a monotone decreasing of nonnegative real numbers. Prove that $\sum_{n=1}^{\infty} a_n/n$ is finite if and only if $\lim_{n\to\infty} a_n = 0$ and

$$
\sum_{n=1}^{\infty} (a_n - a_{n+1}) \ln n < \infty.
$$

In Problems 11829 and 11865, the monotonicity condition of a_n is vital for the conclusion. Here is an counterexample: $a_n = \ln n$ if $n = k^2$, and $a_n = 1/n$ otherwise. Finally, we provide an extension of the necessary part of the proposed problem: Let a_n be a monotone decreasing sequence of real numbers that converges to 0. If the sequence b_n is such that $\sum_{n=1}^{\infty} a_n b_n$ converges, then

$$
\lim_{n\to\infty} \left(\sum_{k=1}^{n} b_k \right) a_n = 0.
$$

Proof. For any $\epsilon > 0$, there exists a $m \in \mathbb{N}$ such that for every $n > m$

$$
|R_n| := \left| \sum_{k=m+1}^{n} a_k b_k \right| < \frac{\epsilon}{4}.
$$

We now estimate $(b_{m+1} + b_{m+2} + \cdots + b_n)a_n$ by using Abel's summation formula.

$$
\begin{aligned}
&|(b_{m+1} + b_{m+2} + \cdots + b_n)a_n| \\
&= a_n \left| \frac{1}{a_{m+1}} a_{m+1} b_{m+1} + \cdots + \frac{1}{a_n} a_n b_n \right| \\
&= a_n \left| \frac{1}{a_{m+1}} R_{m+1} + \frac{1}{a_{m+2}} (R_{m+2} - R_{m+1}) + \cdots + \frac{1}{a_n} (R_n - R_{n-1}) \right| \\
&= a_n \left| \left(\frac{1}{a_{m+1}} - \frac{1}{a_{m+2}} \right) R_{m+1} + \cdots + \left(\frac{1}{a_{n-1}} - \frac{1}{a_n} \right) R_{n-1} + \frac{1}{a_n} R_n \right|
\end{aligned}
$$

$$\leq \frac{\epsilon}{4} a_n \left(\left(\frac{1}{a_{m+2}} - \frac{1}{a_{m+1}} \right) + \cdots + \left(\frac{1}{a_n} - \frac{1}{a_{n-1}} \right) + \frac{1}{a_n} \right)$$

$$= \frac{\epsilon}{4} a_n \left(\frac{2}{a_n} - \frac{1}{a_{m+1}} \right) < \frac{\epsilon}{2}.$$

Since $a_n \to 0$, there exists an $N(\epsilon) > m$ such that for $n > N(\epsilon)$

$$|(b_1 + b_2 + \cdots + b_m)a_n| \leq \frac{\epsilon}{2}.$$

Thus

$$\left| \left(\sum_{k=1}^{n} b_k \right) a_n \right| \leq \left| \left(\sum_{k=1}^{m} b_k \right) a_n \right| + \left| \left(\sum_{k=m+1}^{n} b_k \right) a_n \right| \leq \epsilon.$$

$\square$

In the proposed problem, taking $b_k = 1/k$ yields $a_n H_n \to 0$ or $a_n = o(1/\ln n)$. Without the monotonicity condition on a_n, we challenge the reader to prove the following assertion: Let a_n be a sequence of nonnegative real numbers that converges to 0. Then there exists a sequence of positive numbers b_n such that nb_n is nonincreasing, $\sum_{n=1}^{\infty} b_n$ diverges, and $\sum_{n=1}^{\infty} a_n b_n$ converges.

The following Monthly problem, can be viewed as the complementary of the above problem, offers additional practice.

Problem 12084 (Proposed by G. Stoica, 126(1), 2019). Let $a_1, a_2, \ldots$ be a sequence of nonnegative numbers. Prove that $(1/n)\sum_{k=1}^{n} a_k$ is unbounded if and only if there exists a decreasing sequence $b_1, b_2, \ldots$ such that $\lim_{n\to\infty} b_n = 0, \sum_{n=1}^{\infty} b_n$ is finite, and $\sum_{n=1}^{\infty} a_n b_n$ is infinity. Is the word "decreasing" essential?

2.11   A power series with an exponential tail

Problem 12012 (Proposed by O. Furdui and A. Sîntămărian, 124(10), 2017). Let k be a nonnegative integer. Find the set of real numbers x for which the power series

$$\sum_{n=k}^{\infty} \binom{n}{k} \left(e - 1 - \frac{1}{1!} - \frac{1}{2!} - \cdots - \frac{1}{n!} \right) x^n$$

converges, and determine the sum.

Discussion.
Observe that, once we find the closed form of the sum

$$\sum_{n=k}^{\infty} \left(e - 1 - \frac{1}{1!} - \frac{1}{2!} - \cdots - \frac{1}{n!} \right) x^n, \qquad (2.31)$$

the proposed sum can be derived from the closed form of (2.31) by applying appropriate derivative operators. How can one obtain a closed expression for (2.31)? We either apply Abel's summation formula as we did in Section 2.3 or we use the generating function approach as we did in Section 2.4.

Solution.
Let $S_k(x)$ denote the proposed series. We first show that the power series converges for all real numbers x. Indeed, from Taylor's theorem, there exists $\xi \in (0,1)$ such that

$$\left| \binom{n}{k} \left(e - \sum_{i=1}^{n} \frac{1}{i!} \right) x^n \right| \leq \binom{n}{k} \frac{e^\xi}{(n+1)!} |x|^n \leq \frac{e|x|^k}{k!(n+1)} \cdot \frac{|x|^{n-k}}{(n-k)!}.$$

By the Weierstrass M-test, this implies that $S_k(x)$ converges for all real numbers. In particular, it converges uniformly on any bounded subset of $\mathbb{R}$. Thus, we are free to interchange summation and differentiation.

Next, we claim that

$$S_0(x) := \sum_{n=0}^{\infty} \left(e - 1 - \frac{1}{1!} - \frac{1}{2!} - \cdots - \frac{1}{n!} \right) x^n = \begin{cases} \frac{e - e^x}{1 - x}, & \text{if } x \neq 1, \\ e, & \text{if } x = 1. \end{cases}$$

$$(2.32)$$

Let $a_n = e - 1 - \frac{1}{1!} - \frac{1}{2!} - \cdots - \frac{1}{n!}$ and $b_k = x^k$. If $x \neq 1$, we have $B_k = \sum_{i=0}^{k} b_i = (1 - x^{k+1})/(1 - x)$. Applying the limit version of Abel's summation formula:

$$\sum_{k=0}^{\infty} a_k b_k = \lim_{n \to \infty} B_n a_{n+1} + \sum_{k=0}^{\infty} B_k (a_k - a_{k+1}),$$

we find

$$\begin{aligned}
S_0(x) &= \lim_{n \to \infty} \frac{1 - x^{n+1}}{1 - x} \left(e - 1 - \frac{1}{1!} - \frac{1}{2!} - \cdots - \frac{1}{n!} - \frac{1}{(n+1)!} \right) \\
&\quad + \sum_{k=0}^{\infty} \frac{1 - x^{k+1}}{1 - x} \frac{1}{(k+1)!} \\
&= \frac{1}{1 - x} \left(\sum_{k=0}^{\infty} \frac{1}{(k+1)!} - \sum_{k=0}^{\infty} \frac{x^{k+1}}{(k+1)!} \right) = \frac{e - e^x}{1 - x}.
\end{aligned}$$

The continuity of $S_0(x)$ implies that

$$S_0(1) = \sum_{n=0}^{\infty} \left(e - 1 - \frac{1}{1!} - \frac{1}{2!} - \cdots - \frac{1}{n!} \right) = \lim_{x \to 1} S_0(x) = e.$$

This proves (2.32).
Finally, note that

$$\frac{d^k}{dx^k} x^n = n(n-1) \cdots (n-k+1) x^{n-k} = k! \binom{n}{k} x^{n-k},$$

which is equivalent to

$$\binom{n}{k} x^n = \frac{x^k}{k!} \frac{d^k}{dx^k} x^n.$$

Thus, for $x \neq 1$, term by term differentiation gives

$$S_k(x) = \frac{x^k}{k!} \frac{d^k}{dx^k} S_0(x) = \frac{x^k}{k!} \frac{d^k}{dx^k} \left(\frac{e - e^x}{1 - x} \right). \tag{2.33}$$

Note that

$$\frac{d^j}{dx^j} \left(\frac{1}{1 - x} \right) = \frac{j!}{(1 - x)^{j+1}} \quad \text{for } j \geq 0.$$

Using the Leibniz rule yields

$$\frac{d^k}{dx^k} \left(\frac{e - e^x}{1 - x} \right) = \sum_{j=0}^{k} \binom{k}{j} \frac{d^j}{dx^j} \left(\frac{1}{1 - x} \right) \frac{d^{k-j}}{dx^{k-j}} (e - e^x)$$

$$= \frac{k!e}{(1 - x)^{k+1}} - e^x \sum_{i=0}^{k} \binom{k}{i} \frac{i!}{(1 - x)^{i+1}}.$$

Thus, for $x \neq 1$, (2.33) becomes

$$S_k(x) = \frac{ex^k}{(1 - x)^{k+1}} - x^k e^x \sum_{j=0}^{k} \frac{1}{j!(1 - x)^{k-j+1}}$$

$$= \frac{x^k}{(1 - x)^{k+1}} \left(e - e^x \sum_{j=0}^{k} \frac{(1 - x)^j}{j!} \right).$$

To determine $S_k(1)$, we compute the Taylor series of $e^x - e$ at $x = 1$, which is given by

$$e^x - e = \sum_{i=1}^{\infty} \frac{e}{i!} (x - 1)^i.$$

Thus,

$$\frac{e - e^x}{1 - x} = \sum_{i=1}^{\infty} \frac{e}{i!} (x - 1)^{i-1},$$

which implies that

$$S_k(1) = [(x - 1)^k] \left(\sum_{i=1}^{\infty} \frac{e}{i!} (x - 1)^{i-1} \right) = \frac{e}{(k + 1)!}.$$

There is an alternative derivation based on the method of generating functions. We begin with

$$S_k(x) = e \sum_{n=k}^{\infty} \binom{n}{k} x^n - \sum_{n=k}^{\infty} \binom{n}{k} \left(1 + \frac{1}{1!} + \frac{1}{2!} + \cdots + \frac{1}{n!} \right) x^n$$

$$= e \frac{x^k}{(1 - x)^{k+1}} - \sum_{n=k}^{\infty} \binom{n}{k} \left(1 + \frac{1}{1!} + \frac{1}{2!} + \cdots + \frac{1}{n!} \right) x^n.$$

To find a closed form of the sum in the above equation, we turn to its generating function and find that

$$
\begin{aligned}
G(t) &:= \sum_{k=0}^{\infty} t^k \left(\sum_{n=k}^{\infty} \binom{n}{k} \left(1 + \frac{1}{1!} + \frac{1}{2!} + \cdots + \frac{1}{n!} \right) x^n \right) \\
&= \sum_{n=0}^{\infty} \left(1 + \frac{1}{1!} + \frac{1}{2!} + \cdots + \frac{1}{n!} \right) x^n \sum_{k=0}^{\infty} \binom{n}{k} t^k \\
&= \sum_{n=0}^{\infty} \left(1 + \frac{1}{1!} + \frac{1}{2!} + \cdots + \frac{1}{n!} \right) (x(t+1))^n \\
&= \frac{e^{x(t+1)}}{1 - x(t+1)}.
\end{aligned}
$$

Here in the last equality we have used the well-known generating function property: let $f(t) = \sum_{n=0}^{\infty} a_n t^n$. Then

$$
\sum_{n=0}^{\infty} (a_0 + a_1 + \cdots + a_n) t^n = \frac{f(t)}{1 - t}.
$$

By the Cauchy product, we finally have

$$
[t^k] \, G(t) = \frac{e^x}{1-x} [t^k] \left(\frac{e^{xt}}{1 - \frac{x}{1-x} t} \right) = x^k e^x \sum_{j=0}^{k} \frac{1}{j!(1-x)^{k-j+1}}.
$$

$\square$

Remark. There is an elementary way to establish (2.32). In fact, we have

$$
\begin{aligned}
S_0(x) &= \sum_{n=0}^{\infty} \left(\sum_{j=n+1}^{\infty} \frac{1}{j!} \right) x^n \\
&= \sum_{j=1}^{\infty} \frac{1}{j!} \left(\sum_{n=0}^{j-1} x^n \right) = \sum_{j=1}^{\infty} \frac{1}{j!} \frac{1 - x^j}{1-x} \\
&= \begin{cases} \frac{e - e^x}{1 - x}, & \text{if } x \neq 1, \\ e, & \text{if } x = 1. \end{cases}
\end{aligned}
$$

We now show an extension for $S_0(x)$: Let $f(x) = \sum_{n=0}^{\infty} a_n x^n$ converge on $(-R, R)$. Then

$$
\sum_{n=0}^{\infty} n! a_n \left(e^t - 1 - \frac{t}{1!} - \frac{t^2}{2!} - \cdots - \frac{t^n}{n!} \right) = \int_0^t e^{t-x} f(x) \, dx, \quad (|t| < R).
$$

$$(2.34)$$

Let

$$
y(t) := \sum_{n=0}^{\infty} n! a_n \left(e^t - 1 - \frac{t}{1!} - \frac{t^2}{2!} - \cdots - \frac{t^n}{n!} \right).
$$

Then

$$
\begin{aligned}
y'(t) &= a_0 e^t + \sum_{n=1}^{\infty} n! a_n \left(e^t - 1 - \frac{t}{1!} - \frac{t^2}{2!} - \cdots - \frac{t^{n-1}}{(n-1)!} \right) \\
&= (e^t - 1)a_0 + \sum_{n=1}^{\infty} n! a_n \left(e^t - 1 - \frac{t}{1!} - \frac{t^2}{2!} - \cdots - \frac{t^n}{n!} \right) \\
&\quad + a_0 + \sum_{n=1}^{\infty} n! a_n \frac{t^n}{n!} \\
&= \sum_{n=0}^{\infty} n! a_n \left(e^t - 1 - \frac{t}{1!} - \frac{t^2}{2!} - \cdots - \frac{t^n}{n!} \right) + \sum_{n=0}^{\infty} a_n t^n \\
&= y(t) + f(t).
\end{aligned}
$$

Solving this linear differential equation subject to $y(0) = 0$ yields

$$
y(t) = \int_0^t e^{t-x} f(x)\, dx.
$$

From which taking $f(x) = e^{px}$ gives

$$
\sum_{n=0}^{\infty} p^n \left(e^t - 1 - \frac{t}{1!} - \frac{t^2}{2!} - \cdots - \frac{t^n}{n!} \right) = \frac{e^t - e^{pt}}{1 - p}.
$$

Setting $p = x$ and $t = 1$ we obtain (2.32) again. Furthermore, (2.34) can easily be extended as

$$
\sum_{n=0}^{\infty} n f^{(n)}(0) \left(e^t - 1 - \frac{t}{1!} - \frac{t^2}{2!} - \cdots - \frac{t^n}{n!} \right) = \int_0^t x e^{t-x} f'(x)\, dx.
$$

2.12 An infinite matrix product

Problem 11739 (Proposed by F. Adams, A. Bloch and J. Lagarias, 120(9), 2013).

Let $B(x) = \begin{pmatrix} 1 & x \\ x & 1 \end{pmatrix}$. Consider the infinite matrix product

$$
M(t) = B(2^{-t}) B(3^{-t}) B(5^{-t}) \cdots = \prod_p B(p^{-t}),
$$

where the product runs over all primes, taken in increasing order. Evaluate $M(2)$.

Discussion.

One way to proceed is to use matrix diagonalization. The symmetry of $B(x)$ asserts that $B(x)$ is diagonalizable. Let $B(x) = PD(x)P^{-1}$. Here P is an invertible constant matrix. Then

$$P^{-1}M(t)P = P^{-1}\left(\prod_p D(p^{-t})\right)P.$$

Thus, $M(t)$ has a closed form as long as the infinite product $\prod_p D(p^{-t})$ has a closed form. Another way to proceed is to solve the recurrence for the n-partial finite product.

Solution I.

We prove the following more general result: For all $t > 1$,

$$M(t) = \frac{1}{2}\begin{pmatrix} \frac{\zeta^2(t)+\zeta(2t)}{\zeta(t)\,\zeta(2t)} & \frac{\zeta^2(t)-\zeta(2t)}{\zeta(t)\,\zeta(2t)} \\[2mm] \frac{\zeta^2(t)-\zeta(2t)}{\zeta(t)\,\zeta(2t)} & \frac{\zeta^2(t)+\zeta(2t)}{\zeta(t)\,\zeta(2t)} \end{pmatrix},$$

where $\zeta(s)$ is the Riemann zeta function.

Note that the characteristic equation of B is

$$\det(\lambda I - B) = (\lambda - 1)^2 - x^2 = 0.$$

We find the eigenvalues of B are $1 \pm x$, the orthonormal bases for the eigenspaces are

$$\lambda = 1 - x, \quad \mathbf{v}_1 = \begin{pmatrix} 1/\sqrt{2} \\ -1/\sqrt{2} \end{pmatrix}; \quad \lambda = 1 + x, \quad \mathbf{v}_2 = \begin{pmatrix} 1/\sqrt{2} \\ 1/\sqrt{2} \end{pmatrix}.$$

Let $P = \begin{pmatrix} 1/\sqrt{2} & 1/\sqrt{2} \\ -1/\sqrt{2} & 1/\sqrt{2} \end{pmatrix}$. Then $P^{-1}BP$ is diagonal:

$$P^{-1}BP = \begin{pmatrix} 1 - x & 0 \\ 0 & 1 + x \end{pmatrix}.$$

Moreover,

$$\begin{aligned} P^{-1}M(t)P &= (P^{-1}B(2^{-t})P)\,(P^{-1}B(3^{-t})P)\,(P^{-1}B(5^{-t})P)\cdots \\ &= \prod_p \begin{pmatrix} 1 - p^{-t} & 0 \\ 0 & 1 + p^{-t} \end{pmatrix} \\ &= \begin{pmatrix} \prod_p (1 - p^{-t}) & 0 \\ 0 & \prod_p (1 + p^{-t}) \end{pmatrix}. \end{aligned}$$

Applying the Euler infinite product formula for the Riemann zeta function

$$\zeta(s) = \prod_p (1 - p^{-s})^{-1},$$

we find

$$\prod_p (1 - p^{-t}) \cdot \prod_p (1 + p^{-t}) = \prod_p (1 - p^{-2t}) = \frac{1}{\zeta(2t)},$$

and so

$$P^{-1} M(t) P = \begin{pmatrix} \frac{1}{\zeta(t)} & 0 \\ 0 & \frac{\zeta(t)}{\zeta(2t)} \end{pmatrix}.$$

Finally, we conclude

$$M(t) = P\left(P^{-1} M(t) P\right) P^{-1} = \frac{1}{2} \begin{pmatrix} \frac{\zeta^2(t) + \zeta(2t)}{\zeta(t)\,\zeta(2t)} & \frac{\zeta^2(t) - \zeta(2t)}{\zeta(t)\,\zeta(2t)} \\ \frac{\zeta^2(t) - \zeta(2t)}{\zeta(t)\,\zeta(2t)} & \frac{\zeta^2(t) + \zeta(2t)}{\zeta(t)\,\zeta(2t)} \end{pmatrix}.$$

In particular, when $t = 2$, since $\zeta(2) = \pi^2/6$, $\zeta(4) = \pi^4/90$, we find that

$$M(2) = \frac{3}{2\pi^2} \begin{pmatrix} 7 & 3 \\ 3 & 7 \end{pmatrix}.$$

$\square$

Solution II.
For $n \geq 0$, define

$$M_n(t) = \prod_{k=1}^n B(p_k^{-t}) = \begin{pmatrix} a_n & b_n \\ b_n & a_n \end{pmatrix}.$$

Then $a_0 = 1, b_0 = 0$ and

$$\begin{aligned} a_n &= a_{n-1} + p_n^{-t} b_{n-1}, \\ b_n &= p_n^{-t} a_{n-1} + b_{n-1}. \end{aligned}$$

By iteration, we have

$$a_n + b_n = (a_{n-1} + b_{n-1})(1 + p_n^{-t}) = \prod_{k=1}^n (1 + p_k^{-t});$$

$$a_n - b_n = (a_{n-1} - b_{n-1})(1 - p_n^{-t}) = \prod_{k=1}^n (1 - p_k^{-t}).$$

Since

$$\prod_{k=1}^{n}(1 - p_k^{-t}) \to \frac{1}{\zeta(t)}, \quad \prod_{k=1}^{n}(1 + p_k^{-t}) = \prod_{k=1}^{n}\frac{1 - (p_k^{-t})^2}{1 - p^{-t}} \to \frac{\zeta(t)}{\zeta(2t)},$$

as $n \to \infty$, we find that

$$\lim_{n \to \infty} a_n = \frac{1}{2}\left(\frac{\zeta(t)}{\zeta(2t)} + \frac{1}{\zeta(t)}\right),$$

$$\lim_{n \to \infty} b_n = \frac{1}{2}\left(\frac{\zeta(t)}{\zeta(2t)} - \frac{1}{\zeta(t)}\right),$$

which implies that

$$M(t) = \frac{1}{2}\begin{pmatrix} \frac{\zeta^2(t)+\zeta(2t)}{\zeta(t)\,\zeta(2t)} & \frac{\zeta^2(t)-\zeta(2t)}{\zeta(t)\,\zeta(2t)} \\[2mm] \frac{\zeta^2(t)-\zeta(2t)}{\zeta(t)\,\zeta(2t)} & \frac{\zeta^2(t)+\zeta(2t)}{\zeta(t)\,\zeta(2t)} \end{pmatrix}.$$

In particular, for $t = 2$, we see that

$$\lim_{n \to \infty} a_n = \frac{21}{2\pi^2}, \quad \lim_{n \to \infty} b_n = \frac{9}{2\pi^2}$$

which gives

$$M(2) = \frac{3}{2\pi^2}\begin{pmatrix} 7 & 3 \\ 3 & 7 \end{pmatrix}$$

again. $\qquad\qquad\qquad\qquad\qquad\qquad\qquad\qquad\qquad\qquad\qquad\qquad\qquad\square$

Remark. We collect three problems for additional practice.

1. Let $B(x) = \begin{pmatrix} x & 1 \\ 1 & x \end{pmatrix}$. Find a closed form for

$$P_n = \prod_{k=2}^{n} B(k).$$

2. **Problem 11685** (Proposed by D. Knuth, 120(1), 2013). Prove that

$$\prod_{k=0}^{\infty}\left(1 + \frac{1}{2^{2^k} - 1}\right) = \frac{1}{2} + \sum_{k=0}^{\infty}\frac{1}{\prod_{j=0}^{k-1}(2^{2^j} - 1)}.$$

In other words, prove that

$$(1 + 1)\left(1 + \frac{1}{3}\right)\left(1 + \frac{1}{15}\right)\left(1 + \frac{1}{255}\right)\cdots$$

$$= \frac{1}{2} + 1 + 1 + \frac{1}{3} + \frac{1}{3 \cdot 15} + \frac{1}{3 \cdot 15 \cdot 255} + \cdots.$$

3. **Problem 11883** (Proposed by H. Ohtsuka, 123(1), 2016). For $|q| > 1$, prove that

$$\sum_{k=0}^{\infty} \frac{1}{(q^{2^0} + q)(q^{2^1} + q) \cdots (q^{2^k} + q)} = \frac{1}{q - 1} \prod_{i=0}^{\infty} \frac{1}{q^{1-2^i} + 1}.$$

3

Integrations

Over the years, we have seen a great many elegant and striking integral problems and clever solutions published in the Monthly. Most of them contain mathematical ingenuities. In this chapter, we select 15 of these irresistible integrals. In presenting the combination of approaches required to evaluate these integrals, we have tried to follow the most interesting route to the results and endeavored to highlight connections to other problems and to more advanced topics. It is interesting to see the range of mathematical approaches this chapter exploits. These evaluations provide nice application of infinite series, gamma and beta functions, parametric differentiation and integration, Laplace transforms and contour integration.

3.1 A Lobachevsky integral

Problem 11423 (Proposed by G. Minton, 116(3), 2009). Show that if n and m are positive integers with $n \geq m$ and $n - m$ even, then $\int_0^\infty x^{-m} \sin^n x \, dx$ is a rational multiple of π.

Discussion.
Since there are integer parameters m and n appearing in the problem, we are going to proceed by induction. To build "$P(k+1)$" from "$P(k)$," integration by parts suggests we induct on m.

Solution.
We use induction on m. Let

$$I(n, m) = \int_0^\infty \frac{\sin^n x}{x^m} \, dx.$$

First, for any positive odd number $n = 2k + 1$, we recall that

$$\sin^{2k+1} x = \frac{1}{2^{2k}} \sum_{j=0}^{k} (-1)^{k+j} \binom{2k+1}{j} \sin(2k - 2j + 1)x,$$

and

$$\int_0^\infty \frac{\sin(ax)}{x} \, dx = \frac{\pi}{2} \qquad \text{for } a > 0.$$

Hence

$$I(2k+1,1) = \frac{1}{2^{2k+1}} \sum_{j=0}^{k} (-1)^{k+j} \left(\begin{array}{c} 2k+1 \\ j \end{array} \right) \pi$$

is a rational multiple of π. For $m = 2$, note that integration by parts yields

$$I(n,2) = n \int_0^\infty \frac{\sin^{n-1} x \cos x}{x} \, dx.$$

Using the product to sum formula for sine and cosine, for $n = 2k$, we can expand $\sin^{2k-1} x \cos x$ as

$$\frac{1}{2^{2k-1}} \sum_{j=0}^{k-1} (-1)^{k+j-1} \left(\begin{array}{c} 2k-1 \\ j \end{array} \right) (\sin(2k-2j)x + \sin(2k-2j-2)x)$$

so

$$I(2k,2) = \frac{k}{2^{2k-1}} \left(\sum_{j=0}^{k-2} (-1)^{k+j-1} \left(\begin{array}{c} 2k-1 \\ j \end{array} \right) + \frac{1}{2} \left(\begin{array}{c} 2k-1 \\ k-1 \end{array} \right) \right) \pi$$

is also a rational multiple of π. For $m \geq 2$, integration by parts twice leads to

$$
\begin{aligned}
I(n, m+1) &= -\frac{1}{m} \int_0^\infty \sin^n x \, d \left(\frac{1}{x^m} \right) \\
&= -\frac{n}{m(m-1)} \int_0^\infty \sin^{n-1} x \cos x \, d \left(\frac{1}{x^{m-1}} \right) \\
&= -\frac{n^2}{m(m-1)} I(n, m-1) + \frac{n(n-1)}{m(m-1)} I(n-2, m-1).
\end{aligned}
$$

When $n - (m+1)$ is even and nonnegative, the right-hand side is a rational multiple of π by the induction hypothesis. Therefore, the left-hand side is also such a multiple, which completes the proof. $\square$

Remark. Since

$$\sum_{j=0}^{k} (-1)^{k-j} \left(\begin{array}{c} 2k+1 \\ j \end{array} \right) = \left(\begin{array}{c} 2k \\ k \end{array} \right),$$

this gives a more compact formula

$$I(2k+1,1) = \frac{\pi}{2^{2k+1}} \left(\begin{array}{c} 2k \\ k \end{array} \right).$$

Similarly,

$$I(2k,2) = \frac{\pi}{2^{2k-1}} \left(\begin{array}{c} 2k-2 \\ k-1 \end{array} \right).$$

Note that assumption on $n - m$ even is vital for the conclusion because

$$\int_0^\infty \frac{\sin^3 x}{x^2}\, dx = \frac{3}{4}\ln 3$$

is not a rational multiple of π. In general, similar to the above arguments, when $n \geq m \geq 2$, we find that

$$\int_0^\infty \frac{\sin^{2n} x}{x^{2m-1}}\, dx = \frac{1}{2^{2n-1}} \sum_{k=1}^n (-1)^{m+k} \binom{2n}{n-k} \frac{(2k)^{2m-2}}{(2m-2)!} \ln k;$$

$$\int_0^\infty \frac{\sin^{2n+1} x}{x^{2m}}\, dx = \frac{1}{2^{2n}} \sum_{k=0}^n (-1)^{m+k} \binom{2n+1}{n-k} \frac{(2k+1)^{2m-1}}{(2m-1)!} \ln(2k+1).$$

Based on the Monthly Editorial Notes, Lobachevsky studied $I(n,m)$ as early as 1842. Recently, a similar problem appeared as
Mathematics Magazine Problem 2020 (Proposed by J. Sorel, 90(2), 2017). Find all natural numbers n such that the integral

$$I_n := \int_0^1 x^n \arctan x\, dx$$

is a rational number.

By induction it was shown [84] that I_n is a rational number precisely when $n = 4k + 3$ with $k = 0, 1, 2, \ldots$. A much more challenging problem appeared in the Monthly in 1967:
Advanced Problem 5529 (Proposed by D. S. Mitrinovic, 74(8), 1967). Evaluate

$$I_n := \int_{-\infty}^\infty \prod_{j=1}^n \frac{\sin k_j(x - a_j)}{x - a_j}\, dx$$

with k_j, a_j, $j = 1, 2, \ldots, n$ real numbers.

The next year a solution was published in the form

$$I_n = \left(\prod_{j=2}^n \frac{\sin k_j(a_1 - a_j)}{a_1 - a_j} \right) \pi \tag{3.1}$$

under the assumption that $k_1 \geq k_2 \geq \ldots k_n \geq 0$. But this solution, as Klamkin (1970) pointed out, can not be true since it is not symmetric in the parameters while I_n is. Djoković and Glasser showed that Formula (3.1) holds true under the additional restrictions: $k_1 \geq k_2 + k_3 + \ldots + k_n$ and all of the k_j are positive. It appeared no simple general fix for (3.1). As a unsolved question, this problem is listed in Monthly several times and has disappeared in the later 1980's. Recently, Borwein et. al. in a remarkable paper [18] completely set the case $n = 3$ and exhibit some more general structure of I_n.

For the special case

$$J_n := \int_{-\infty}^{\infty} \text{sinc}x \cdot \text{sinc}\left(\frac{x}{3}\right) \cdots \text{sinc}\left(\frac{x}{2n+1}\right) dx$$

where $\text{sinc}x := \sin x / x$. Using *Mathematica*, we obtain

$$J_0 = J_1 = \cdots = J_6 = \pi,$$

but

$$J_7 = \int_{-\infty}^{\infty} \text{sinc}x \cdot \text{sinc}\left(\frac{x}{3}\right) \cdots \text{sinc}\left(\frac{x}{15}\right) dx$$

$$= \frac{467807924713440738696537864469}{467807924720320453655260875000}\pi < \pi.$$

Based on these curious results, [18] gives the following generalization:

First Bite Theorem. If $\sum_{k=1}^{n-1} a_k \leq a_0 < \sum_{k=1}^{n} a_k$, then

$$\int_{-\infty}^{\infty} \prod_{k=0}^{n} \text{sinc}(a_k x)\, dx = \frac{1}{a_0}\left(1 - \frac{(a_1 + a_2 + \cdots + a_n - a_0)^n}{2^{n-1} n! a_1 a_2 \cdots a_n}\right)\pi.$$

Finally, we end this section with six related problems for additional practice.

1. Let n and m be positive integers with $0 \leq m < n$. Show that

$$\int_0^{\infty} \frac{\sin^n x}{x^n} \cos(mx)\, dx$$

    is a rational multiple of π.

2. Let n and m be positive integers such that $n \geq m > p$, where $p = (n-m)(\text{mod } 2)$. Let $\lfloor x \rfloor$ be the floor function. Show that

$$\int_0^{\infty} \frac{\sin^n x}{x^m}\, dx = \frac{(-1)^{\lfloor (n-m)/2 \rfloor} \pi^{1-p}}{2^{n-p}(m-1)!}$$

$$\sum_{k=0}^{\lfloor n/2 \rfloor - p} (-1)^k \binom{n}{k}(n-2k)^{m-1} \ln^p(n-2k).$$

3. Let S_n be an series analog of J_n and define by

$$S_n := \sum_{k=1}^{\infty} \text{sinc}^n(k) \quad \text{for all } n \geq 1.$$

    (a) Show that

$$S_n = -\frac{1}{2} + r_n \pi,$$

    where r_n is a rational number, $1 \leq n \leq 6$.

(b) Show that the pattern of answers in (a) breaks at $n = 7$. With this insight, can you find a general formula of S_n which is analog to the *First Bite Theorem*?

4. Let

$$C_n = \int_0^\infty \frac{\sin(5x)}{x} \prod_{k=1}^n \cos\left(\frac{x}{k}\right) dx.$$

Show that $C_n = \pi/2$ for all $1 \le n \le 82$, but $C_{83} < \pi/2$. Can you reveal the principle behind these results? Similar to the *First Bite Theorem* above, can you find a formula for the following more general integral:

$$\int_0^\infty \frac{\sin(mx)}{x} \prod_{k=1}^n \cos(a_k x) \, dx?$$

5. For positive integer m, evaluate

$$\int_0^\infty \frac{\sin(mx)}{x} J_0^m(x) \, dx,$$

where $J_0(x)$ is the Bessel function of the first kind of order zero.

6. Calculate

$$\pi_1 := \int_0^\infty \prod_{n=1}^\infty \cos\left(\frac{x}{n}\right) dx \quad \text{and}$$

$$\pi_2 := \int_0^\infty \cos(2x) \prod_{n=1}^\infty \cos\left(\frac{x}{n}\right) dx.$$

It is known that

$$\pi_1 < \frac{\pi}{4} \quad \text{and} \quad \pi_2 < \frac{\pi}{8}.$$

Although π_2 is within 10^{-42} of $\pi/8$, a remarkable approximation, we do not have concise closed form expressions for π_1 and π_2 yet. As [10] pointed out, even asking for the numerical value is rather challenging because the oscillatory behavior of $\prod_{n=1}^\infty \cos\left(\frac{x}{n}\right)$.

3.2   Two log gamma integrals

Problem 11329 (Proposed by T. Amdeberhan and V. Moll, 114(10), 2007). Let $f(t) = 2^{-t} \ln \Gamma(t)$, where Γ denotes the classical *gamma function*, and let γ be Euler's constant. Derive the following integral identities:

$$\int_0^\infty f(t) \, dt = 2 \int_0^1 f(t) \, dt - \frac{\gamma + \ln \ln 2}{\ln 2},$$

$$\int_0^\infty t \, f(t) \, dt = 2 \int_0^1 (t+1) f(t) \, dt - \frac{(\gamma + \ln \ln 2)(1 + 2\ln 2) - 1}{\ln^2 2}.$$

Discussion.
Because of the complexity of $f(t)$, the usual integration techniques do not
work on these two integrals. To evaluate them, we will transform the integrals
into some known integrals using a functional equation.

Solution.
From

$$\int_0^\infty e^{-x} \ln x \, dx = \Gamma'(1) = -\gamma,$$

it follows that for $\alpha > 0$,

$$\int_0^\infty e^{-\alpha t} \ln t \, dt = -\frac{\gamma + \ln \alpha}{\alpha} \quad \text{and} \quad \int_0^\infty t e^{-\alpha t} \ln t \, dt = \frac{1 - (\gamma + \ln \alpha)}{\alpha^2}.$$

$$(3.2)$$

Here the second integral in (3.2) is obtained by differentiating the first with
respect to α. In particular, set $\alpha = \ln 2$. Using $\Gamma(t+1) = t\Gamma(t)$, we see that f
satisfies the desired functional equation

$$2f(t+1) = f(t) + 2^{-t} \ln t = f(t) + e^{-\alpha t} \ln t.$$

Hence

$$\int_0^\infty 2f(t) \, dt - \int_0^1 2f(t) \, dt \quad = \quad \int_1^\infty 2f(t) \, dt = \int_0^\infty 2f(t+1) \, dt$$

$$= \quad \int_0^\infty f(t) \, dt + \int_0^\infty e^{-\alpha t} \ln t \, dt,$$

and therefore

$$\int_0^\infty f(t) \, dt = 2 \int_0^1 f(t) \, dt - \frac{\gamma + \ln \alpha}{\alpha}. \qquad (3.3)$$

Similarly,

$$\int_0^\infty 2(t-1)f(t) \, dt - \int_0^1 2(t-1)f(t) \, dt = \int_1^\infty 2(t-1)f(t) \, dt$$

$$= \int_0^\infty 2t f(t+1) \, dt = \int_0^\infty t f(t) \, dt + \int_0^\infty t e^{-\alpha t} \ln t \, dt,$$

which yields

$$\int_0^\infty 2(t-1)f(t) \, dt = \int_0^1 2(t-1)f(t) \, dt + \int_0^\infty t f(t) \, dt + \frac{1 - (\gamma + \ln \alpha)}{\alpha^2}.$$

Combining this with (3.3) shows that

$$\int_0^\infty t f(t) \, dt = 2 \int_0^1 (t+1)f(t) \, dt - \frac{(\gamma + \ln \alpha)(1 + 2\alpha) - 1}{\alpha^2}.$$

$$\square$$

Remark: It is interesting to reveal a little bit of the background of this problem. The story began with curiosity about the Laplace transform of the digamma function,

$$L(\alpha) := \int_0^\infty e^{-\alpha t} \psi(t+1)\, dt.$$

The digamma function itself has been studied extensively and many of its properties and identities are listed in ([49], pp. 952-955). However, an explicit formula of $L(\alpha)$ is absent from the literature and tabulations of the digamma function.

The study of definite integrals, where the integrand is a combination of powers, logarithms and trigonometric functions, was initiated by Euler. Some famous results include

$$\int_0^1 x \ln(2\cos x)\, dx = -\frac{7}{16}\zeta(3) \quad \text{and} \quad \int_0^1 x^2 \ln(2\cos x)\, dx = -\frac{\pi}{4}\zeta(3).$$

In 2009, Oloa [73] introduced the Euler-type integral

$$M(\alpha) := \frac{\pi}{4} \int_0^{\pi/2} \frac{x^2\, dx}{x^2 + \ln^2(2e^{-\alpha}\cos x)}.$$

By applying the expansion

$$\frac{x^2}{x^2 + \ln^2(2\cos x)} = x\sin 2x + \sum_{n=1}^\infty (-1)^{n-1}\left(\frac{a_n}{n!} - \frac{a_{n+1}}{(n+1)!}\right) x\sin(2nx)$$

with

$$a_n := \int_0^1 (t)_n\, dt = \int_0^1 t(t+1)\cdots(t+n-1)\, dt,$$

he determined the special value

$$M(0) = \frac{\pi}{4} \int_0^{\pi/2} \frac{x^2\, dx}{x^2 + \ln^2(2\cos x)} = \frac{1}{2}(1 - \gamma + \ln(2\pi)).$$

Surprisingly, guided by numerical experimentation, Glasser and Manna [46] demonstrated the following relationship between $M(\alpha)$ and $L(\alpha)$: for $\alpha > \ln 2$,

$$M(\alpha) = \frac{\gamma}{\alpha} + L(\alpha) = \frac{\gamma}{\alpha} - \gamma - \ln\alpha + \alpha\int_0^\infty e^{-\alpha t}\ln\Gamma(t)\, dt,$$

where γ is Euler's constant. Their derivations offer a remarkable applications of the gamma function, digamma function and complex integrals.

The graph of $M(\alpha)$ generated by *Mathematica* is shown in Figure 3.1. It has a cusp at $\alpha = \ln 2$. The proposers of this problem investigated another branch of $M(\alpha)$ where $0 < \alpha < \ln 2$ in [7] and found

$$M(\alpha) = \frac{\gamma}{\alpha} + \frac{\alpha + \ln(1 - e^{-\alpha}) - \gamma - \ln\alpha}{1 - e^{-\alpha}} + \frac{\alpha}{1 - e^{-\alpha}}\int_0^1 e^{-\alpha t}\ln\Gamma(t)\, dt.$$

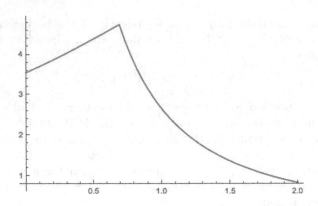

FIGURE 3.1
The graph of $M(\alpha)$

This proposed problem indeed is Lemma 2.4 in the paper, which is used to confirm the continuity of $M(\alpha)$ and the jump of 4 for $M'(\alpha)$ at $\alpha = \ln 2$.

Now, we end this section with five problems for additional practice. The last one gives an explicit expression of $L(\alpha)$ (Due to Dixit [35]).

1. Show that $\int_0^1 t \ln \Gamma(t) \, dx = \frac{1}{4} \ln(2\pi) - \ln A$, where A is the Glaisher-Kinkelin constant.

2. Show that

$$\int_0^1 \ln^2 \Gamma(t) \, dt + \int_0^1 \ln \Gamma(t) \ln \Gamma(1-t) \, dt = \frac{1}{2} \ln^2(2\pi) + \frac{1}{24} \pi^2.$$

3. Show that

$$\int_0^\infty \left(\psi(t+1) - \frac{1}{2(t+1)} - \ln t \right) dt = \frac{1}{2} \ln 2\pi.$$

4. Evaluate

$$\int_0^1 e^{-\alpha t} \ln \Gamma(t) \, dt \quad \text{and} \quad \int_0^1 t e^{-\alpha t} \ln \Gamma(t) \, dt.$$

5. Let $\alpha > 0$. Show that

$$L(\alpha) = \left(\frac{1}{e^\alpha - 1} - \frac{1}{\alpha} + 1 \right) \ln \frac{2\pi}{\alpha} + 2\alpha \sum_{n=1}^\infty \frac{\ln n}{\alpha^2 + 4n^2\pi^2}$$

$$+ \frac{1}{4} \left(\psi\left(\frac{i\alpha}{2\pi} \right) + \psi\left(-\frac{i\alpha}{2\pi} \right) \right) - \frac{\gamma + \ln \alpha}{\alpha}.$$

3.3   Short gamma products with simple values

Problem 11426 (Proposed by M. L. Glasser, 116(4), 2009). Find

$$\frac{\Gamma(1/14)\,\Gamma(9/14)\,\Gamma(11/14)}{\Gamma(3/14)\,\Gamma(5/14)\,\Gamma(13/14)},$$

where Γ denotes the usual gamma function, given by $\Gamma(z) = \int_0^\infty t^{z-1}e^{-t}\,dt$.

Discussion.

Recall the well-known *Gauss multiplication formula*, a long gamma product but with simple values:

$$\prod_{k=1}^{n-1} \Gamma\left(\frac{k}{n}\right) = \frac{(2\pi)^{(n-1)/2}}{\sqrt{n}}.$$

There is another less-known formula due to Sándor and Tóth [81]: for $n \geq 2, n \neq p^r$, a prime power, then

$$\prod_{\gcd(k,n)=1} \Gamma\left(\frac{k}{n}\right) = (2\pi)^{\phi(n)/2},$$

where $\phi(n)$ is the Euler totient function, the number of integers between 1 and n that are relatively prime to n. This singles out some factors from the Gauss product but remains simple values. For example, for $n = 14$, a product of six factors gives

$$\Gamma(1/14)\,\Gamma(3/14)\,\Gamma(5/14)\Gamma(9/14)\,\Gamma(11/14)\,\Gamma(13/14) = (2\pi)^3.$$

This problem suggests an even shorter gamma product still with simple values:

$$\Gamma(1/14)\,\Gamma(9/14)\,\Gamma(11/14) = 4\pi^{3/2},$$

which is not a consequence of either of the product formulas above. We will proceed with this problem by using the *Legendre duplication formula*, and will see that a large collection of short products exist which exhibit simple values for $n = 2m$ with m odd.

Solution.

We show the desired ratio is 2. Indeed, rewrite the Legendre duplication formula as

$$\Gamma(2z)\,\Gamma(1/2) = 2^{2z-1}\,\Gamma(z)\,\Gamma(z+1/2).$$

Setting $z = 1/14, 9/14$ and $11/14$ respectively, we find that

$$\begin{aligned}
\Gamma(1/7)\,\Gamma(1/2) &= 2^{-6/7}\,\Gamma(1/14)\,\Gamma(4/7); \\
\Gamma(9/7)\,\Gamma(1/2) &= 2^{2/7}\,\Gamma(9/14)\,\Gamma(8/7); \\
\Gamma(11/7)\,\Gamma(1/2) &= 2^{4/7}\,\Gamma(11/14)\,\Gamma(9/7).
\end{aligned}$$

Applying $\Gamma(1 + z) = z\Gamma(z)$ yields $\Gamma(8/7) = \Gamma(1/7)/7$ and $\Gamma(11/7) = 4\Gamma(4/7)/7$. Therefore the product of the three equations above leads to

$$\Gamma(1/14)\Gamma(9/14)\Gamma(11/14) = 4\left(\Gamma(1/2)\right)^3.$$

Similarly, we have

$$\Gamma(3/14)\Gamma(5/14)\Gamma(13/14) = 2\left(\Gamma(1/2)\right)^3.$$

The ratio of 2 now follows from combining these two products.   □

Remark. Recall the definition of the *Legendre symbol* $\left(\frac{a}{p}\right)$: Let p be an odd prime. Then

$$\left(\frac{a}{p}\right) = \begin{cases} 1 & \text{if } a \text{ is a quadratic residue of } p \text{ and } a \not\equiv 0\,(\text{mod } p), \\ -1 & \text{if } a \text{ is a quadratic nonresidue of } p, \\ 0 & a \equiv 0\,(\text{mod } p). \end{cases}$$

Using this notation, we offer the following generalization: Let p be a prime with $p = 7\,(\text{mod } 8)$. Then

$$\prod_{a \in O(2p)} \left(\Gamma\left(\frac{a}{2p}\right)\right)^{\left(\frac{a}{p}\right)} = 2^{\sum_{k=1}^{(p-1)/2}\left(\frac{k}{p}\right)}, \tag{3.4}$$

where $O(n)$ is the set of odd positive integers less than n. For example, let $p = 7$, since

$$\left(\frac{1}{7}\right) = \left(\frac{9}{7}\right) = \left(\frac{11}{7}\right) = 1, \quad \left(\frac{7}{7}\right) = 0, \quad \left(\frac{3}{7}\right) = \left(\frac{5}{7}\right) = \left(\frac{13}{7}\right) = -1,$$

(3.4) immediately deduces that

$$\frac{\Gamma(1/14)\,\Gamma(9/14)\,\Gamma(11/14)}{\Gamma(3/14)\,\Gamma(5/14)\,\Gamma(13/14)} = 2^{\left(\frac{1}{7}\right)+\left(\frac{2}{7}\right)+\left(\frac{3}{7}\right)} = 2.$$

This extension (3.4) is also obtained by Tauraso (http://www.mat.uniroma2.it/~tauraso/AMM/AMM11426.pdf). We offer a composite solution as follows: Using $p = 7\,(\text{mod } 8)$, we have

$$\left(\frac{2p-a}{p}\right) = \left(\frac{-a}{p}\right) = \left(\frac{-1}{p}\right) \cdot \left(\frac{a}{p}\right) = (-1)^{(p-1)/2}\left(\frac{a}{p}\right) = -\left(\frac{a}{p}\right).$$

Let the left-hand side of (3.4) be $P(p)$. Then

$$P(p) = \prod_{a \in O(p)} \left(\frac{\Gamma(a/2p)}{\Gamma((2p-a)/2p)}\right)^{\left(\frac{a}{p}\right)}.$$

Applying the Euler reflection formula and the Legendre duplication formula of the gamma function, if $x \neq 1, 1/2$, we have

$$
\begin{aligned}
\frac{\Gamma(x)}{\Gamma(1-x)} &= \Gamma^2(x)\frac{\sin(\pi x)}{\pi} \\
&= 2^{1-4x}\frac{\Gamma^2(2x)}{\Gamma^2(x+1/2)}\frac{\sin(\pi(1-2x))}{\sin(\pi(1-2x)/2)} \\
&= 2^{1-4x}\frac{\Gamma^2(2x)}{\Gamma^2(x+1/2)}\frac{\Gamma(x+1/2)\Gamma(1/2-x)}{\Gamma(2x)\Gamma(1-2x)} \\
&= 2^{1-4x}\frac{\Gamma(2x)\Gamma(1/2-x)}{\Gamma(x+1/2)\Gamma(1-2x)}.
\end{aligned}
$$

Substituting $x = a/2p$ yields

$$
P(p) = \prod_{a \in O(p)} 2^{(1-\frac{2a}{p})(\frac{a}{p})}\left(\frac{\Gamma(a/p)\Gamma((p-a)/2p)}{\Gamma((p-a)/p)\Gamma((p+a)/2p)}\right)^{(\frac{a}{p})}.
$$

Next, by the assumption on p, we have

$$
\left(\frac{(p\pm a)/2}{p}\right) = \left(\frac{2}{p}\right)^{-1}\left(\frac{p\pm a}{p}\right) = (-1)^{-(p^2-1)/8}\left(\frac{\pm a}{p}\right) = \pm\left(\frac{a}{p}\right).
$$

Therefore,

$$
\begin{aligned}
P(p) &= \prod_{a \in O(p)} 2^{(1-\frac{2a}{p})(\frac{a}{p})}\left(\frac{\Gamma(\frac{a}{p})(a/p)\Gamma(\frac{p-a}{p})((p-a)/p)}{\Gamma(\frac{(p+a)/2}{p})((p+a)/2p)\Gamma(\frac{(p-a)/2}{p})((p-a)/2p)}\right) \\
&= \prod_{a \in O(p)} 2^{(1-\frac{2a}{p})(\frac{a}{p})}\left(\frac{\prod_{k=1}^{p-1}\Gamma(\frac{k}{p})(k/p)}{\prod_{k=1}^{p-1}\Gamma(\frac{k}{p})(k/p)}\right) \\
&= 2^{\sum_{a \in O(p)}(1-\frac{2a}{p})(\frac{a}{p})}.
\end{aligned}
$$

Note that

$$
\sum_{a \in O(p)}\left(\frac{a}{p}\right) = \sum_{k=1}^{(p-1)/2}\left(\frac{p-2k}{p}\right) = -\sum_{k=1}^{(p-1)/2}\left(\frac{k}{p}\right)
$$

and

$$
\begin{aligned}
\sum_{a \in O(p)}\frac{a}{p}\left(\frac{a}{p}\right) &= \sum_{k=1}^{p-1}\frac{k}{p}\left(\frac{k}{p}\right) - \sum_{k=1}^{(p-1)/2}\frac{2k}{p}\left(\frac{2k}{p}\right) \\
&= \sum_{k=(p+1)/2}^{p-1}\frac{k}{p}\left(\frac{k}{p}\right) - \sum_{k=1}^{p-1}\frac{2k}{p}\left(\frac{2k}{p}\right)
\end{aligned}
$$

$$= \sum_{k=1}^{(p-1)/2} \frac{p-k}{p} \left(\frac{p-k}{p}\right) - \sum_{k=1}^{(p-1)/2} \frac{k}{p} \left(\frac{k}{p}\right)$$

$$= - \sum_{k=1}^{(p-1)/2} \left(\frac{k}{p}\right).$$

We finally obtain

$$\sum_{a \in O(p)} \left(1 - \frac{2a}{p}\right) \left(\frac{a}{p}\right) = - \sum_{k=1}^{(p-1)/2} \left(\frac{k}{p}\right) + 2 \sum_{k=1}^{(p-1)/2} \left(\frac{k}{p}\right) = \sum_{k=1}^{(p-1)/2} \left(\frac{k}{p}\right).$$

This proves (3.4) as claimed.

The gamma function is probably one of the most ubiquitous functions in analysis. Recently Borwein and Corless [22] made an informative survey of papers on the gamma function published in the Monthly. Through those articles over one century, they provided a amazed cross-section of mathematics: Analysis, geometry, statistics, combinatorics, logic and number theory.

We now end this section to compile two problems for additional practice:

1. (Due to Zucker, personal communication) Let $n = 2^m - 1$ with $m > 1$. Prove that

$$\Gamma\left(\frac{1}{2n}\right) \prod_{k=1}^{m-1} \Gamma\left(\frac{2^k + n}{2n}\right) = 2^{m-1} \pi^{m/2}.$$

2. (Due to Nijenhuis [72]) Let n be an odd integer, and let A be the cyclic subgroup of $\Phi(2n)$ generated by $n + 2$, or any one of its cosets. Let $v(n)$ denote the cardinality of A, and $b(a)$ the number of elements of A that are larger than n. Prove that

$$\prod_{x \in A} \Gamma\left(\frac{x}{2n}\right) = 2^{b(a)} \pi^{v(n)/2}.$$

Comment. Notice that the number of sets A may not be unique. For example, when $n = 31$, we have six sets of A as follows:

$$\{1, 33, 35, 39, 47\}, \quad \{3, 17, 37, 43, 55\}, \quad \{5, 9, 41, 49, 51\},$$

$$\{7, 19, 25, 45, 59\}, \quad \{11, 13, 21, 53, 57\}, \quad \{15, 23, 27, 29, 61\}.$$

Among of them, each set, multiplying the member by 33 $(\text{mod } 62)$, gives the next set, in circle order. For example, applying the formula to the first set with $v = 5$ and $b = 4$ yields

$$\Gamma\left(\frac{1}{62}\right) \Gamma\left(\frac{33}{62}\right) \Gamma\left(\frac{35}{62}\right) \Gamma\left(\frac{39}{62}\right) \Gamma\left(\frac{47}{62}\right) = 2^4 \pi^{5/2}.$$

3.4   Evaluate an integral by Feynman's way

Problem 11966 (Proposed by C. I. Vălean, March, 2017). Prove that

$$\int_0^1 \frac{x \ln(1+x)}{1+x^2}\, dx = \frac{\pi^2}{96} + \frac{(\ln 2)^2}{8}.$$

Discussion.
Let $x = \tan\theta$. The proposed integral becomes

$$\int_0^{\pi/4} \tan\theta \ln(1 + \tan\theta)\, d\theta,$$

which does not make things easy. Integration by parts also seems hard to proceed. So we try to apply Richard Feynman's "a different box of tools" [40] — that evaluate integrals by differentiation and integration with respect to a parameter. In this way, we are able to transform the transcendental integrands into rational functions.

Solution I — By parametric differentiation.
Let

$$I(p) = \int_0^1 \frac{x \ln(1+px)}{1+x^2}\, dx.$$

The Leibniz rule yields

$$I'(p) = \int_0^1 \frac{x^2}{(1+x^2)(1+px)}\, dx.$$

In view of the partial fraction decomposition

$$\frac{x^2}{(1+x^2)(1+px)} = \frac{1}{1+p^2}\left(\frac{1}{1+px} + \frac{-1+px}{1+x^2}\right),$$

integration with respect to x gives

$$I'(p) = \frac{\ln(1+p)}{p(1+p^2)} - \frac{\pi}{4}\frac{1}{1+p^2} + \frac{\ln 2}{2}\frac{p}{1+p^2}.$$

Since $I(0) = 0$, we find the desired integral

$$\begin{aligned}
I(1) &= I(0) + \int_0^1 I'(p)\, dp \\
&= \int_0^1 \left(\frac{\ln(1+p)}{p(1+p^2)} - \frac{\pi}{4}\frac{1}{1+p^2} + \frac{\ln 2}{2}\frac{p}{1+p^2}\right) dp \\
&= \int_0^1 \frac{\ln(1+p)}{p(1+p^2)}\, dp - \frac{\pi^2}{16} + \frac{(\ln 2)^2}{4}
\end{aligned}$$

$$= \int_0^1 \frac{(1+p^2-p^2)\ln(1+p)}{p(1+p^2)}\,dp - \frac{\pi^2}{16} + \frac{(\ln 2)^2}{4}$$

$$= \int_0^1 \frac{\ln(1+p)}{p}\,dp - \int_0^1 \frac{p\ln(1+p)}{1+p^2}\,dp - \frac{\pi^2}{16} + \frac{(\ln 2)^2}{4}$$

$$= \int_0^1 \frac{\ln(1+p)}{p}\,dp - I(1) - \frac{\pi^2}{16} + \frac{(\ln 2)^2}{4}.$$

Recall that $\ln(1+x) = \sum_{n=1}^\infty (-1)^{n+1}x^n/n$. Then

$$\int_0^1 \frac{\ln(1+p)}{p}\,dp = \sum_{n=1}^\infty \int_0^1 \frac{(-1)^{n+1}p^{n-1}}{n} = \sum_{n=1}^\infty \frac{(-1)^{n+1}}{n^2} = \frac{\pi^2}{12}.$$

This yields

$$I(1) = \frac{1}{2}\left(\frac{\pi^2}{48} + \frac{(\ln 2)^2}{4}\right) = \frac{\pi^2}{96} + \frac{(\ln 2)^2}{8}$$

as claimed.   □

Solution II — By parametric integration.
Let the proposed integral be I. Notice that

$$\ln(1+x) = \int_0^1 \frac{x\,dt}{1+xt}.$$

We have

$$\begin{aligned}
I &= \int_0^1 \int_0^1 \frac{x^2}{(1+x^2)(1+xt)}\,dt\,dx \\
&= \int_0^1 \int_0^1 \frac{x^2}{(1+x^2)(1+xt)}\,dx\,dt \\
&= \int_0^1 \frac{1}{1+t^2}\int_0^1 \left(\frac{tx-1}{1+x^2} + \frac{1}{1+tx}\right)dx\,dt \\
&= \int_0^1 \frac{1}{1+t^2}\left(\frac{t\ln 2}{2} - \frac{\pi}{4} + \frac{\ln(1+t)}{t}\right)dt \\
&= \frac{\ln 2}{2}\int_0^1 \frac{t\,dt}{1+t^2} - \frac{\pi}{4}\int_0^1 \frac{dt}{1+t^2} + \int_0^1 \frac{\ln(1+t)}{t(1+t^2)}\,dt \\
&= \frac{(\ln 2)^2}{4} - \frac{\pi^2}{16} + \int_0^1 \frac{\ln(1+t)}{t}\,dt - \int_0^1 \frac{t\ln(1+t)}{1+t^2}\,dt \\
&= \frac{(\ln 2)^2}{4} - \frac{\pi^2}{16} + \int_0^1 \frac{\ln(1+t)}{t}\,dt - I.
\end{aligned}$$

Solving this for I we find that

$$I = \frac{(\ln 2)^2}{8} - \frac{\pi^2}{32} + \frac{1}{2}\int_0^1 \frac{\ln(1+t)}{t}\,dt = \frac{(\ln 2)^2}{8} - \frac{\pi^2}{32} + \frac{1}{2}\frac{\pi^2}{12} = \frac{\pi^2}{96} + \frac{(\ln 2)^2}{8}.$$

□

Remark. Parametric differentiation and integration often provides a straightforward method to evaluate difficult integrals which conventionally require the more sophisticated method of contour integration. Sometimes it may lead to new proofs of classical results. For more examples, please refer to Chapter 19 in [29]. Here we present possibly one of the best solutions to Euler's Basel problem [36] via parametric differentiation, which is due to Muzaffar [69]. His solution begins with

$$I(x) = \int_0^{\pi/2} \sin^{-1}(x \sin \theta)\, d\theta, \qquad x \in (0, 1).$$

Differentiating under the integral sign gives

$$I'(x) = \int_0^{\pi/2} \frac{\sin \theta}{\sqrt{1 - x^2 \sin^2 \theta}}\, d\theta = \int_0^{\pi/2} \frac{\sin \theta}{\sqrt{(1 - x^2) + x^2 \cos^2 \theta}}\, d\theta.$$

Substituting $u = \cos \theta$ yields

$$\begin{aligned}
I'(x) &= \int_0^1 \frac{1}{\sqrt{(1 - x^2) + x^2 u^2}}\, du \\
&= \frac{1}{x} \ln \left(u + \sqrt{(1 - x^2)/x^2 + u^2} \right) \Big|_0^1 \\
&= \frac{1}{2x} \ln \left(\frac{1 + x}{1 - x} \right) = \sum_{n=1}^{\infty} \frac{x^{2n}}{2n + 1}.
\end{aligned}$$

Hence,

$$I(x) = I(0) + \int_0^x I'(t)\, dt = \sum_{n=1}^{\infty} \frac{x^{2n+1}}{(2n + 1)^2} = \frac{1}{2}(\text{Li}_2(x) - \text{Li}_2(-x)),$$

where $\text{Li}_2(x)$ is the dilogarithm function. Setting $x = 1$ we find

$$\begin{aligned}
\frac{1}{2}(\text{Li}_2(1) - \text{Li}_2(-1)) &= \frac{3}{4} \sum_{n=1}^{\infty} \frac{1}{n^2} = \int_0^{\pi/2} \sin^{-1}(\sin \theta)\, d\theta \\
&= \int_0^1 \frac{\sin^{-1}(t)}{\sqrt{1 - t^2}}\, dt = \frac{1}{2}(\sin^{-2} t)^2 \Big|_0^1 = \frac{\pi^2}{8},
\end{aligned}$$

which implies that $\zeta(2) = \pi^2/6$ immediately. As a byproduct, integrating

$$\frac{\sin^{-1}(t)}{\sqrt{1 - t^2}} = \sum_{n=1}^{\infty} \frac{(2t)^{2n-1}}{n\binom{2n}{n}},$$

from 0 to 1/2, we recover another Euler's beautiful formula

$$\zeta(2) = 3 \sum_{n=1}^{\infty} \frac{1}{n^2 \binom{2n}{n}}.$$

We now collect six additional problems for your practice.

1. **Another proof of** $\zeta(2) = \pi^2/6$. Let $I(x) = \int_0^{\pi/2} \tan^{-1}(x \tan \theta) \, d\theta$ for $x \in (0, 1)$. Show that

$$I(x) = \frac{1}{2} \left(\ln x \ln \left(\frac{1+x}{1-x} \right) + \text{Li}_2(x) - \text{Li}_2(-x) \right).$$

2. **A Classical Problem.** Show that

$$\int_0^\infty \ln x e^{-x^2} \, dx = -\frac{\sqrt{\pi}}{4} (2 \ln 2 + \gamma).$$

3. **Putnam Problem 2005-A5.** Evaluate

$$\int_0^1 \frac{\ln(1+x)}{1+x^2} \, dx.$$

4. **Problem 10884** (Proposed by Z. Ahmed, 108(6), 2001). Evaluate

$$\int_0^1 \frac{\arctan \sqrt{2+x^2}}{(1+x^2)\sqrt{2+x^2}} \, dx.$$

   Hint: Use

$$\frac{\arctan u}{u} = \int_0^1 \frac{dy}{1+u^2 y^2}.$$

5. **Problem 11101** (Proposed by E. F. Skelton, 111(7), 2004). Show that

$$\int_0^\infty a \arctan \left(\frac{b}{\sqrt{a^2+x^2}} \right) \frac{dx}{\sqrt{a^2+x^2}} = \frac{a\pi}{2} (\ln(b+\sqrt{a^2+b^2}) - \ln a)$$

   for $a, b > 0$. *Hint:* Use

$$\arctan \left(\frac{b}{\sqrt{a^2+x^2}} \right) \frac{a}{\sqrt{a^2+x^2}} = \int_0^b \frac{a \, du}{a^2+x^2+u^2}.$$

6. **Problem 11113** (Proposed by I. Sofair, 111(9), 2004). Evaluate

$$I_k(a, b) = \int_0^\infty \int_0^\infty \frac{e^{-k\sqrt{x^2+y^2}} \sin(ax) \sin(bx)}{xy\sqrt{x^2+y^2}} \, dx dy$$

   in closed form for $a, b, k > 0$. *Hint:* Use $\sin(ax) = x \int_0^a \cos(xt) \, dt$.

3.5 Three ways to evaluate a log-sine integral

Problem 11639 (Proposed by O. Kouba, 119(4), 2012). Evaluate

$$\int_0^{\pi/2} (\ln(2\sin x))^2 \, dx.$$

Discussion.
In general, it is difficult to determine whether or not an integral can be evaluated precisely. Thanks to *Mathematica*, we see this problem has the exact value $\pi^3/24$. Once we "know" the answer analytically, it is a fairly simple matter to "prove" it. Indeed, this problem can be done in at least three different ways. The first method is based on the Fourier series of $\ln(2\sin(x/2))$; the second is a direct calculation which involves the Euler beta function; the final solution relies on a contour integration. It is interesting to see how wide a range of mathematical topics these solutions exploit.

Solution I — By Fourier series of $\ln(2\sin(x/2))$.
Recall that

$$-\ln\left(2\sin\frac{x}{2}\right) = \sum_{n=1}^\infty \frac{\cos nx}{n}. \qquad (3.5)$$

Then

$$\ln^2\left(2\sin\frac{x}{2}\right) = \sum_{n,m=1}^\infty \frac{\cos nx \cos mx}{nm}.$$

The orthogonality of $\cos nx$ on $[0,\pi]$ implies that

$$\begin{aligned}
\int_0^\pi \ln^2\left(2\sin\frac{x}{2}\right) dx &= \sum_{n,m=1}^\infty \frac{1}{nm} \int_0^\pi \cos nx \cos mx \, dx \\
&= \sum_{n=m=1}^\infty \frac{1}{nm} \int_0^\pi \cos nx \cos mx \, dx \\
&= \frac{\pi}{2} \sum_{n=1}^\infty \frac{1}{n^2} = \frac{\pi^3}{12}.
\end{aligned}$$

Let $t = x/2$. Then

$$\int_0^\pi \ln^2\left(2\sin\frac{x}{2}\right) dx = 2\int_0^{\pi/2} \ln^2(2\sin t) \, dt.$$

Hence

$$\int_0^{\pi/2} (\ln(2\sin x))^2 \, dx = \frac{1}{2}\int_0^\pi \ln^2(2\sin t) \, dt = \frac{\pi^3}{24}.$$

☐

Solution II — By the Euler beta function.

Denote the integral by I. Observe that

$$I = \int_0^{\pi/2} (\ln 2 + \ln \sin x)^2 \, dx$$

$$= \frac{\ln^2 2}{2} \pi + 2 \ln 2 \int_0^{\pi/2} \ln(\sin x) \, dx + \int_0^{\pi/2} \ln^2(\sin x) \, dx.$$

It is well-known that

$$\int_0^{\pi/2} \ln(\sin x) \, dx = -\frac{\ln 2}{2} \pi.$$

Hence

$$I = -\frac{\ln^2 2}{2} \pi + \int_0^{\pi/2} \ln^2(\sin x) \, dx.$$

Now it suffices to show that

$$\int_0^{\pi/2} \ln^2(\sin x) \, dx = \frac{\ln^2 2}{2} \pi + \frac{1}{24} \pi^3.$$

Let $t = \sin^2 x$. Then

$$\int_0^{\pi/2} \ln^2(\sin x) \, dx = \frac{1}{2} \int_0^1 \frac{\ln^2(\sqrt{t})}{\sqrt{t}\sqrt{1-t}} \, dt$$

$$= \frac{1}{8} \int_0^1 t^{1/2-1}(1-t)^{1/2-1} \ln^2 t \, dt$$

On the other hand, recall

$$B(x, y) = \int_0^1 t^{x-1}(1-t)^{y-1} \, dt = \frac{\Gamma(x)\Gamma(y)}{\Gamma(x+y)}.$$

Applying the Leibniz rule yields

$$\int_0^{\pi/2} \ln^2(\sin x) \, dx = \frac{1}{8} \frac{d^2}{dx^2} B(x+1/2, 1/2)\big|_{x=0}.$$

From $\Gamma(1/2) = \sqrt{\pi}$, the Legendre duplication formula gives

$$B(x+1/2, 1/2) = \frac{\Gamma(x+1/2)\Gamma(1/2)}{\Gamma(x+1)} = \frac{(2x)\Gamma(2x)\pi}{4^x(x\Gamma(x))^2}.$$

Since

$$x\Gamma(x) = e^{-\gamma x} \prod_{n=1}^{\infty} \left(1 + \frac{x}{n}\right)^{-1} e^{x/n}$$

$$= \left(1 - \gamma x + \frac{1}{2}(\gamma x)^2 + o(x^2)\right) \prod_{n=1}^{\infty} \left(1 + \frac{x^2}{2n^2} + o(x^2)\right)$$

$$= 1 - \gamma x + \left(\frac{1}{2}\gamma^2 + \frac{1}{2}\sum_{n=1}^{\infty}\frac{1}{n^2}\right)x^2 + o(x^2)$$

$$= 1 - \gamma x + \left(\frac{1}{2}\gamma^2 + \frac{\pi^2}{12}\right)x^2 + o(x^2),$$

it follows

$$B(x + 1/2, 1/2) = \pi - (2\pi \ln 2)x + \left(2\pi \ln^2 2 + \frac{\pi^3}{6}\right)x^2 + o(x^2).$$

Therefore,

$$\frac{d^2}{dx^2}B(x + 1/2, 1/2)\Big|_{x=0} = 4\pi \ln^2 2 + \frac{1}{3}\pi^3,$$

and so

$$\int_0^{\pi/2} \ln^2(\sin x)\,dx = \frac{\ln^2 2}{2}\pi + \frac{1}{24}\pi^3.$$

□

Solution III — By contour integration.

Let $z = re^{i\theta}$. Then, on the unit circle

$$\ln(1 - z) = \ln(2\sin(\theta/2)) + \frac{1}{2}i(\theta - \pi). \tag{3.6}$$

Integrating $\ln^2(1 - z)/iz$ along the contour shown in Figure 3.2, since there is no singularity on the contour, the residue theorem yields

$$\oint \frac{\ln^2(1 - z)}{iz}\,dz = 0. \tag{3.7}$$

On the small arc, as $\epsilon \to 0$, we have

$$\left|\oint \frac{\ln^2(1 - z)}{iz}\,dz\right| = \left|\oint \frac{(1 - z)\ln^2(1 - z)}{iz(1 - z)}\,dz\right|$$

$$\leq \max_{|1-z|=\epsilon} |(1 - z)\ln^2(1 - z)/z|\,\frac{2\pi\epsilon}{\epsilon} \to 0.$$

Thus, by (3.6) and $dz = ie^{i\theta}d\theta = izd\theta$, (3.7) becomes

$$\int_0^{2\pi} \left(\ln(2\sin(\theta/2)) + \frac{1}{2}i(\theta - \pi)\right)^2 d\theta = 0.$$

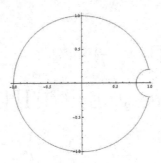

FIGURE 3.2
The contour of integration

The real part gives

$$\int_0^{2\pi} \ln^2(2\sin(\theta/2))\, d\theta = \frac{1}{4}\int_0^{2\pi} (\theta-\pi)^2\, d\theta = \frac{\pi^3}{6}.$$

The substitution $\theta = 2\pi - t$ yields

$$\int_\pi^{2\pi} \ln^2(2\sin(\theta/2))\, d\theta = \int_0^\pi \ln^2(2\sin(t/2))\, dt,$$

and so

$$\int_0^\pi \ln^2(2\sin(\theta/2))\, d\theta = \frac{\pi^3}{12}.$$

Replacing $\theta/2$ by x yields

$$\int_0^{\pi/2} (\ln(2\sin x))^2\, dx = \frac{\pi^3}{24}$$

as claimed.   □

Remark. The Fourier series (3.5) was due to Euler. Here is a short elementary proof:

$$\sum_{n=1}^{\infty} \frac{\cos(nx)}{n} = \sum_{n=1}^{\infty} \frac{e^{nxi} + e^{-nxi}}{2n}$$

$$= -\frac{1}{2}\left(\ln(1-e^{xi}) + \ln(1-e^{-xi})\right)$$

$$= -\frac{1}{2}\ln(2-2\cos x)$$

$$= -\frac{1}{2}\ln\left(4\sin^2\frac{x}{2}\right)$$

$$= -\ln\left(2\sin\frac{x}{2}\right).$$

In Solution I, as Jean-Pierre Grivaux did in the published solution, another path to evaluate this integral is via Parseval's identity:

$$\frac{1}{2\pi} \int_0^{2\pi} \ln^2\left(2\sin\frac{x}{2}\right) dx = \frac{1}{2} \sum_{n=1}^{\infty} \frac{1}{n^2} = \frac{\pi^2}{12}.$$

Solution II may seem unnecessarily lengthy, however, it enables us to investigate the more general family of integrals

$$I(p,q) = \int_0^{\pi/2} \ln^p \sin x \ln^q \cos x \, dx, \qquad (p, q \geq 0).$$

Indeed, along the same lines as in Solution II, we have

$$
\begin{aligned}
I(p,q) &= \frac{1}{2^{p+q+1}} \frac{\partial^{p+q}}{\partial \alpha^p \partial \beta^q} \left(\frac{\Gamma(\alpha+1/2)\Gamma(\beta+1/2)}{\Gamma(\alpha+\beta+1)} \right)\Bigg|_{\alpha=\beta=0} \\
&= \frac{\pi}{2^{p+q+1}} \frac{\partial^{p+q}}{\partial \alpha^p \partial \beta^q} \left(\frac{2^{-2(\alpha+\beta)}\Gamma(2\alpha+1)\Gamma(2\beta+1)}{\Gamma(\alpha+\beta+1)\Gamma(\alpha+1)\Gamma(\beta+1)} \right)\Bigg|_{\alpha=\beta=0}.
\end{aligned}
$$

In particular, we have

$$\int_0^{\pi/2} \ln \sin x \ln \cos x \, dx = \frac{\ln^2 2}{2}\pi - \frac{1}{48}\pi^3,$$

$$\int_0^{\pi/2} \ln^2 \sin x \ln \cos x \, dx = -\frac{\ln^3 2}{2}\pi + \frac{1}{8}\pi\zeta(3),$$

and

$$\int_0^{\pi/2} \ln^2 \sin x \ln^2 \cos x \, dx = \frac{\pi}{2}\left(\frac{1}{160}\pi^4 + \ln^4 2 - \zeta(3)\ln 2\right).$$

For $q = 0$, we can also find $I(p,0)$ as we did in Solution III. Integrating $\ln^p(1-z)$ on the contour in Figure 3.2 and taking the real part yields

$$\sum_{k=0}^{[p/2]} (-1)^k \binom{p}{2k} \int_0^{\pi/2} x^{2k} \ln^{p-2k}(2\sin x) \, dx = 0.$$

This implies the special values

$$\int_0^{\pi/2} \ln^3 \sin x \, dx = -\frac{\ln 2}{8}\pi^3 - \frac{\ln^3 2}{2}\pi - \frac{3}{4}\pi\zeta(3);$$

$$\int_0^{\pi/2} \ln^4 \sin x \, dx = \frac{19}{480}\pi^5 + \frac{\ln^2 2}{4}\pi^3 + \frac{\ln^4 2}{2}\pi + 3\ln 2\pi\zeta(3).$$

Recently, Borwein and Straub [23] studied another generalized log-sine integral in the form

$$\mathrm{Ls}_n^{(k)}(\theta) := -\int_0^\theta x^k \ln^{n-1-k}(2\sin(x/2)) \, dx,$$

which appeared in the ϵ-expansion of various Feynman diagrams in quantum mechanics. Explicit evaluation at π and multiples are obtained. For example,

$$\int_0^{\pi/3} \ln^2(2\sin(x/2))\,dx = \frac{7}{108}\pi^3; \qquad \int_0^{\pi/3} x\ln^2(2\sin(x/2))\,dx = \frac{17}{6480}\pi^4.$$

Finally, we end this section by compiling some additional problems for practice.

1. Show that $\int_0^{\pi} \ln^2(\tan(x/4))\,dx = \pi^3/4$.

2. Let $\alpha, \beta > -1$ and integers $p, q \geq 0$. Prove that

$$\int_0^{\pi/2} \ln^p \sin x \ln^q \cos x (\sin x)^{2\alpha+1}(\cos x)^{2\beta+1}\,dx$$

$$= \frac{1}{2^{p+q+1}} \frac{\partial^{p+q}}{\partial\alpha^p\partial\beta^q} \left(\frac{\Gamma(\alpha+1)\Gamma(\beta+1)}{\Gamma(\alpha+\beta+2)} \right).$$

3. Let $\alpha, \beta > -1$. Prove that

$$\int_0^{\pi/2} \ln \sin x (\sin x)^{2\alpha+1}(\cos x)^{2\beta+1}\,dx$$

$$= \frac{\Gamma(\alpha+1)\Gamma(\beta+1)}{4\Gamma(\alpha+\beta+2)}(\psi(\alpha+1) - \psi(\alpha+\beta+2)),$$

where ψ is the digamma function.

4. For $a, b, p, q > 0$, show that

$$\int_0^{\pi/2} \ln(a^2 \cos^2 x + b^2 \sin^2 x)\,dx = \pi \ln\left(\frac{a+b}{2}\right);$$

$$\int_0^{\pi} \frac{\ln(a^2 \cos^2 x + b^2 \sin^2 x)}{p^2 \cos^2 x + q^2 \sin^2 x}\,dx = \frac{2\pi}{pq} \ln\left(\frac{aq+bp}{p+q}\right).$$

5. Let

$$F(x) := 2^{-2x} \frac{(2x)\Gamma(2x)}{x\Gamma(x)}.$$

Show that, for $|x| < 1/2$,

$$\ln F(x) = -(\gamma + 2\ln 2)x + \sum_{k=2}^{\infty} (-1)^k \zeta(k) \frac{2^k - 1}{k} x^k,$$

where ζ is the Riemann zeta function.

6. Let $a_n = \int_0^{\pi} \ln^n(\sin x)\,dx$. Show that (Due to Beumer [13])

    (a) $(n-1)a_n = \ln 2\, a_{n-1} + \sum_{k=1}^{n-1}(1 - 2^{-k})\zeta(k+1)a_{n-k-1}$.

(b) $\sum_{n=0}^{\infty} a_n \frac{x^n}{n!} = \frac{\sqrt{\pi}}{2} \frac{\Gamma((x+1)/2)}{\Gamma(x/2+1)}$.

7. Let $a_n = \int_0^{\pi} \ln^n(\sin x)\, dx$.

   (a) Define
   $$D(s) = \sum_{k=0}^{\infty} \frac{(2k-1)!!}{(2k)!!} \cdot \frac{1}{(2k+1)^s}.$$
   Show that $a_n = (-1)^n n! D(n+1)$.

   (b) Let
   $$\sigma_k = \frac{1}{1^k} - \frac{1}{2^k} + \frac{1}{3^k} - \frac{1}{4^k} + \cdots = (1 - 2^{1-k})\zeta(k).$$

   Show that
   $$a_n = (-1)^n \frac{\pi}{2}
   \begin{vmatrix}
   \sigma_1 & -1 & 0 & 0 & \cdots & 0 \\
   \sigma_2 & \sigma_1 & -2 & 0 & \cdots & 0 \\
   \sigma_3 & \sigma_2 & \sigma_1 & -3 & \cdots & 0 \\
   \cdots & \cdots & \cdots & \cdots & \cdots & \cdots \\
   \sigma_n & \sigma_{n-1} & \sigma_{n-2} & \sigma_{n-3} & \cdots & \sigma_1
   \end{vmatrix}$$

8. **Problem 11418** (Proposed by G. Lamb, 116(3), 2009). Find
   $$\int_{-\infty}^{\infty} \frac{t^2 \operatorname{sech}^2 t}{a - \tanh t}\, dt$$
   for complex a with $|a| > 1$.

9. **Problem 11629** (Proposed by O. Oloa, 119(3), 2012). Let
   $$f(\sigma) = \int_0^1 x^{\sigma} \left(\frac{1}{\ln x} + \frac{1}{1-x} \right)^2 dx.$$
   (a) Show that $f(0) = \ln(2\pi) - 3/2$.
   (b) Find a closed form expression for $f(\sigma)$ for $\sigma > 0$.

10. **Problem 12051** (Proposed by P. Ribeiro, 125(6), 2018). Prove
   $$\sum_{n=0}^{\infty} \binom{2n}{n} \frac{1}{4^n (2n+1)^3} = \frac{\pi^3}{48} + \frac{\pi}{4} \ln^2 2.$$

3.6 A log-product integral

Problem 11993 (Proposed by C. I. Vălean, 124(7), 2017). Prove that
$$\int_0^1 \frac{\log(1-x)(\log(1+x))^2}{x}\, dx = -\frac{\pi^4}{240}.$$

Discussion.
Since we are able to evaluate a number of integrals involving a single logarithmic function, we now aim to reduce the number of logarithms with different arguments in the integrand. To this end, invoking the logarithmic properties, we call the algebraic identity

$$(a+b)^3 + (a-b)^3 = 2a^3 + 6ab^2.$$

Solution.
Let the proposed integral be I. Appealing to the algebraic identity

$$6ab^2 = (a+b)^3 + (a-b)^3 - 2a^3,$$

with $a = \log(1-x)$ and $b = \log(1+x)$, we rewrite the proposed integral as

$$I = \frac{1}{6}\int_0^1 \left(\log^3(1-x^2) + \log^3\left(\frac{1-x}{1+x}\right) - 2\log^3(1-x) \right) \frac{dx}{x}. \quad (3.8)$$

Using the substitution $x^2 = t$, we have

$$\int_0^1 \log^3(1-x^2)\,\frac{dx}{x} = \frac{1}{2}\int_0^1 \log^3(1-t)\,\frac{dt}{t}$$

$$= \frac{1}{2}\int_0^1 \frac{\log^3 u}{1-u}\,du \quad (\text{let } u = 1-t)$$

$$= \frac{1}{2}\sum_{n=0}^\infty \int_0^1 u^n \log^3 u\,du$$

$$= -3\sum_{n=0}^\infty \frac{1}{(n+1)^4} = -3\zeta(4),$$

where ζ is the Riemann zeta function. As a byproduct, we also get

$$\int_0^1 \log^3(1-t)\,\frac{dt}{t} = -6\zeta(4).$$

Similarly, using the substitution $x = \frac{1-t}{1+t}$, we have

$$\int_0^1 \log^3\left(\frac{1-x}{1+x}\right)\frac{dx}{x} = 2\int_0^1 \frac{\log^3 t}{1-t^2}\,dt$$

$$= 2\sum_{n=0}^\infty \int_0^1 u^{2n} \log^3 u\,du$$

$$= -12\sum_{n=0}^\infty \frac{1}{(2n+1)^4} = -\frac{45}{4}\zeta(4),$$

where we have used the fact that

$$\sum_{n=0}^{\infty} \frac{1}{(2n+1)^4} = (1-2^{-4})\sum_{n=1}^{\infty}\frac{1}{n^4} = \frac{15}{16}\zeta(4).$$

Plugging these results into (3.8), since $\zeta(4) = \pi^4/90$, we find that

$$I = \frac{1}{6}\left(-3\zeta(4) - \frac{45}{4}\zeta(4) + 12\zeta(4)\right) = -\frac{3}{8}\zeta(4) = -\frac{\pi^4}{240}$$

as desired. $\qquad\qquad\qquad\qquad\qquad\qquad\qquad\qquad\qquad\qquad\qquad\qquad\Box$

Remark. The solution clearly presents a useful approach to evaluate integrals involving the product of logarithms. We collect a few more problems for your practice.

1. Show that

$$\int_0^1 \frac{\log^2 x \log^2(1+x)}{x}\, dx = 2\zeta(2)\zeta(3) - \frac{29}{8}\zeta(5).$$

2. Show that

$$\int_0^1 \frac{\log^n x \log^2(1+x)}{x}\, dx = 2(-1)^n n!$$

$$\left(\sum_{k=1}^{\infty}(-1)^k \frac{H_k}{k^{n+2}} + (1 - 2^{-n-2})\zeta(n+3)\right).$$

Can you find a closed form for

$$\int_0^1 \frac{\log^n x \log^m(1+x)}{x}\, dx$$

for $m, n \in \mathbb{N}$?

3. Prove that

$$\int_0^1 \frac{\log x \log(1-x)\log^2(1+x)}{x}\, dx = \frac{7}{8}\zeta(2)\zeta(3) - \frac{25}{16}\zeta(5).$$

3.7 A Putnam problem beyond

Problem 11234 (Proposed by J. Brennan and R. Ehrenborg, 113(6), 2006). Let $a_1, \ldots, a_n$ and $b_1, \ldots, b_{n-1}$ be real numbers, with $a_1 < b_1 < a_2 \ldots a_{n-1} < b_{n-1} < a_n$. Let h be an integrable function from $\mathbb{R}$ to $\mathbb{R}$. Show that

$$\int_{-\infty}^{\infty} h\left(\frac{(x-a_1)\cdots(x-a_n)}{(x-b_1)\cdots(x-b_{n-1})}\right)dx = \int_{-\infty}^{\infty} h(x)\, dx.$$

Discussion

Let $a_1 = -1, a_2 = 1$ and $b_1 = 0$. The proposed problem becomes **Putnam Problem 1968-B4** [4]: Prove that

$$\int_{-\infty}^{\infty} h\left(x - \frac{1}{x}\right) dx = \int_{-\infty}^{\infty} h(x)\, dx.$$

This result goes back as far as Cauchy and can be proved as follows. Let $I = \int_{-\infty}^{\infty} f(x - 1/x)\, dx$. Applying the substitution $u = x - 1/x$ yields two branches:

$$x_{\pm} = \frac{1}{2}\left(u \pm \sqrt{u^2 + 4}\right), \qquad u \in \mathbb{R}.$$

Since

$$\frac{dx_-}{du} + \frac{dx_+}{du} = \frac{1}{2}\left(1 - \frac{u}{\sqrt{u^2+4}} + 1 + \frac{u}{\sqrt{u^2+4}}\right) = 1,$$

it follows that

$$\begin{aligned}
I &= \int_{-\infty}^{0} f(u)\, dx_- + \int_{0}^{\infty} f(u)\, dx_+ \\
&= \int_{-\infty}^{\infty} f(u)\left(\frac{dx_-}{du} + \frac{dx_+}{du}\right) du \\
&= \int_{-\infty}^{\infty} f(u)\, du.
\end{aligned}$$

We demonstrate the above argument goes through for the general case. It is remarkable that

$$\int_{-\infty}^{\infty} f(u)\, dx = \int_{-\infty}^{\infty} f(x)\, dx$$

remains true when

$$u = x + C - \sum_{k=1}^{n-1} \frac{\alpha_k}{x - b_k} \tag{3.9}$$

with constant C and all the $\alpha_k > 0$ for $1 \le k \le n - 1$.

Solution.

We will instead prove the above generalization and then verify that

$$\frac{(x - a_1) \cdots (x - a_n)}{(x - b_1) \cdots (x - b_{n-1})} = x + C - \sum_{k=1}^{n-1} \frac{\alpha_k}{x - b_k} \tag{3.10}$$

satisfies $\alpha_k > 0$ for $1 \le k \le n - 1$. To this end, without loss of generality, we assume that $b_1 < b_2 < \cdots < b_{n-1}$. On each of the following intervals

$$(-\infty, b_1), (b_1, b_2), \cdots, (b_{n-1}, \infty)$$

the function u given by (3.9) is continuous and has limits $-\infty$ and ∞ at the left and right endpoints, respectively. Thus the range of u on each of these intervals is $\mathbb{R}$. Furthermore, from (3.9) we have

$$x^n - \left(u - C + \sum_{k=1}^{n-1} b_k \right) x^{n-1} + \cdots = 0.$$

This implies that its solution has n branches satisfying

$$x_1 + x_2 + \cdots + x_n = u - C + \sum_{k=1}^{n-1} b_k.$$

Hence

$$\frac{d}{du}(x_1 + x_2 + \cdots + x_n) = 1.$$

Thus the proposed integral

$$I : = \int_{-\infty}^{b_1} f(u)\, dx_1 + \int_{b_1}^{b_2} f(u)\, dx_2 + \cdots \int_{b_{n-1}}^{\infty} f(u)\, dx_n$$

$$= \int_{-\infty}^{\infty} f(u)\, \frac{d}{du}(x_1 + x_2 + \cdots + x_n)\, du$$

$$= \int_{-\infty}^{\infty} f(u)\, du.$$

Now we verify that $\alpha_k > 0$ for $1 \le k \le n-1$. Multiplying (3.10) by $x - b_k$ on both sides, then letting $x = b_k$ yields

$$-\alpha_k = \frac{(b_k - a_1)\cdots(b_k - a_k)(b_k - a_{k+1})\cdots(b_k - a_n)}{(b_k - b_1)\cdots(b_k - b_{k-1})(b_k - b_{k+1})\cdots(b_k - b_{n-1})}.$$

In the above ratio, by assumption, there are $n - (k+1) + 1 = n - k$ negative factors in the numerator and $(n-1) - (k+1) + 1 = n - k - 1$ negative factors in the denominator. It follows that

$$\alpha_k = \frac{(b_k - a_1)\cdots(b_k - a_k)(a_{k+1} - b_k)\cdots(a_n - b_k)}{(b_k - b_1)\cdots(b_k - b_{k-1})(b_{k+1} - b_k)\cdots(b_{n-1} - b_k)} > 0,$$

for $1 \le k \le n-1$.

$\square$

Remark. This problem displays a principle of invariant integral value. For example, for $b > 0$,

$$\int_0^{\infty} \exp(-x^2 - b^2/x^2)\, dx = \frac{1}{2e^{2b}} \int_{-\infty}^{\infty} \exp(-(x - b/x)^2)\, dx$$

$$= \frac{1}{2e^{2b}} \int_{-\infty}^{\infty} \exp(-x^2)\, dx = \frac{\sqrt{\pi}}{2e^{2b}}.$$

Moreover, let $f(x) = 1/(x^2 + c^2)^\alpha, \alpha > 1/2$. Then

$$\int_{-\infty}^{\infty} f(x)\, dx = \frac{2}{c^{2\alpha-1}} \int_0^\infty \left(\frac{1}{1+u^2}\right)^\alpha du$$

$$= \frac{1}{c^{2\alpha-1}} \int_0^\infty \frac{t^{-1/2}}{(1+t)^\alpha}\, dt = \frac{1}{c^{2\alpha-1}} B(\alpha - 1/2, 1/2),$$

where $B(x, y)$ is the Euler beta function.

Applying this invariant principle also greatly simplifies the original proof of the following formula in [14]:

$$\int_{-\infty}^{\infty} \left(\frac{x^2}{x^4 + 2ax^2 + 1}\right)^\alpha dx \; = \; \int_{-\infty}^{\infty} \left(\frac{1}{(x - 1/x)^2 + 2(a+1)}\right)^\alpha dx$$

$$= \; \int_{-\infty}^{\infty} f(x)\, dx \quad (\text{with } c = \sqrt{2(a+1)})$$

$$= \; \frac{1}{(2(a+1))^{\alpha-1/2}} B(\alpha - 1/2, 1/2).$$

This master formula, together with

$$\int_0^\infty \left(\frac{x^2}{x^4 + 2ax^2 + 1}\right)^\alpha dx = \int_0^\infty \left(\frac{x^2}{x^4 + 2ax^2 + 1}\right)^\alpha \frac{x^2 + 1}{x^b + 1} \frac{dx}{x^2}$$

provides a method for unifying a large class of integrals. See [14] for details.

Notice that the positivity of α_k in (3.10) is necessary. Here is a counterexample: Let $u = x + 1/x$, then

$$\int_{-\infty}^{\infty} \exp[-(x + 1/x)^2]\, dx = \sqrt{\pi}e^{-4} \neq \int_{-\infty}^{\infty} \exp(-x^2)\, dx = \sqrt{\pi}.$$

The following Liouville formula has been misquoted in several tables of integrals (For example, see [49], 3.257, p. 346):

$$\int_0^\infty [(ax + b/x)^2 + c]^{-p-1}\, dx = \frac{\sqrt{\pi}\Gamma(p + 1/2)}{2ac^{p+1/2}\Gamma(p+1)}.$$

This expression is true only for $b < 0$. Indeed, for $b > 0, 4ab + c > 0$, we have

$$\int_0^\infty [(ax + b/x)^2 + c]^{-p-1}\, dx = \frac{1}{2} \int_{-\infty}^{\infty} \left(\frac{x^2}{a^2x^4 + (2ab + c)x^2 + b^2}\right)^{p+1} dx$$

$$= \frac{B(p + 1/2, 1/2)}{2a(4ab + c)^{p+1/2}}.$$

It is interesting to ask for what other function $u(x)$ will the integral of $f(u)$ be the same as the integral of $f(x)$. Glasser suggested that the invariant principle is valid if (3.10) is replaced by

$$u = x - \sum_{k=1}^{\infty} a_k \cot[(x - b_k)^{-1}].$$

We leave it to the reader to work out the details. Now, we end this section with four problems for additional practice.

1. For all $a, b \in \mathbb{R}$, show that

$$\int_{-\infty}^{\infty} \frac{(2x + a + b)^2}{(x + a)^2 (x + b)^2 + (2x + a + b)^2} \, dx = 2\pi.$$

2. Let $u = \alpha x - \beta/x$ with $\alpha, \beta > 0$. For any continuous function F, show that

$$\int_0^{\infty} \frac{F(u)}{1 + x^2} \, dx = (\alpha + \beta) \int_0^{\infty} \frac{F(x)}{x^2 + (\alpha + \beta)^2} \, dx.$$

Use this result to evaluate

$$\int_0^{\infty} \frac{J_0(|\alpha x - \beta/x|)}{1 + x^2} \, dx,$$

where $J_0(x)$ is the Bessel function of the first kind of zero order.

3. Let $u = \sum_{k=1}^{n} a_k/(x + b_k)$. Show that

$$\int_{-\infty}^{\infty} f(u) \, dx = \sum_{k=1}^{n} a_k \int_{-\infty}^{\infty} f(1/x) \, dx.$$

4. Let $u = x/(e^x - 1)$ or $x/(1 - e^{-x})$. For $\alpha > 0$, show that

$$\int_0^{\infty} \left(u e^{-u} \right)^{\alpha} dx = \int_0^{\infty} \left(u e^{-u} \right)^{\alpha} du.$$

3.8   A surface integral with many faces

Problem 11277 (Proposed by P. De, 114(3), 2007). Find

$$\int_{\phi=0}^{\pi/2} \int_{\theta=0}^{\pi/2} \frac{\log(2 - \sin\theta \cos\phi) \sin\theta}{2 - 2\sin\theta \cos\phi + \sin^2\theta \cos^2\phi} \, d\theta d\phi.$$

Discussion.
You may recognize that the proposed double integral is a surface integral over the first octant of the unit sphere in spherical coordinates. Thus, in rectangular coordinates, this integral becomes

$$I = \int_0^1 \int_0^{\sqrt{1-x^2}} \frac{\ln(2 - x)}{1 + (1 - x)^2} \frac{1}{\sqrt{1 - x^2 - y^2}} \, dy dx.$$

On the other hand, we can calculate this integral based on a lemma stated below. This lemma enables us to convert a class of double integrals into single integrals, which leads to a straightforward evaluation of I. Here we give two solutions based on the above discussions. Surprisingly, both approaches end with the calculation of the same integral

$$\int_0^1 \frac{\ln(1+x)}{1+x^2}\, dx.$$

Solution I.

Since I is a surface integral over the first octant of the unit sphere in spherical coordinates, converting it to rectangular coordinates yields

$$I = \int_0^1 \int_0^{\sqrt{1-x^2}} \frac{\ln(2-x)}{1+(1-x)^2} \frac{1}{\sqrt{1-x^2-y^2}}\, dy dx.$$

For the inner integral, we have

$$\int_0^{\sqrt{1-x^2}} \frac{1}{\sqrt{1-x^2-y^2}}\, dy = \sin^{-1}\left(\frac{y}{\sqrt{1-x^2}}\right)\Bigg|_0^{\sqrt{1-x^2}} = \frac{\pi}{2}.$$

Hence,

$$I = \frac{\pi}{2} \int_0^1 \frac{\ln(2-x)}{1+(1-x)^2}\, dx = \frac{\pi}{2} \int_0^1 \frac{\ln(1+x)}{1+x^2}\, dx.$$

There are a number of ingenious ways to evaluate this integral. One way is to take advantage of "symmetry." Let $x = \tan\alpha$. We obtain

$$\int_0^1 \frac{\ln(1+x)}{1+x^2}\, dx = \int_0^{\pi/4} \ln(1+\tan\alpha)\, d\alpha = \int_0^{\pi/4} \ln\left(\frac{\sqrt{2}\cos(\pi/4-\alpha)}{\cos\alpha}\right)\, d\alpha$$

$$= \frac{\pi}{8}\ln 2 + \int_0^{\pi/4} \ln\cos(\pi/4-\alpha)\, d\alpha - \int_0^{\pi/4} \ln\cos\alpha\, d\alpha$$

$$= \frac{\pi}{8}\ln 2,$$

where the last two integrals are equal by symmetry about $\alpha = \pi/8$. This immediately deduces

$$I = \frac{\pi}{2} \cdot \frac{\pi}{8}\ln 2 = \frac{\pi^2}{16}\ln 2.$$

$\square$

Solution II.

We begin with a lemma.

Lemma. *If f is integrable on $[0,1]$, then*

$$\int_0^{\pi/2} \int_0^{\pi/2} f(\sin\theta\cos\phi)\sin\theta\, d\phi d\theta = \frac{\pi}{2} \int_0^1 f(x)\, dx. \qquad (3.11)$$

Proof. Let S be the unit sphere in the first octant. The double integral can then be viewed as a surface integral over S with the angle θ measured from the z-axis. Thus, $x = \sin\theta\cos\phi, y = \sin\theta\sin\phi, z = \cos\theta$ and

$$\|\mathbf{T}_\theta \times \mathbf{T}_\phi\| = |\sin\theta|.$$

This yields

$$\int_0^{\pi/2}\int_0^{\pi/2} f(\sin\theta\cos\phi)\sin\theta\,d\phi d\theta = \iint_S f(x)\,dS.$$

On the other hand, viewing the x-axis as the polar axis and ϕ as the angle, we have

$$\iint_S f(x)\,dS = \int_0^{\pi/2}\int_0^1 f(x)dx d\phi = \frac{\pi}{2}\int_0^1 f(x)\,dx,$$

which proves the lemma.

Replacing θ by $\pi/2 - \theta$ in (3.11), we also find

$$\int_0^{\pi/2}\int_0^{\pi/2} f(\cos\theta\cos\phi)\cos\theta\,d\theta d\phi = \frac{\pi}{2}\int_0^1 f(x)\,dx.$$

To use the lemma to evaluate I, let $f(x) = \ln(2-x)/(2-2x+x^2)$ in (3.11). Then

$$I = \frac{\pi}{2}\int_0^1 \frac{\ln(2-x)}{2-2x+x^2}\,dx = \frac{\pi}{2}\int_0^1 \frac{\ln(1+x)}{1+x^2}\,dx.$$

In view of

$$\int_0^1 \frac{\ln(1+x)}{1+x^2}\,dx = \frac{\pi}{8}\ln 2,$$

again, we find that

$$I = \frac{\pi^2}{16}\ln 2.$$

$\square$

Remark. (3.11) can be proved without using the surface integral. To this end, let $x = \sin\theta\cos\phi$ in the inner integral. Then

$$\int_0^{\pi/2} f(\sin\theta\cos\phi)\sin\theta\,d\phi = \int_0^{\sin\theta} \frac{\sin\theta}{\sqrt{\sin^2\theta - x^2}} f(x)\,dx.$$

Using $u = \sin\theta$ in the outer integral gives

$$\int_0^{\pi/2}\int_0^{\pi/2} f(\sin\theta\cos\phi)\sin\theta\,d\phi d\theta$$

$$= \int_0^1\int_0^u \frac{u}{\sqrt{u^2 - x^2}\sqrt{1-u^2}} f(x)dx du$$

$$= \int_0^1 \left(\int_x^1 \frac{u}{\sqrt{u^2 - x^2}\sqrt{1 - u^2}} \, du \right) f(x) dx \quad \text{(switch the integral order)}$$

$$= \frac{1}{2} \int_0^1 \left(\int_{x^2}^1 \frac{1}{\sqrt{v - x^2}\sqrt{1 - v}} \, dv \right) f(x) dx \quad \text{(set } v = u^2\text{)}.$$

Notice that applying the substitution $v = (1 + x^2)/2 + s(1 - x^2)/2$ yields

$$\int_{x^2}^1 \frac{1}{\sqrt{v - x^2}\sqrt{1 - v}} \, dv = \int_{-1}^1 \frac{ds}{\sqrt{1 - s^2}} = \pi.$$

This proves (3.11) again. Along the same lines, (3.11) can be generalized

$$\int_0^\pi \int_0^{2\pi} f(a \sin\phi \cos\theta + b \sin\phi \sin\theta + c \cos\phi) \sin\phi \, d\theta d\phi$$

$$= 2\pi \int_{-1}^1 f(x\sqrt{a^2 + b^2 + c^2}) \, dx$$

and

$$\iiint_B f(ax + by + cz) \, dV = \pi \int_{-1}^1 (1 - x^2) f(x\sqrt{a^2 + b^2 + c^2}) \, dx,$$

where B is the unit ball in $\mathbb{R}^3$. We leave the proofs to the reader.

It is also worth noting that this proposed problem provides a prefect example of discovering mathematical truths through the use of computers. Bailey and Borwein evaluated the numerical value of I to exceedingly high precision and found the exact value $\pi^2 \ln 2/16$ via the *Inverse Symbolic Calculator*, an online numerical constant recognition tool available at `http://wayback.cecm.sfu.ca/projects/ISC/ISCmain.html`

To establish the result rigorously, let $A(\theta, \phi) = 1 - \sin\theta \cos\phi$. They then established an analytical evaluation via double series expansions and Wallis's formula:

$$
\begin{aligned}
I_2 &= \int_0^{\pi/2} \int_0^{\pi/2} \frac{\ln(1 + A)}{1 + A^2} \sin\theta \, d\theta d\phi \\
&= \int_0^{\pi/2} \int_0^{\pi/2} \left(\sum_{k=0}^\infty (-1)^k \frac{A^{k+1}}{k+1} \right) \cdot \left(\sum_{m=0}^\infty (-1)^m A^{2m} \right) \sin\theta \, d\theta d\phi \\
&= \sum_{m=0}^\infty \sum_{k=0}^\infty \frac{(-1)^{m+k}}{k+1} \int_0^{\pi/2} \int_0^{\pi/2} A^{2m+k+1} \sin\theta \, d\theta d\phi \\
&= \frac{\pi}{2} \sum_{m=1}^\infty \sum_{k=0}^\infty \frac{(-1)^{m+k+1}}{(k+1)(2m+k)},
\end{aligned}
$$

where they used

$$\int_0^{\pi/2} \int_0^{\pi/2} A^k(\theta, \phi) \sin\theta \, d\theta d\phi = \frac{\pi}{2k+1}.$$

Now, it suffices to show that

$$\sum_{m=1}^{\infty}\sum_{k=0}^{\infty} \frac{(-1)^{m+k+1}}{(k+1)(2m+k)} = \frac{\pi}{8}\ln 2.$$

Note that

$$\sum_{m=1}^{\infty}\sum_{k=0}^{\infty} \frac{(-1)^{m+k+1}}{(k+1)(2m+k)} = \sum_{m=1}^{\infty}\sum_{k=0}^{\infty} (-1)^{m+k+1} \int_0^1 x^k\,dx \cdot \int_0^1 y^{2m+k-1}\,dy.$$

Invoking the geometric series twice gives

$$\sum_{m=1}^{\infty}\sum_{k=0}^{\infty} \frac{(-1)^{m+k+1}}{(k+1)(2m+k)} = \int_0^1\int_0^1 \frac{y}{(1+y^2)(1+xy)}\,dx\,dy.$$

Integrating the inner integral with respect to x yields

$$\sum_{m=1}^{\infty}\sum_{k=0}^{\infty} \frac{(-1)^{m+k+1}}{(k+1)(2m+k)} = \int_0^1 \frac{\ln(1+y)}{1+y^2}\,dy = \frac{\pi}{8}\ln 2$$

as desired.

Experimental mathematics is a newly developed approach to mathematical discovery. Modern computers make this easier and have extended our study and research range. Recently Stenger [89] stated that "Experimental math has been very successful in mathematical research, but there are a couple of reasons why it might be even more successful in helping to solve Monthly problems." To illustrate this, he looked at three particular methods and applied them to solve six Monthly problems. Note that Experimental math is primarily heuristic. It does guide us to an answer, but we still have to prove it.

Here we provide two problems from *SIAM Review* for additional practice.

1. (Due to A. H. Nuttall) For $\alpha > 0$, show that

$$\int_0^\pi \left(\frac{\sin x}{x}\exp(x\cot x)\right)^\alpha dx = \frac{\pi\alpha^\alpha}{\Gamma(1+\alpha)}.$$

The above result has been confirmed numerically to 15 decimal places for numerous values of $\alpha \in [0, 150]$.

2. (Due to H. P. Robinson) Does

$$\int_0^\infty (\zeta(x) - 1)\,dx = \lim_{n\to\infty}\left(\sum_{k=2}^n \frac{1}{\ln k} - \int_0^n \frac{dx}{\ln x}\right)?$$

It was found that both sides match to at least 43 decimal places.

3.9   Evaluate a definite integral by the gamma function

Problem 11564 (Proposed by A. Stadler, 118(4), 2011). Prove that

$$\int_0^\infty \frac{e^{-x}(1 - e^{-6x})}{x(1 + e^{-2x} + e^{-4x} + e^{-6x} + e^{-8x})}\, dx = \ln\left(\frac{3 + \sqrt{5}}{2}\right).$$

Discussion.
The usual integration techniques studied in the first year calculus will not work on this integral directly. Using the substitution $t = e^{-x}$ yields

$$-\int_0^1 \frac{1 - t^6}{(1 + t^2 + t^4 + t^6 + t^8)\ln t}\, dt = -\int_0^1 \frac{(1 - t^2)(1 - t^6)}{(1 - t^{10})\ln t}\, dt.$$

This suggests we consider the parametric integral

$$\int_0^1 \frac{(1 - t^p)(1 - t^6)}{(1 - t^{10})\ln t}\, dt.$$

The key insight is to use parametric differentiation to remove the $\ln t$ from the denominator.

Solution.
We will prove a more general result: if $p \geq 0, \alpha > 0$ and $\beta = p + \alpha + 2$, then

$$\int_0^\infty \frac{e^{-x}(1 - e^{-px})(1 - e^{-\alpha x})}{x(1 - e^{-\beta x})}\, dx = -\int_0^1 \frac{(1 - t^p)(1 - t^\alpha)}{(1 - t^\beta)\ln t}\, dt$$
$$= \ln\left(\frac{\sin((p + 1)\pi/\beta)}{\sin(\pi/\beta)}\right). \qquad (3.12)$$

Once (3.12) is confirmed, the desired result follows upon letting $p = 2, \alpha = 6, \beta = 10$. Hence,

$$\int_0^\infty \frac{e^{-x}(1 - e^{-6x})}{x(1 + e^{-2x} + e^{-4x} + e^{-6x} + e^{-8x})}\, dx = \ln\left(\frac{\sin(3\pi/10)}{\sin(\pi/10)}\right) = \ln\left(\frac{3 + \sqrt{5}}{2}\right).$$

To prove (3.12), let

$$I(p) = -\int_0^1 \frac{(1 - t^p)(1 - t^\alpha)}{(1 - t^\beta)\ln t}\, dt.$$

We compute

$$
\begin{aligned}
I'(p) &= \int_0^1 \frac{t^p(1-t^\alpha)}{(1-t^\beta)}\, dt \\
&= \int_0^1 t^p(1-t^\alpha) \sum_{n=0}^\infty t^{n\beta}\, dt \\
&= \sum_{n=0}^\infty \int_0^1 t^{p+n\beta}(1-t^\alpha)\, dt \\
&= \sum_{n=0}^\infty \left(\frac{1}{p+1+n\beta} - \frac{1}{p+\alpha+1+n\beta} \right).
\end{aligned}
$$

Recall that

$$
-\frac{\Gamma'(x)}{\Gamma(x)} = \gamma + \frac{1}{x} + \sum_{n=1}^\infty \left(\frac{1}{n+x} - \frac{1}{n} \right).
$$

This yields that

$$
\begin{aligned}
I'(p) &= \frac{1}{\beta} \sum_{n=0}^\infty \left(\frac{1}{n+(p+1)/\beta} - \frac{1}{n+(p+\alpha+1)/\beta} \right) \\
&= \frac{1}{\beta} \left(\frac{\Gamma'((p+\alpha+1)/\beta)}{\Gamma((p+\alpha+1)/\beta)} - \frac{\Gamma'((p+1)/\beta)}{\Gamma((p+1)/\beta)} \right).
\end{aligned}
$$

Hence,

$$
I(p) = \ln \left(\frac{\Gamma((p+\alpha+1)/\beta)}{\Gamma((p+1)/\beta)} \right) + C.
$$

Since $I(0) = 0$, it follows that

$$
C = -\ln \left(\frac{\Gamma((\alpha+1)/\beta)}{\Gamma(1/\beta)} \right),
$$

and so

$$
I(p) = \ln \left(\frac{\Gamma((p+\alpha+1)/\beta)\Gamma(1/\beta)}{\Gamma((p+1)/\beta)\Gamma((\alpha+1)/\beta)} \right). \tag{3.13}
$$

By the Euler reflection formula

$$
\Gamma(x)\Gamma(1-x) = \frac{\pi}{\sin \pi x},
$$

and $\beta = p + \alpha + 2$, we have

$$
\Gamma((p+1)/\beta)\Gamma((\alpha+1)/\beta) = \frac{\pi}{\sin((p+1)\pi/\beta)},
$$

$$
\Gamma((p+\alpha+1)/\beta)\Gamma(1/\beta) = \frac{\pi}{\sin(\pi/\beta)}.
$$

Substituting them into (3.13) yields (3.12).   □

Remark. The approach used to prove (3.12) enables us to evaluate a class of integral of this type, which were initially considered by Euler. The following problems offer additional practice.

1. Let $p, q > -1$ and $p + q > -1$. Then

$$\int_0^\infty \frac{e^{-x}(1 - e^{-px})(1 - e^{-qx})}{x}\, dx = \int_0^1 (1 - x^p)(1 - x^q)\frac{dx}{\ln x}$$

$$= \ln \frac{p + q + 1}{(p+1)(q+1)}.$$

2. Let $a, b, c > 0$ and $d > a + b + c$. Then

$$\int_0^\infty \frac{(1 - e^{-ax})(1 - e^{-bx})e^{-cx}}{x(1 - e^{-dx})}\, dx = \ln \left(\frac{\Gamma(c/d)\Gamma((a+b+c)/d)}{\Gamma((a+c)/d)\Gamma((b+c)/d)} \right).$$

3. Let $p, q > -1$ and $p + \alpha, q + \alpha > -1$. Then

$$\int_0^1 \frac{x^p - x^q}{1 - x}\frac{1 - x^\alpha}{\ln x}\, dx = \ln \left(\frac{\Gamma(q+1)\Gamma(p+\alpha+1)}{\Gamma(p+1)\Gamma(q+\alpha+1)} \right).$$

4. Let $p, q, \alpha > 0$. Then

$$\int_0^1 \frac{x^{p-1} - x^{q-1}}{1 + x^\alpha}\frac{dx}{\ln x}\, dx = \ln \left(\frac{\Gamma((p+\alpha)/2\alpha)\Gamma(q/2\alpha)}{\Gamma((q+\alpha)/2\alpha)\Gamma(p/2\alpha)} \right).$$

5. Let $p, q > 0$. Then

$$\int_0^1 \frac{x^{p-1} - x^{q-1}}{1 - x^{2n}}\frac{1 - x^2}{\ln x}\, dx = \ln \left(\frac{\Gamma((p+2)/2n)\Gamma(q/2n)}{\Gamma((q+2)/2n)\Gamma(p/2n)} \right).$$

6. Let $0 < p, q < \alpha$. Then

$$\int_0^\infty \frac{x^{p-1} - x^{q-1}}{1 - x^\alpha}\frac{dx}{\ln x}\, dx = \ln \left(\frac{\sin(p\pi/\alpha)}{\sin(q\pi/\alpha)} \right);$$

$$\int_0^\infty \frac{x^{p-1} - x^{q-1}}{1 + x^\alpha}\frac{dx}{\ln x}\, dx = \ln \left[\tan \left(\frac{p\pi}{2\alpha} \right) \cot \left(\frac{q\pi}{2\alpha} \right) \right].$$

7. **Problem 5035** (Proposed by Y. Matsuoka, 69(6), 1962). Let α be a fixed positive number. Prove that, as $n \to \infty$,

$$\int_0^{\pi/2} t^\alpha \cos^{2n} t\, dt = \frac{1}{2}\Gamma \left(\frac{\alpha+1}{2} \right) /n^{(\alpha+1)/2}$$

$$- \frac{1}{12}\Gamma \left(\frac{\alpha+5}{2} \right) /n^{(\alpha+3)/2} + O(1/n^{(\alpha+5)/2}).$$

Hint: Consider the Lagrange reversion of a power series: $t^\alpha = a_1 \sin^\alpha t + a_2 \sin^{\alpha+2} t + a_3 \sin^{\alpha+4} t + \cdots$.

8. **Problem 12184** (Proposed by P. Perfetti, 127(5), 2020). Prove

$$\int_1^\infty \frac{\ln(x^4 - 2x^2 + 2)}{x\sqrt{x^2 - 1}}\, dx = \pi \ln(2 + \sqrt{2}).$$

3.10 Digamma via a double integral

Problem 11937 (Proposed by J. C. Sampedro, 123(9), 2016). Let s be a complex number not a zero of the gamma function $\Gamma(s)$. Prove

$$\int_0^1 \int_0^1 \frac{(xy)^{s-1} - y}{(1 - xy)\ln(xy)}\, dx\, dy = \frac{\Gamma'(s)}{\Gamma(s)}.$$

Discussion.
By the Euler reflection formula, we have

$$\Gamma(z)\Gamma(1 - z)\sin \pi z = \pi.$$

This implies that $\Gamma(z) \neq 0$ for all $z \in \mathbb{C}$. To ensure the proposed integral converges, we must assume that $\mathrm{Re}(s) > 0$. To show the convergence of the double integral for $\mathrm{Re}(s) > 0$, we rewrite it as

$$-\int_0^1 \int_0^1 \frac{1 - (xy)^{s-1}}{1 - xy}\frac{dx\, dy}{\ln(xy)} + \int_0^1 \int_0^1 \frac{1 - y}{1 - xy}\frac{dx\, dy}{\ln(xy)}.$$

Notice that $(1 - (xy)^{s-1})/(1 - xy)$ has a finite limit as $xy \to 1$, both functions x^{s-1} and y^{s-1} are integrable at 0, and $\int_0^1 \int_0^1 \frac{dx\, dy}{|\ln(xy)|} < \infty$. Thus the first integral converges absolutely. Since $0 \le \frac{1-y}{1-xy} \le 1$ for $x, y \in (0, 1)$, the second integral converges absolutely as well.

There are a number of well-known series and integral representations for the digamma function $\psi(s) = \Gamma'(s)/\Gamma(s)$. This prompts us to work on transforming the double integral into either a series or an integral representation of the digamma function. We give one example of each in the following two solutions.

Solution I — by series.
Recall that

$$\frac{\Gamma'(s)}{\Gamma(s)} = -\gamma - \frac{1}{x} - \sum_{n=1}^\infty \left(\frac{1}{s+n} - \frac{1}{n}\right) = -\gamma - \sum_{n=1}^\infty \left(\frac{1}{s+n-1} - \frac{1}{n}\right),$$

where γ is Euler's constant. We now show the double integral precisely equals to the series above. Denote the double integral by S. Expanding $1/(1 - xy)$ into a geometric series yields

$$S = \int_0^1 \int_0^1 \frac{(xy)^{s-1} - y}{\ln(xy)} \sum_{n=0}^{\infty} (xy)^n \, dxdy$$

$$= \sum_{n=0}^{\infty} \int_0^1 \int_0^1 \frac{(xy)^{s-1} - y}{\ln(xy)} (xy)^n \, dxdy.$$

Here interchanging summation and integration is justified by the terms having the same sign. Since for $0 < xy < 1$,

$$\frac{(xy)^k}{\ln(xy)} = -\int_k^{\infty} (xy)^t \, dt,$$

we find that

$$S = -\sum_{n=0}^{\infty} \int_0^1 \int_0^1 \left(\int_{n+s-1}^{\infty} (xy)^t \, dt - \int_n^{\infty} y(xy)^t \, dt \right) dxdy$$

$$= -\sum_{n=0}^{\infty} \left(\int_{n+s-1}^{\infty} \int_0^1 \int_0^1 (xy)^t \, dxdydt - \int_n^{\infty} \int_0^1 \int_0^1 y(xy)^t \, dxdydt \right)$$

$$= -\sum_{n=0}^{\infty} \left(\int_{n+s-1}^{\infty} \frac{dt}{(t+1)^2} - \int_n^{\infty} \frac{dt}{(t+1)(t+2)} \right)$$

$$= -\sum_{n=0}^{\infty} \left(\frac{1}{n+s} - \ln\left(\frac{n+2}{n+1}\right) \right)$$

$$= -\sum_{n=1}^{\infty} \left(\frac{1}{n+s-1} - \ln\left(\frac{n+1}{n}\right) \right).$$

Since the integrands in the first equality are nonnegative, we are able to reverse the order of integration. By the definition

$$\gamma = \lim_{n\to\infty} \left(1 + \frac{1}{2} + \cdots + \frac{1}{n} - \ln n \right),$$

we have

$$\gamma = \sum_{n=1}^{\infty} \left(\frac{1}{n} - \ln\left(\frac{n+1}{n}\right) \right),$$

and conclude that

$$S = -\gamma - \sum_{n=1}^{\infty} \left(\frac{1}{s+n-1} - \frac{1}{n} \right) = \frac{\Gamma'(s)}{\Gamma(s)}.$$

□

Solution II — by integral representation.

The integral representation we have in mind is

$$\frac{\Gamma'(s)}{\Gamma(s)} = -\int_0^1 \left(\frac{t^{s-1}}{1-t} + \frac{1}{\ln t} \right) dt. \tag{3.14}$$

That is a well-known formula of the digamma function ([49], 4.281.4, p. 583).

We show a more general result: Let s be a complex number such that $\mathrm{Re}(s) > 0$ and let $n \in \mathbb{N}$. Then

$$\int_0^1 \int_0^1 \frac{(xy)^{s-1} - y^n}{(1-xy)\ln(xy)} \, dx dy = -\frac{\ln n!}{n} - \int_0^1 \left(\frac{t^{s-1}}{1-t} + \frac{1}{\ln t} \right) dt. \tag{3.15}$$

Let $t = xy$. Then $x = t/y$ and $dx = dt/y$. We compute

$$\int_0^1 \int_0^1 \frac{(xy)^{s-1} - y^n}{(1-xy)\ln(xy)} \, dx dy$$

$$= \int_0^1 \int_0^y \frac{t^{s-1} - y^n}{y(1-t)\ln t} \, dt dy$$

$$= \int_0^1 \frac{1}{(1-t)\ln t} \left(\int_t^1 \left(\frac{t^{s-1}}{y} - y^{n-1} \right) dy \right) dt$$

$$= -\int_0^1 \frac{1}{(1-t)\ln t} \left(t^{s-1} \ln t + \frac{1}{n}(1-t^n) \right) dt$$

$$= -\int_0^1 \left(\frac{t^{s-1}}{1-t} + \frac{1}{\ln t} \right) dt - \int_0^1 \left(\frac{1-t^n}{n(1-t)} - 1 \right) \frac{dt}{\ln t}.$$

Since

$$\frac{1-t^n}{n(1-t)} - 1 = \frac{1}{n} \sum_{k=0}^{n-1} t^k - 1 = \frac{1}{n} \sum_{k=0}^{n-1} (t^k - 1),$$

we find that

$$\int_0^1 \left(\frac{1-t^n}{n(1-t)} - 1 \right) \frac{dt}{\ln t} = \frac{1}{n} \int_0^1 \sum_{k=0}^{n-1} \frac{t^k - 1}{\ln t} \, dt = \frac{1}{n} \sum_{k=0}^{n-1} \ln(k+1) = \frac{1}{n} \ln(n!).$$

This, together with (3.14), confirms (3.15) as claimed. □

Remark. Here we give an elementary proof of (3.14). In view of

$$\ln x = \int_0^\infty \frac{e^{-t} - e^{-xt}}{t} \, dt,$$

we have

$$\Gamma'(s) = \int_0^\infty e^{-x} x^{s-1} \ln x \, dx$$

$$= \int_0^\infty e^{-x} x^{s-1} \left(\int_0^\infty \frac{e^{-t} - e^{-xt}}{t} \, dt \right) dx$$

$$= \int_0^\infty \left(e^{-t} \int_0^\infty x^{s-1} e^{-x} - \int_0^\infty x^{s-1} e^{-x(1+t)} \, dx \right) \frac{dt}{t}$$

$$= \Gamma(s) \int_0^\infty \left(e^{-t} - \frac{1}{(1+t)^s} \right) \frac{dt}{t}.$$

Hence,

$$\frac{\Gamma'(s)}{\Gamma(s)} = \int_0^\infty \left(\frac{e^{-t}}{t} - \frac{1}{t(1+t)^s} \right) dt.$$

The substitution $t \to e^t - 1$ leads to

$$\frac{\Gamma'(s)}{\Gamma(s)} = \int_0^\infty \left(\frac{e^{-t}}{t} - \frac{e^{-st}}{1 - e^{-t}} \right) dt.$$

This, with the substitution $t \to e^{-t}$, yields (3.14).

We collect a few problems for additional practice.

1. Let $s > -1$. Show that

$$\int_0^1 \int_0^1 \frac{(-\ln(xy))^s}{1 - xy} \, dx dy = \Gamma(s+2)\zeta(s+2).$$

2. (Hadjicostas-Chapman formula [27]). Let $\mathrm{Re}(s) > -2$. Show that

$$\int_0^1 \int_0^1 \frac{(1-x)[-\ln(xy)]^s}{1 - xy} \, dx dy = \Gamma(s+2) \left(\zeta(s+2) - \frac{1}{s+1} \right).$$

3. Let $p, q > 0$ and $p \neq q$. Show that

$$\int_0^1 \int_0^1 \frac{x^{p-1} y^{q-1}}{(1 + xy)(-\ln(xy))} \, dy dx = \frac{1}{p-q} \ln \left(\frac{\Gamma(q/2)\Gamma((p+1)/2)}{\Gamma(p/2)\Gamma((q+1)/2)} \right).$$

   What happens if $p = q$?

4. Let $s > 0$. Show that

$$\int_0^1 \int_0^1 \frac{(xy)^{s-1}(1-x)}{(1 - xy)(-\ln(xy))} \, dx dy = \ln s - \frac{\Gamma'(s)}{\Gamma(s)}.$$

   In particular, for $s = 1$, this yields Sondow's formula [83]

$$\gamma = \int_0^1 \int_0^1 \frac{1-x}{(1 - xy)(-\ln(xy))} \, dx dy.$$

5. **Problem 11322** (Proposed by J. Sondow, 114(9), 2007). Let N be a positive integer. Prove that

$$\int_{x=0}^1 \int_{y=0}^1 \frac{(x(1-x)y(1-y))^N}{(1 - xy)(-\ln(xy))} \, dy dx$$

$$= \sum_{n=N+1}^\infty \int_{t=n}^\infty \left(\frac{N!}{t(t+1) \cdots (t+N)} \right)^2 dt.$$

6. **Problem 11331** (Proposed by R. Bagby, 114(10), 2007). Show that if k is a positive integer, then

$$\int_0^\infty \left(\frac{\ln(1+t)}{t} \right)^{k+1} dt = (k+1) \sum_{j=1}^k a_j \zeta(j+1),$$

where ζ denotes the Riemann zeta function and a_j is the coefficient of x^j in $x \prod_{n=1}^{k-1} (1 - nx)$.
Hint: The factor $(k+1)$ before the series was missing in the original statement. Show the proposed integral is equivalent to

$$\int_0^1 \frac{x^{k-1}(-\ln x)^{k+1}}{(1-x)^{k+1}} dx.$$

7. **SIAM Review Problem 85-22** (Proposed by E. O. George and C. C. Rousseau). Evaluate

$$M_n(s) = n(n-1) \int_0^1 \int_0^u \left(\frac{uv}{(1-u)(1-v)} \right)^{s/2} (u-v)^{n-2} \, dv du.$$

Here $M_n(s)$ is the moment generating function for the mid-range of a random n-sample from the logistic distribution.

3.11 Another double integral

Problem 11650 (Proposed by M. Becker, 119(6), 2012). Evaluate

$$\int_{x=0}^\infty \int_{y=x}^\infty e^{-(x-y)^2} \sin^2(x^2 + y^2) \frac{x^2 - y^2}{(x^2 + y^2)^2} dy dx.$$

Discussion.
Let I be the proposed integral. To reformulate the problem into an equivalent but manageable one, we switch to polar coordinates and obtain

$$
\begin{aligned}
I &= \int_0^\infty \int_{\pi/4}^{\pi/2} e^{-r^2(1-\sin 2\theta)} \sin^2(r^2) \frac{\cos 2\theta}{r} d\theta dr \\
&= \int_0^\infty \frac{e^{-r^2} \sin^2(r^2)}{r} \int_{\pi/4}^{\pi/2} e^{r^2 \sin 2\theta} \cos 2\theta d\theta dr \\
&= \int_0^\infty \frac{e^{-r^2} \sin^2(r^2)}{r} \frac{1}{2r^2} e^{r^2 \sin 2\theta} \Big|_{\pi/4}^{\pi/2} dr \\
&= \frac{1}{2} \int_0^\infty \frac{\sin^2(r^2)}{r^3} (e^{-r^2} - 1) \, dr.
\end{aligned}
$$

Furthermore, the change of variable $t = r^2$ yields

$$I = \frac{1}{4} \int_0^\infty \frac{\sin^2 t}{t^2} (e^{-t} - 1)\, dt. \tag{3.16}$$

Thus it suffices to evaluate (3.16). This can be done in a variety of ways. We present three different calculations as below.

Solution I.
We calculate (3.16) by parametric differentiation. It is well-known that

$$\int_0^\infty \frac{\sin^2 t}{t^2}\, dt = \frac{\pi}{2}.$$

To compute the integral $\int_0^\infty e^{-t} \sin^2 t / t^2\, dt$, we introduce the parametric integral

$$F(\alpha) = \int_0^\infty \frac{\sin^2(\alpha t)}{t^2} e^{-t}\, dt.$$

By the Leibniz rule, we have

$$F'(\alpha) = \int_0^\infty \frac{\sin(2\alpha t)}{t} e^{-t}\, dt;$$

$$F''(\alpha) = 2 \int_0^\infty e^{-t} \cos(2\alpha t)\, dt,$$

where the justifications of differentiating under the integrals are ensured by the uniformly convergence on α. Integrating by parts leads to

$$F''(\alpha) = \frac{2}{1 + 4\alpha^2}.$$

Appealing to $F(0) = F'(0) = 0$, integrating $F''(\alpha)$ twice yields

$$F(\alpha) = \alpha \arctan(2\alpha) - \frac{1}{4} \ln(1 + 4\alpha^2).$$

Hence,

$$F(1) = \int_0^\infty \frac{\sin^2 t}{t^2} e^{-t}\, dt = \arctan 2 - \frac{1}{4} \ln 5.$$

In summary, we conclude

$$I = \frac{1}{4} \arctan 2 - \frac{1}{16} \ln 5 - \frac{1}{8} \pi.$$

Notice that $\arctan 2 + \arctan(1/2) = \pi/2$. We find a more compact form:

$$I = -\frac{1}{16} \ln 5 - \frac{1}{4} \arctan\left(\frac{1}{2}\right).$$

□

Solution II.
We calculate (3.16) by parametric integration. Rewrite

$$I = \frac{1}{4} \int_0^\infty \frac{e^{-t} - 1}{t} \frac{\sin^2 t}{t} \, dt.$$

Since

$$\frac{e^{-t} - 1}{t} = -\int_0^1 e^{-st} \, ds \quad \text{and} \quad \sin^2 t = \frac{1}{2}(1 - \cos(2t)),$$

it follows that

$$I = -\frac{1}{8} \int_0^1 \int_0^\infty \frac{e^{-st}}{t}(1 - \cos(2t)) \, dt ds.$$

For the inner integral, using $1/t = \int_0^\infty e^{-tu} \, du$, for $s > 0$, we have

$$
\begin{aligned}
\int_0^\infty \frac{e^{-st}}{t}(1 - \cos(2t)) \, dt &= \int_0^\infty \left(\int_0^\infty e^{-(s+u)t}(1 - \cos(2t)) \, dt \right) du \\
&= \int_0^\infty \left(\frac{1}{s+u} - \frac{s+u}{4 + (s+u)^2} \right) du \\
&= \ln \left(\frac{\sqrt{4 + s^2}}{s} \right).
\end{aligned}
$$

The interchange of order of integration is justified by nonnegativity of the integrands. Integrating by parts gives

$$I = -\frac{1}{8} \int_0^1 \ln \left(\frac{\sqrt{4 + s^2}}{s} \right) ds = -\frac{1}{16} \ln 5 - \frac{1}{4} \arctan \left(\frac{1}{2} \right).$$

□

Solution III.
We calculate (3.16) by the Laplace transform. To this end, we show that

$$\mathcal{L}\left[\frac{\sin^2 t}{t^2} \right](s) = \int_0^\infty e^{-st} \frac{\sin^2 t}{t^2} \, dt = \arctan \frac{2}{s} - \frac{s}{4} \ln \frac{s^2 + 4}{s^2}. \tag{3.17}$$

Once (3.17) is established, we have

$$I = \frac{1}{4} \left(\mathcal{L}\left[\frac{\sin^2 t}{t^2} \right](1) - \mathcal{L}\left[\frac{\sin^2 t}{t^2} \right](0) \right) = \frac{1}{4} \arctan 2 - \frac{1}{16} \ln 5 - \frac{1}{8} \pi.$$

We now prove (3.17) by using parametric differentiation. Let

$$L(p) = \mathcal{L}\left[\frac{\sin^2(pt)}{t^2} \right](s).$$

Then

$$L''(p) = \frac{d^2}{dp^2}\left(\mathcal{L}\left[\frac{\sin^2(pt)}{t^2}\right](s)\right) = \frac{d}{dp}\left(\mathcal{L}\left[\frac{\sin(2pt)}{t}\right](s)\right)$$

$$= 2\mathcal{L}[\cos(2pt)](s) = \frac{2s}{4p^2 + s^2}.$$

Since $L(0) = L'(0) = 0$, we integrate this equation twice to get

$$L(p) = a\arctan\left(\frac{2p}{s}\right) - \frac{s}{4}\ln\left(\frac{s^2 + 4p^2}{s^2}\right).$$

(3.17) now follows from setting $p = 1$.   □

Remarks. It is interesting to see that this problem can be solved by *Mathematica*. Indeed, after passing to the polar coordinates, the command

```
Integrate[Sin[r^2]^2*(Exp[-r^2] - 1)/(2 r^3), {r, 0, Infinity}]
```

produces the output

```
1/16 (-2 \[Pi] + 4 ArcTan[2] - Log[5])
```

The approach by the Laplace transform seems go through for much board sense. To illustrate this method further, we present another solution to the following Monthly problem based on the Laplace transform.

Problem 11322 (Proposed by J. Sondow, 114(9), 2007). Let N be a positive integer. Prove that

$$\int_0^1\int_0^1 \frac{(x(1-x)y(1-y))^N}{(1-xy)(-\ln(xy))}\,dy dx = \sum_{n=N+1}^{\infty}\int_n^{\infty}\left(\frac{N!}{t(t+1)\cdots(t+N)}\right)^2 dt.$$

Let

$$f_N(s) = \begin{cases} (1-e^{-s})^N & \text{for } s \geq 0 \\ 0 & \text{for } s < 0. \end{cases}$$

First, by induction, we show that

$$F_N(p) := \mathcal{L}[f_n] = \int_0^{\infty} f_N(s)e^{-ps}\,ds = \frac{N!}{p(p+1)\cdots(p+N)}. \qquad (3.18)$$

Clearly, (3.18) hold for $N = 0$. In general, for $p > 0$, we have

$$F_N(p) - F_{N+1}(p) = \int_0^{\infty} f_N(s)e^{-s}e^{-ps}\,ds = \int_0^{\infty} f_N(s)e^{-(p+1)s}\,ds = F_N(p+1),$$

and so $F_{N+1}(p) = F_N(p) - F_N(p+1)$. By the induction hypothesis, we find that

$$F_{N+1}(p) = \frac{N!}{p(p+1)\cdots(p+N)} - \frac{N!}{(p+1)(p+2)\cdots(p+N+1)}$$

$$= \frac{(N+1)!}{p(p+1)\cdots(p+N+1)}.$$

Thus (3.18) holds for all integers $N \geq 0$.

Invoking the convolution properties of the Laplace transform, we have

$$\mathcal{L}[f_N * f_N](p) = F_N^2(p) = \left(\frac{N!}{p(p+1)\cdots(p+N)}\right)^2.$$

Integrating with respect to the parameter p gives

$$\int_0^\infty \frac{e^{-ps}}{s} f_N * f_N(s)\, ds = \int_p^\infty \left(\frac{N!}{u(u+1)\cdots(u+N)}\right)^2 du.$$

On the other hand, invoking Fubini's theorem, we find

$$
\begin{aligned}
\int_0^\infty \frac{e^{-ps}}{s} f_N * f_N(s)\, ds &= \int_0^\infty \int_0^\infty \frac{e^{-ps}}{s} f_N(t) f_N(s-t)\, dt\, ds \\
&= \int_0^\infty \int_0^\infty \frac{e^{-ps}}{s} f_N(t) f_N(s-t)\, ds\, dt \\
&= \int_0^\infty \int_0^\infty \frac{e^{-p(u+t)}}{s} f_N(t) f_N(u)\, du\, dt \\
&= \int_0^1 \int_0^1 \frac{(xy)^{p-1}(1-x)^N(1-y)^N}{-\ln(xy)}\, dy\, dx \\
&\qquad (\text{let } x = e^{-t}, y = e^{-u}).
\end{aligned}
$$

Hence, for $p > 0$, we conclude that

$$\int_0^1 \int_0^1 \frac{(xy)^{p-1}(1-x)^N(1-y)^N}{-\ln(xy)}\, dy\, dx = \int_p^\infty \left(\frac{N!}{u(u+1)\cdots(u+N)}\right)^2 du.$$

Summing both sides for p from $N+1$ to infinity yields

$$
\begin{aligned}
\int_0^1 \int_0^1 & \left(\sum_{p=N+1}^\infty (xy)^{p-1}\right) \frac{(1-x)^N(1-y)^N}{-\ln(xy)}\, dy\, dx \\
&= \sum_{p=N+1}^\infty \int_p^\infty \left(\frac{N!}{u(u+1)\cdots(u+N)}\right)^2 du,
\end{aligned}
$$

where the interchange the order of integration and summation is justified by the positivity of integrands. Appealing to the geometric series, this proves

$$\int_0^1 \int_0^1 \frac{(x(1-x)y(1-y))^N}{(1-xy)(-\ln(xy))}\, dy\, dx = \sum_{p=N+1}^\infty \int_p^\infty \left(\frac{N!}{u(u+1)\cdots(u+N)}\right)^2 du$$

as desired.

We end this section with eight problems for additional practice.

1. Let

$$f(s) = \int_0^1 x^s \left(\frac{1}{\ln x} + \frac{1}{1-x} \right)^2 dx.$$

For $s > 0$, show that

$$f(s) = s\psi(s+1) + \ln \left(\frac{2\pi(1+s)^{s+1}e^{-2(s+3/4)}}{\Gamma^2(1+s)} \right),$$

where ψ is the digamma function.

Hint: Let $g(t) = (1/(1-e^{-t}) - 1/t)^2$. Then

$$L(s) := \mathcal{L}[g](s) = \int_0^\infty g(t)e^{-st}\,dt = \int_0^1 x^{s-1}\left(\frac{1}{\ln x} + \frac{1}{1-x} \right)^2 dx = f(s-$$

You can recover $L(s)$ from showing that

$$L''(s) = \frac{1}{s} + \int_0^\infty \left(\frac{t^2 e^{-st}}{(1-e^{-t})^2} - \frac{2te^{-st}}{1-e^{-t}} \right) dt = \frac{1}{s} + (s-1)\psi''(s).$$

2. For $|a| < 1/2$, evaluate (Due to M. L. Glasser)

$$\int_0^\infty \sin\left(\frac{t}{2}\right) \cos(at - \sin t) \frac{e^{\cos t}}{t}\,dt.$$

3. **SIAM Review Problem 94-14** (Proposed by N. Ortner). For $\alpha > 0$, evaluate

$$\int_0^\infty \frac{x \arctan(\alpha/x)}{1 + x^2 + \sqrt{1+x^2}}\,dx.$$

4. **Problem 11275** (Proposed by M. Becker, 114(2), 2007). Find

$$\int_{y=0}^\infty \int_{x=y}^\infty \frac{(x-y)^2 \ln((x+y)/(x-y))}{xy \sinh(x+y)}\,dx\,dy.$$

Hint: By changing variables show that the double integral is equal to

$$-\frac{\pi^2}{2} \int_0^1 \frac{u^2 \ln u}{1-u^2}\,du.$$

5. **Problem 11953** (Proposed by C. V. Vălean, 124(1), 2017). Evaluate

$$\int_0^\infty \int_0^\infty \frac{\sin x \sin y \sin(x+y)}{xy(x+y)}\,dx\,dy.$$

Hint. Let $\operatorname{sinc}(x) = \sin x / x$. The Fourier transform gives $\operatorname{sinc}(x) = \mathcal{F}(h)/2$, where h is the indicator function of the interval $[-1, 1]$. Applying the Plancherel theorem

$$\int_{-\infty}^{\infty} \mathcal{F}(f)(x)\overline{\mathcal{F}}(g)(x)\, dx = 2\pi \int_{-\infty}^{\infty} f(t)\bar{g}(t)\, dt,$$

shows that

$$\int_{-\infty}^{\infty} \operatorname{sinc}(x)\operatorname{sinc}(x + y)\, dx = \pi\operatorname{sinc}(y).$$

6. **Problem 12070** (Proposed by C. V. Vălean, 125(9), 2018). Prove

$$\int_0^{\pi/4} \int_0^{\pi/4} \frac{\cos x \cos y(y \sin y \cos x - x \sin x \cos y)}{\cos(2x) - \cos(2y)}\, dx dy$$
$$= \frac{7\zeta(3) + 4\pi \ln 2}{64}.$$

7. **SIAM Review Problem 65-5** (Proposed by N. Mullineux and J. R. Reed). Show that

$$\int_{-\infty}^{\infty} \int_0^{\infty} \frac{e^{-y} \cos(xy)}{\mu y \sqrt{\alpha^2 + y^2} + \alpha^2 + y^2}\, dy dx = \frac{\pi}{\alpha^2}.$$

8. **SIAM Review Problem 77-3** (A define integral of Bohr). Find the exact value of

$$K = \int_0^{\infty} F(x)(F'(x) - \ln x)\, dx,$$

where

$$F(x) = \int_{-\infty}^{\infty} \frac{\cos(xy)}{(1 + y^2)^{3/2}}\, dy.$$

3.12 An integral with log and arctangent

Problem 12054 (Proposed by C. I. Vălean, 125(6), 2018). Prove

$$\int_0^1 \frac{\arctan x}{x} \ln\left(\frac{1 + x^2}{(1 - x)^2}\right) dx = \frac{\pi^3}{16}.$$

Discussion.

The substitution $x = \tan\theta$ or $x = (1-u)/(1+u)$ doesn't seem to lead anywhere, so we try to evaluate this integral directly by transforming the integrand into a power series.

Solution.

First, we show that

$$\arctan x \ln(1+x^2) = 2\sum_{n=1}^{\infty} \frac{(-1)^{n-1}H_{2n}}{2n+1} x^{2n+1}, \tag{3.19}$$

where H_n is the nth harmonic number. To see this, recall that

$$\arctan x = \sum_{n=0}^{\infty} \frac{(-1)^n}{2n+1} x^{2n+1} \quad \text{and} \quad \ln(1+x^2) = \sum_{n=1}^{\infty} \frac{(-1)^{n-1}}{n} x^{2n}.$$

The Cauchy product of these two series gives

$$\begin{aligned}
\arctan x \ln(1+x^2) &= \sum_{n=1}^{\infty}(-1)^{n-1}\left(2\sum_{k=1}^{n}\frac{1}{2k}\cdot\frac{1}{2n+1-2k}\right)x^{2n+1}\\
&= 2\sum_{n=1}^{\infty}(-1)^{n-1}\frac{1}{2n+1}\sum_{k=1}^{n}\left(\frac{1}{2k}+\frac{1}{2n+1-2k}\right)x^{2n+1}\\
&\quad \times(\text{using partial fractions})\\
&= 2\sum_{n=1}^{\infty}\frac{(-1)^{n-1}H_{2n}}{2n+1}x^{2n+1},
\end{aligned}$$

which proves (3.19). Hence,

$$\begin{aligned}
\int_0^1 \frac{\arctan x}{x}\ln(1+x^2)\,dx &= \int_0^1\left(2\sum_{n=1}^{\infty}\frac{(-1)^{n-1}H_{2n}}{2n+1}x^{2n}\right)dx\\
&= 2\sum_{n=1}^{\infty}\frac{(-1)^{n-1}H_{2n}}{(2n+1)^2}. \tag{3.20}
\end{aligned}$$

The interchange of the order of summation and integration is justified as follows: For $x \in [0,1]$, using the estimate of the alternating series remainder yields

$$\left|\sum_{n=N}^{\infty}\frac{(-1)^{n-1}H_{2n}}{2n+1}x^{2n}\right| \leq \frac{H_{2(N+1)}}{2N+3} \to 0, \quad \text{as } N \to \infty.$$

Next, we have

$$\int_0^1 \frac{\arctan x}{x}\ln(1-x)^2\,dx = 2\sum_{n=0}^{\infty}\int_0^1\frac{(-1)^n}{2n+1}x^{2n}\ln(1-x)\,dx.$$

Since

$$\int_0^1 x^{2n} \ln(1-x)\, dx = -\int_0^1 \left(\sum_{k=1}^{\infty} \frac{x^{2n+k}}{k} \right) dx = -\sum_{k=1}^{\infty} \frac{1}{k} \int_0^1 x^{2n+k}\, dx$$

$$= -\sum_{k=1}^{\infty} \frac{1}{k(2n+k+1)} = -\frac{1}{2n+1} \sum_{k=1}^{\infty}$$

$$\times \left(\frac{1}{k} - \frac{1}{2n+k+1} \right)$$

$$= -\frac{1}{2n+1} H_{2n+1},$$

where the positivity of integrands justifies the interchange of the order of summation and integration, we have

$$\int_0^1 \frac{\arctan x}{x} \ln(1-x)^2\, dx = -2 \sum_{n=0}^{\infty} \frac{(-1)^n H_{2n+1}}{(2n+1)^2}. \tag{3.21}$$

Finally, combining (3.20) and (3.21), we find that

$$\int_0^1 \frac{\arctan x}{x} \ln\left(\frac{1+x^2}{(1-x)^2} \right) dx = 2 \sum_{n=1}^{\infty} \frac{(-1)^{n-1} H_{2n}}{(2n+1)^2} + 2 \sum_{n=0}^{\infty} \frac{(-1)^n H_{2n+1}}{(2n+1)^2}$$

$$= 2 \sum_{n=0}^{\infty} \frac{(-1)^n (H_{2n+1} - H_{2n})}{(2n+1)^2} \quad (\text{let } H_0 = 0)$$

$$= 2 \sum_{n=0}^{\infty} \frac{(-1)^n}{(2n+1)^3} = \frac{\pi^3}{16},$$

where we have used the well-known result

$$\sum_{n=0}^{\infty} \frac{(-1)^n}{(2n+1)^3} = \frac{\pi^3}{32}.$$

$\square$

Remark. It is interesting to reveal one unsuccessful journey to evaluate this integral by the substitution $x = \tan\theta$. In this case, we have

$$\int_0^1 \frac{\arctan x}{x} \ln\left(\frac{1+x^2}{(1-x)^2} \right) dx = -\int_0^{\pi/4} \frac{\theta}{\sin\theta \cos\theta} \ln(1 - \sin(2\theta))\, d\theta.$$

The change of variable $2\theta \to \theta$ leads to

$$\int_0^1 \frac{\arctan x}{x} \ln\left(\frac{1+x^2}{(1-x)^2} \right) dx = -\frac{1}{2} \int_0^{\pi/2} \frac{\theta}{\sin\theta} \ln(1 - \sin\theta)\, d\theta.$$

To compute

$$\int_0^{\pi/2} \frac{\theta}{\sin\theta} \ln(1 - \sin\theta)\, d\theta, \tag{3.22}$$

we initially introduced the parametric integral

$$J(p) := \int_0^{\pi/2} \frac{\theta}{\sin\theta} \ln(1 - p\sin\theta)\, d\theta.$$

The Leibniz rule gives

$$J'(p) = \int_0^{\pi/2} \frac{\theta}{1 - p\sin\theta}\, d\theta.$$

This integral may look simple but it is really hard to calculate. Even *Mathematica* does not provide a useful result. Then we turn to the power series of $\ln(1-x)$ and obtain

$$\int_0^{\pi/2} \frac{\theta}{\sin\theta} \ln(1 - \sin\theta)\, d\theta = \sum_{n=1}^{\infty} \frac{1}{n} \int_0^{\pi/2} \theta \sin^{n-1}\theta\, d\theta.$$

For $n \geq 1$, let

$$S_n = \int_0^{\pi/2} \theta \sin^{n-1}\theta\, d\theta.$$

Clearly, $S_1 = \pi^2/8, S_2 = 1$. For $n \geq 3$, integrating by parts leads to the following reduction formula

$$S_n = \frac{1}{(n-1)^2} + \frac{n-2}{n-1} S_{n-2}.$$

For $k \geq 1$, this implies that

$$\begin{aligned}
S_{2k-1} &= \frac{1}{[2(k-1)]^2} + \frac{2k-3}{2(k-1)[2(k-2)]^2} + \cdots + \frac{(2k-3)(2k-5)\cdots 1}{2(k-1)2(k-2)\cdots 2} \frac{\pi^2}{8} \\
&= \sum_{n=0}^{k-2} \frac{(2k-3)(2k-5)\cdots[2(k-n)-1]}{2(k-1)2(k-2)\cdots[2(k-n-1)]} \cdot \frac{1}{2(k-n-1)} \\
&\quad + \frac{(2k-3)!!}{[2(k-1)]!!} \frac{\pi^2}{8};
\end{aligned}$$

$$\begin{aligned}
S_{2k} &= \frac{1}{(2k-1)^2} + \frac{2(k-1)}{(2k-1)(2k-3)^2} + \cdots + \frac{2(k-1)\,2(k-2)\cdots 2}{(2k-1)(2k-3)\cdots 3} \\
&= \sum_{n=0}^{k-1} \frac{2(k-1)2(k-2)\cdots 2(k-n)}{(2k-1)(2k-3)\cdots[2(k-n)+1]} \cdot \frac{1}{2(k-n)+1}.
\end{aligned}$$

However, we are unable to find the closed forms of the following series

$$\sum_{k=1}^{\infty} \frac{1}{2k} S_{2k} \quad \text{and} \quad \sum_{k=1}^{\infty} \frac{1}{2k-1} S_{2k-1}.$$

In summary, although we have

$$\int_0^\pi \frac{\theta}{\sin\theta} \ln(1 - \sin\theta)\, d\theta = \frac{\pi}{2} \int_0^\pi \frac{1}{\sin\theta} \ln(1 - \sin\theta)\, d\theta = -\frac{3}{8}\pi^3,$$

currently we are unable to evaluate (3.22) in an elementary way.

We now end this section with ten problems for additional practice.

1. **Problem 10884** (Proposed by Z. Ahmed, 108(6), 2001). Evaluate

$$\int_0^1 \frac{\arctan\sqrt{2 + x^2}}{(1 + x^2)\sqrt{2 + x^2}}\, dx.$$

2. (**Coxeter's Integral**) Evaluate

$$\int_0^{\pi/2} \arccos\left(\frac{\cos\theta}{1 + 2\cos\theta}\right) dx.$$

Hint: One way to proceed is first to find

$$\int_0^1 \frac{\arctan\sqrt{p + x^2}}{(1 + x^2)\sqrt{p + x^2}}\, dx$$

for $p > 1$, then convert Coxeter's integral into one special case of the above integral by using

$$\arccos t = 2\arctan\sqrt{\frac{1 - t}{1 + t}}.$$

3. **Problem 11152** (Proposed by M. Ivan, 112(5), 2005). Evaluate

$$\int_0^1 \frac{\ln(\cos(\pi x/2))}{x(1 + x)}\, dx.$$

4. **Problem 11961** (Proposed by M. Berindeanu, 124(2), 2017). Evaluate

$$\int_0^{\pi/2} \frac{\sin x}{1 + \sqrt{\sin x}}\, dx.$$

5. For $0 < \alpha, \beta < 1$, show that

$$\int_\alpha^1 \int_\beta^1 \frac{dx dy}{(x + y)^2 + (1 + xy)^2} = \frac{1}{2} F\left(\frac{(1 - \alpha)(1 - \beta)}{(1 + \alpha)(1 + \beta)}\right),$$

where

$$F(x) = \int_0^x \frac{\arctan t}{t}\, dt.$$

6. **Problem 4826** (Proposed by M. S. Klamkin and L. A. Shepp, 66(1), 1959). Let $F(x) = \int_0^x \frac{\arctan t}{t}\,dt$. Express $F(1)$ in terms of $F(2 - \sqrt{3})$, then obtain a more rapidly converging expansion. Indeed, this problem can be traced back to Ramanujan. He established the astonishing identity

$$\frac{\sin(2x)}{1^2} + \frac{\sin(6x)}{3^2} + \frac{\sin(10x)}{5^2} + \cdots = F(\tan x) - x\ln(\tan x)$$

by showing their derivatives are equal.

7. **Problem 4865** (Proposed by L. Lewin, 66(8), 1959). Let $F(x) = \int_0^x \frac{\arctan t}{t}\,dt$. Prove that

$$6F(1) - 4F(1/2) - 2F(1/3) - F(3/4) = \pi \ln 2.$$

8. Define $\Gamma(s, x) = \int_x^\infty t^{s-1}e^{-t}\,dt$. Find

$$\int_0^1 \int_0^1 \ln \Gamma(s, x)\,ds\,dx.$$

9. Show that

$$\int_0^1 \frac{x^{2n}\ln x}{\sqrt{1-x^2}}\,dx = \frac{\binom{2n}{n}\pi}{2^{2n+1}} \left(\sum_{k=1}^{2n} \frac{(-1)^{k-1}}{k} - \ln 2 \right),$$

$$\int_0^1 \frac{x^{2n+1}\ln x}{\sqrt{1-x^2}}\,dx = \frac{(2n)!!}{(2n+1)!!} \left(\ln 2 + \sum_{k=1}^{2n+1} \frac{(-1)^{k-1}}{k} - \ln 2 \right).$$

10. **SIAM Review Problem 84-8** (Proposed by J. A. Morrison). Prove directly that

$$\int_0^1 \left(\frac{2}{\pi}\arctan\left(\frac{2}{\pi}\arctan(1/x) + \frac{1}{\pi}\ln\left(\frac{1+x}{1-x}\right) \right) - \frac{1}{2} \right) \frac{dx}{x}$$
$$= \frac{1}{2}\ln\left(\frac{\pi}{2\sqrt{2}} \right).$$

3.13   Another integral with log and arctangent

Problem 12158 (Proposed by H. Grandmontagne, 127(1), 2020). Prove

$$\int_0^1 \frac{(\ln(x))^2 \arctan(x)}{1+x}\,dx = \frac{21\pi\zeta(3)}{64} - \frac{\pi^2 G}{24} - \frac{\pi^3 \ln(2)}{32}$$

where $\zeta(3)$ is Apéry's constant $\sum_{k=1}^{\infty} 1/k^3$ and $G = \sum_{k=0}^{\infty} (-1)^k/(2k+1)^2$ is Catalan's constant.

Discussion.

Recall that in the previous section, when the standard substitution $x = \tan\theta$ does not work, we turn to transforming the integrand into a power series. Since

$$\arctan x = \sum_{n=0}^{\infty} \frac{(-1)^n}{2n+1} x^{2n+1},$$

the Cauchy product of series gives

$$
\begin{aligned}
\frac{\arctan x}{1-x} &= x + x^2 + \left(1 - \frac{1}{3}\right) x^3 + \left(1 - \frac{1}{3}\right) x^4 \\
&\quad + \left(1 - \frac{1}{3} + \frac{1}{5}\right) x^5 + \left(1 - \frac{1}{3} + \frac{1}{5}\right) x^6 + \cdots \\
&= \sum_{n=0}^{\infty} \left(\sum_{k=0}^{n} \frac{(-1)^k}{2k+1}\right) x^{2n+1} + \sum_{n=1}^{\infty} \left(\sum_{k=0}^{n-1} \frac{(-1)^k}{2k+1}\right) x^{2n}.
\end{aligned}
$$

Replacing x by $-x$ yields

$$\frac{\arctan x}{1+x} = \sum_{n=0}^{\infty} \left(\sum_{k=0}^{n} \frac{(-1)^k}{2k+1}\right) x^{2n+1} - \sum_{n=1}^{\infty} \left(\sum_{k=0}^{n-1} \frac{(-1)^k}{2k+1}\right) x^{2n}.$$

Since

$$\int_0^1 (\ln x)^2 x^k \, dx = \frac{2}{(k+1)^3} \qquad \text{for all } k \geq 0,$$

the proposed integral is successfully reduced to the following series:

$$
\begin{aligned}
\int_0^1 \frac{(\ln(x))^2 \arctan(x)}{1+x} \, dx &= \frac{1}{4} \sum_{n=0}^{\infty} \frac{1}{(n+1)^3} \left(\sum_{k=0}^{n} \frac{(-1)^k}{2k+1}\right) \\
&\quad - 2 \sum_{n=1}^{\infty} \frac{1}{(2n+1)^3} \left(\sum_{k=0}^{n-1} \frac{(-1)^k}{2k+1}\right). \qquad (3.23)
\end{aligned}
$$

However, this reformulation provides little insight because both double series can be viewed as generalizations of Euler sums which have not been studied in details yet. In the following, using integration by parts, we convert the proposed integral into a double integral. The presence of symmetry enables us to reduce the amount of calculation in arriving at the desired answer. At the end of this section, we will revisit the double series in (3.23).

Solution.

Rewrite the proposed integral as

$$\int_0^1 \frac{(\ln(x))^2 \arctan(x)}{1+x} \, dx = \int_0^1 \arctan(x) d\left(\int_0^x \frac{(\ln(t))^2}{1+t} \, dt\right).$$

Let

$$f(x) = \int_0^x \frac{(\ln(t))^2}{1+t} \, dt.$$

Then

$$f(1) = \int_0^1 \frac{(\ln(t))^2}{1+t} \, dy = \sum_{n=0}^\infty (-1)^n \int_0^1 (\ln(t))^2 t^n \, dt = \sum_{n=0}^\infty \frac{2(-1)^n}{(n+1)^3} = \frac{3\zeta(3)}{2}.$$

Applying the fundamental theorem of calculus, we have

$$\int_0^1 (f(x) \arctan(x))' \, dx = f(x) \arctan(x)|_0^1 = \frac{3\zeta(3)}{2} \cdot \frac{\pi}{4} = \frac{3\pi\zeta(3)}{8}.$$

On the other hand, since

$$\int_0^1 (f(x) \arctan(x))' \, dx = \int_0^1 \frac{f(x)}{1+x^2} \, dx + \int_0^1 \frac{(\ln(x))^2 \arctan(x)}{1+x} \, dx,$$

we find that

$$\int_0^1 \frac{(\ln(x))^2 \arctan(x)}{1+x} \, dx = \frac{3\pi\zeta(3)}{8} - \int_0^1 \frac{f(x)}{1+x^2} \, dx. \tag{3.24}$$

Let

$$I := \int_0^1 \frac{f(x)}{1+x^2} \, dx.$$

Using the substitution $t = xy$, we have

$$f(x) = \int_0^1 \frac{x(\ln(xy))^2}{1+xy} \, dy.$$

Hence,

$$I = \int_0^1 \int_0^1 \frac{x[(\ln(x))^2 + 2\ln(x)\ln(y) + (\ln(y))^2]}{(1+x^2)(1+xy)} \, dy dx = I_1 + I_2 + I_3. \tag{3.25}$$

First, we have

$$\begin{aligned}
I_1 &= \int_0^1 \int_0^1 \frac{x(\ln(x))^2}{(1+x^2)(1+xy)} \, dy dx \\
&= \int_0^1 \left(\frac{(\ln(x))^2}{(1+x^2)} \int_0^1 \frac{x \, dy}{1+xy} \right) dx \\
&= \int_0^1 \left(\frac{(\ln(x))^2}{(1+x^2)} \ln(1+xy)|_{y=0}^{y=1} \right) dx \\
&= \int_0^1 \frac{(\ln(x))^2 \ln(1+x)}{(1+x^2)} \, dx.
\end{aligned}$$

Next, we have

$$
\begin{aligned}
I_3 &= \int_0^1 \int_0^1 \frac{x(\ln(y))^2}{(1+x^2)(1+xy)} \, dy \, dx \\
&= \int_0^1 \left((\ln(y))^2 \int_0^1 \frac{x \, dx}{(1+xy)(1+x^2)} \right) dy \\
&= \int_0^1 \frac{\pi y + 2\ln 2 - 4\ln(1+y)}{4(1+y^2)} (\ln(y))^2 \, dy \\
&= \frac{\pi}{4} \int_0^1 \frac{y(\ln(y))^2}{1+y^2} \, dy + \frac{\ln(2)}{2} \int_0^1 \frac{(\ln(y))^2}{1+y^2} \, dy - I_1.
\end{aligned}
$$

Since

$$
\begin{aligned}
\int_0^1 \frac{y(\ln(y))^2}{1+y^2} \, dy &= \sum_{n=0}^{\infty} (-1)^n \int_0^1 (\ln(y))^2 y^{2n+1} \, dy \\
&= \sum_{n=0}^{\infty} \frac{2(-1)^n}{(2n+2)^3} = \frac{1}{4} \sum_{n=0}^{\infty} \frac{(-1)^n}{(n+1)^3} \\
&= \frac{3\zeta(3)}{16}; \quad \text{and} \\
\int_0^1 \frac{(\ln(y))^2}{1+y^2} \, dy &= \sum_{n=0}^{\infty} (-1)^n \int_0^1 (\ln(y))^2 y^{2n} \, dy \\
&= 2 \sum_{n=0}^{\infty} \frac{(-1)^n}{(2n+1)^3} = 2\beta(3) = \frac{\pi^3}{16},
\end{aligned}
$$

where $\beta(x)$ is the Dirichlet beta function defined by $\beta(x) = \sum_{n=0}^{\infty} (-1)^n/(2n+1)^x$, we find that

$$
I_1 + I_3 = \frac{3\pi\zeta(3)}{64} + \frac{\pi^3 \ln(2)}{32}. \tag{3.26}
$$

Finally, by partial fractions, we have

$$
\frac{x}{(1+xy)(1+x^2)} = \frac{1}{1+y^2} \left(\frac{x+y}{1+x^2} - \frac{y}{1+xy} \right).
$$

Thus, in view of the symmetry, we have

$$
\begin{aligned}
I_2 &= 2 \int_0^1 \int_0^1 \frac{x \ln(x) \ln(y)}{(1+x^2)(1+xy)} \, dy \, dx \\
&= 2 \int_0^1 \int_0^1 \frac{1}{1+y^2} \left(\frac{x+y}{1+x^2} - \frac{y}{1+xy} \right) \ln(x) \ln(y) \, dy \, dx
\end{aligned}
$$

$$= 2 \int_0^1 \int_0^1 \frac{x+y}{(1+x^2)(1+y^2)} \ln(x)\ln(y)\,dxdy$$

$$-2 \int_0^1 \int_0^1 \frac{y\ln(x)\ln(y)}{(1+y^2)(1+xy)}\,dydx$$

$$= 2 \int_0^1 \int_0^1 \frac{x+y}{(1+x^2)(1+y^2)} \ln(x)\ln(y)\,dxdy - I_2,$$

and so

$$I_2 = \int_0^1 \int_0^1 \frac{x+y}{(1+x^2)(1+y^2)} \ln(x)\ln(y)\,dxdy.$$

In view of symmetry again, we have

$$I_2 = 2 \int_0^1 \int_0^1 \frac{x}{(1+x^2)(1+y^2)} \ln(x)\ln(y)\,dxdy$$

$$= 2 \int_0^1 \frac{x\ln(x)}{1+x^2}\,dx \cdot \int_0^1 \frac{\ln(y)}{1+y^2}\,dy.$$

Since

$$\int_0^1 \frac{x\ln(x)}{1+x^2}\,dx = \sum_{n=0}^{\infty} (-1)^n \int_0^1 (\ln(x))x^{2n+1}\,dx$$

$$= \sum_{n=0}^{\infty} \frac{(-1)^{n+1}}{4(2n+1)^2} = -\frac{\pi^2}{48}; \quad \text{and}$$

$$\int_0^1 \frac{\ln(y)}{1+y^2}\,dy = \sum_{k=0}^{\infty} (-1)^k \int_0^2 \ln(y)y^{2k}\,dy$$

$$= \sum_{k=0}^{\infty} \frac{(-1)^{k+1}}{(2k+1)^2} = -G,$$

we find that

$$I_2 = 2 \cdot \left(-\frac{\pi^2}{48}\right) \cdot (-G) = \frac{\pi^2 G}{24}. \tag{3.27}$$

Combining (3.25), (3.26) and (3.27) into (3.24), we conclude

$$\int_0^1 \frac{(\ln(x))^2 \arctan(x)}{1+x}\,dx = \frac{3\pi\zeta(3)}{8} - I = \frac{21\pi\zeta(3)}{64} - \frac{\pi^2 G}{24} - \frac{\pi^3\ln(2)}{32}$$

as claimed.   □

Remark. The key ideas used in the solution are

- to introduce $f(x)$ that leads to the double integral (3.25);
- to evaluate the double integral I_2 via symmetry.

We now go back to the double series in (3.23). For the first series, we rewrite it as

$$S_1 := \sum_{n=0}^{\infty} \frac{1}{(n+1)^3} \left(\sum_{k=0}^{n} \frac{(-1)^k}{2k+1} \right) = \sum_{n=1}^{\infty} \frac{1}{n^3} \left(\sum_{k=0}^{n-1} \frac{(-1)^k}{2k+1} \right).$$

By Abel's summation formula, we have

$$S_1(n) := \left(\sum_{k=1}^{n} \frac{1}{k^3} \right) \left(\sum_{k=0}^{n-1} \frac{(-1)^k}{2k+1} \right) - \sum_{k=1}^{n-1} \frac{(-1)^k}{2k+1} \left(\sum_{i=1}^{k} \frac{1}{i^3} \right).$$

Taking $n \to \infty$ gives

$$S_1 = \frac{\pi \zeta(3)}{4} - \sum_{k=1}^{\infty} \frac{(-1)^k}{2k+1} \left(\sum_{i=1}^{k} \frac{1}{i^3} \right).$$

To evaluate

$$I := \sum_{k=1}^{\infty} \frac{(-1)^k}{2k+1} \left(\sum_{i=1}^{k} \frac{1}{i^3} \right),$$

recall that

$$\sum_{k=1}^{\infty} H_3^{(k)} x^k = \frac{\text{Li}_3(x)}{1-x}, \tag{3.28}$$

where $\text{Li}_n(x)$ is the polylogarithm function. Replacing x by $-x^2$ in (3.28) and then integrating yields

$$\begin{aligned}
I &= \int_0^1 \frac{\text{Li}_3(-x^2)}{1+x^2} \, dx \\
&= \int_0^{\infty} \frac{\text{Li}_3(-x^2)}{1+x^2} \, dx - \int_1^{\infty} \frac{\text{Li}_3(-x^2)}{1+x^2} \, dx \\
&= \int_0^{\infty} \frac{\text{Li}_3(-x^2)}{1+x^2} \, dx - \int_0^1 \frac{\text{Li}_3(-1/x^2)}{1+x^2} \, dx.
\end{aligned}$$

This implies that

$$2I = \int_0^{\infty} \frac{\text{Li}_3(-x^2)}{1+x^2} \, dx + \int_0^1 \frac{\text{Li}_3(-x^2) - \text{Li}_3(-1/x^2)}{1+x^2} \, dx.$$

By the identity $\text{Li}_3(-x) - \text{Li}_3(-1/x) = -\zeta(2) \ln x - \ln^3 x/6$, we find

$$\begin{aligned}
2I &= \int_0^{\infty} \frac{\text{Li}_3(-x^2)}{1+x^2} \, dx + \int_0^1 \frac{-2\zeta(2) \ln x - 4\ln^3 x/3}{1+x^2} \, dx \\
&= \int_0^{\infty} \frac{\text{Li}_3(-x^2)}{1+x^2} \, dx - 2\zeta(2) \int_0^1 \frac{\ln x}{1+x^2} \, dx - \frac{4}{3} \int_0^1 \frac{\ln^3 x}{1+x^2} \, dx.
\end{aligned}$$

For the first integral, in view of $\text{Li}_3(x) = \frac{1}{2} \int_0^1 \frac{x \ln^2 y}{1 - xy} \, dy$, exchanging the order of the integration, we have

$$
\begin{aligned}
\int_0^\infty \frac{\text{Li}_3(-x^2)}{1 + x^2} \, dx &= \int_0^1 \ln^2 y \left(-\frac{1}{2} \int_0^\infty \frac{x^2}{(1 + x^2)(1 + x^2 y)} \, dx \right) dy \\
&= \int_0^1 \ln^2 y \left(-\frac{\pi}{4} \frac{1}{y + \sqrt{y}} \right) = -2\pi \int_0^1 \frac{\ln^2 t}{1 + t} \, dt \\
&= -2\pi \sum_{k=0}^\infty \int_0^1 (-1)^k t^k \ln^2 t \, dt = -3\pi \zeta(3).
\end{aligned}
$$

Moreover, we have

$$
\int_0^1 \frac{\ln x}{1 + x^2} \, dx = -G,
$$

$$
\begin{aligned}
\int_0^1 \frac{\ln^3 x}{1 + x^2} \, dx &= \sum_{k=0}^\infty \int_0^1 (-1)^k x^{2k} \ln^3 x \, dx \\
&= -\sum_{k=0}^\infty \frac{(-1)^k 6}{(2k + 1)^4} = -6\beta(4),
\end{aligned}
$$

where $\beta(x)$ is the Dirichlet beta function. In summary, we obtain

$$
I = G\zeta(2) + 4\beta(4) - \frac{3\pi\zeta(3)}{2}.
$$

Therefore, we finally find

$$
S_1 = \frac{7\pi\zeta(3)}{4} - G\zeta(2) - 4\beta(4).
$$

We challenge the reader to establish the following identity for the second series in (3.23):

$$
\sum_{n=1}^\infty \frac{1}{(2n + 1)^3} \left(\sum_{k=0}^{n-1} \frac{(-1)^k}{2k + 1} \right) = \frac{7\pi\zeta(3)}{128} + \frac{\pi^3 \ln 2}{64} - \frac{1}{2}\beta(4).
$$

It is well-known that integrals of products of logarithmic and polylogarithmic functions are associated with Euler sums. For example, since

$$
\int_0^1 x^{j-1} (\ln x)^{m-1} \, dx = (-1)^{m-1} \frac{(m-1)!}{j^m} \qquad \text{for } j \geq 1, m \geq 2,
$$

we have

$$
\sum_{j=1}^k \frac{1}{j^m} = \frac{(-1)^{m-1}}{(m-1)!} \int_0^1 (\ln x)^{m-1} \frac{1 - x^k}{1 - x} \, dx.
$$

Hence

$$S(m,n) \;:=\; \sum_{k=1}^{\infty} \frac{1}{k^n}\left(\sum_{j=1}^{k}\frac{1}{j^m}\right)$$

$$= \frac{(-1)^{m-1}}{(m-1)!}\int_0^1 \frac{(\ln x)^{m-1}}{1-x}\sum_{k=1}^{\infty}\frac{1-x^k}{k^n}\,dx$$

$$= \zeta(m)\zeta(n) - \frac{(-1)^{m-1}}{(m-1)!}\int_0^1 \frac{(\ln x)^{m-1}\mathrm{Li}_n(x)}{1-x}\,dx,$$

where we have used the well-known identity

$$\int_0^1 \frac{(\ln x)^{m-1}}{1-x}\,dx = (-1)^{m-1}\Gamma(m)\zeta(m) = (-1)^{m-1}(m-1)!\zeta(m).$$

We now end this section to collect some related problems for your practice. In particular, the first four are nice applications of the approach used in the solution. Let G be the Catalan constant.

1. Prove

$$\int_0^1 \frac{\ln(x)\arctan(x)}{1+x}\,dx = \frac{G\ln 2}{2} - \frac{\pi^3}{64}.$$

2. Prove

$$\int_0^1 \frac{x\arctan(x)\ln(1-x^2)}{1+x^2}\,dx = G\ln 2 - \frac{\pi^3}{48} - \frac{\pi\ln^2 2}{8}.$$

3. Prove

$$\int_0^1 \frac{\arctan^2(x)\ln(1+x)}{1+x^2}\,dx = \frac{\pi^3\ln 2}{384} + \frac{21\pi\zeta(3)}{256} - \frac{3\zeta(2)G}{16}.$$

4. Prove

$$\int_0^1 \frac{(\ln x)^2\arctan(x)}{x(1+x^2)}\,dx = \frac{\pi^3\ln 2}{16} - \frac{7\pi\zeta(3)}{32} + \beta(4),$$

   where $\beta(x)$ is the Dirichlet beta function defined by $\beta(x) = \sum_{n=0}^{\infty}(-1)^n/(2n+1)^x$.

5. Show that

$$\int_0^1 \frac{\ln x\,\mathrm{Li}_2(x)}{1-x}\,dx = -\frac{3}{4}\zeta(4).$$

6. Show that

$$\int_0^1 \frac{\ln^n x\,\mathrm{Li}_{n+1}(x)}{1-x}\,dx = \frac{(-1)^{n+1}\,n!}{2}[\zeta(2n+2) - \zeta^2(n+1)].$$

7. Show that

$$\int_0^1 \frac{\ln x \text{Li}_2(-x)}{1+x^2} \, dx = \sum_{n=1}^{\infty} \frac{(-1)^{n+1}}{n^2} \sum_{k=0}^{\infty} \frac{(-1)^k}{(2k+n+1)^2} = \frac{\pi^2 G}{48}.$$

8. Show that

$$\int_0^1 \frac{\ln^2 x \text{Li}_3(x)}{1+x} \, dx = \zeta(6) - \frac{9\zeta(3)^2}{16}.$$

In general, show that

$$\int_0^1 \frac{\ln^n x \text{Li}_{n+1}(x)}{1+x} \, dx = \frac{(-1)^n n!}{2} [\zeta(2n+2) - (1-2^{-n})^2 \zeta^2(n+1)].$$

9. Show that

$$\int_0^1 \frac{\text{Li}_3(-x)}{1+x^2} \, dx = \frac{\pi^2 G}{12} - \frac{3\pi\zeta(3)}{128} - \beta(4).$$

10. Show that

$$\int_0^1 \frac{\ln^n x \text{Li}_{n+1}(-x)}{1+x^2} \, dx = (-1)^{n+1} \frac{n!}{2^{n+1}} (1-2^{-n})\zeta(n+1)\beta(n+1).$$

11. Show that

(a) $\sum_{n=1}^{\infty} \frac{(-1)^n}{n^2} \left(\sum_{k=1}^{2n-1} \frac{(-1)^k}{k} \right) = \pi G - \frac{27}{16} \zeta(3).$

(b) $\sum_{n=1}^{\infty} \frac{(-1)^{n+1}}{n^2} H_{2n-1} = \pi G - \frac{29}{16} \zeta(3).$

(c) $\sum_{n=1}^{\infty} \frac{(-1)^{n+1}}{n^2} \left(\sum_{k=0}^{n-1} \frac{1}{2k+1} \right) = \pi G - \frac{7}{4} \zeta(3).$

12. Let

$$A(m, n) = \sum_{k=1}^{\infty} \frac{(-1)^{k+1} H_k^{(m)}}{k^n}. \tag{3.29}$$

Show that

(a) $A(m, n) = \frac{(-1)^n}{\Gamma(n)} \int_0^1 \frac{(\ln x)^{n-1} \text{Li}_m(-x)}{x(1+x)} \, dx.$

(b) $A(1, 1) = \frac{1}{2}\zeta(2) - \frac{1}{2}(\ln 2)^2.$

(c) $A(2, 1) = \zeta(3) - \frac{1}{2} \ln 2\zeta(2); A(1, 2) = \frac{5}{8}\zeta(3).$

(d) $A(3, 1) = \frac{19}{16}\zeta(4) - \frac{3}{4} \ln 2\zeta(3).$

(e) $A(4, 1) = 2\zeta(5) - \frac{7}{8} \ln 2\zeta(4) - \frac{3}{8}\zeta(2)\zeta(3); A(1, 4) = \frac{59}{32}\zeta(5) - \frac{1}{2}\zeta(2)\zeta(3).$

(f) Let $m \geq 2$. Then

$$A(m, 1) = \frac{(-1)^m}{\Gamma(m)} \int_0^1 \frac{(\ln x)^{m-1} \ln \left(\frac{1+x}{2} \right)}{1-x} \, dx$$

$$= \frac{m}{2}\zeta(m+1) - \ln 2\eta(m) - \frac{1}{2}\sum_{k=1}^{m-2} \eta(k+1)\eta(m-k),$$

where $\eta(s)$ is called the *Dirichlet eta function* which is defined by

$$\eta(x) = \sum_{k=1}^{\infty} \frac{(-1)^{k+1}}{k^x} = (1 - 2^{1-x})\zeta(x).$$

(g) If n is even, then

$$A(1,n) = \frac{1+n}{2}\eta(1+n) - \frac{1}{2}\zeta(1+n) - \sum_{k=1}^{n/2-1} \eta(2k)\zeta(1+n-2k).$$

13. Let $A(m,n)$ be defined as (3.29). Let $\mathrm{Li}_n(x)$ be the polylogarithm function. Show that

(a) $A(1,3) = \frac{11}{4}\zeta(4) - \frac{7}{4}\ln 2\zeta(3) + \frac{1}{2}(\ln 2)^2\zeta(2) - \frac{1}{12}(\ln 2)^4 - 2\mathrm{Li}_4(1/2)$.

(b) $A(2,2) = -\frac{51}{16}\zeta(4) + \frac{7}{2}\ln 2\zeta(3) - (\ln 2)^2\zeta(2) + \frac{1}{6}(\ln 2)^4 + 4\mathrm{Li}_4(1/2)$.

(c) $A(2,3) = -\frac{5}{8}\zeta(2)\zeta(3) - \frac{11}{32}\zeta(5)$.

(d) **Open question:** Determine the closed form of $A(m, 2n+1)$ for $m \geq 1, n \geq 2$.

14. Recall the Witten zeta function (see Section 2.1)

$$\mathcal{W}(r,s,t) = \sum_{n,m=1} \frac{1}{n^r m^s (n+m)^t}.$$

(a) For $r, s > 1$ and $t > 0$, show that

$$\mathcal{W}(r,s,t) = \frac{1}{\Gamma(t)} \int_0^1 \mathrm{Li}_r(x)\mathrm{Li}_s(x)\frac{(-\ln x)^{t-1}}{x}\, dx.$$

(b) Show that $\mathcal{W}(s+1,s,1) = \zeta^2(s+1)/2$ for $s \geq 1$.

(c) Show that $\mathcal{W}(2,2,1) = 2\zeta(2)\zeta(3) - 3\zeta(5)$.

(d) Introduce the *multivariate zeta function* $\zeta(r,s)$ defined by

$$\zeta(r,s) := \sum_{n>m} \frac{1}{n^r m^s}.$$

Can you represent $\mathcal{W}(r,s,t)$ in terms of the multivariate zeta function values?

3.14 An orthonormal function sequence

Problem 11850 (Proposed by Z. Ahmed, 122(6), 2015). Let

$$A_n(x) = \sqrt{\frac{2}{\pi}}\frac{1}{n!}(1+x^2)^{n/2}\frac{d^n}{dx^n}\left(\frac{1}{1+x^2}\right).$$

Prove that $\int_{-\infty}^{\infty} A_m(x)A_n(x)\,dx = \delta(m,n)$ for nonnegative integers m and n. Here, $\delta(m,n) = 1$ if $m = n$, and otherwise $\delta(m,n) = 0$.

Discussion.
In **Problem 10777** (107(1), 2000), the current proposer asked to evaluate

$$I(m,n) := \int_0^\infty \frac{d^m}{dx^m}\left(\frac{1}{1+x^2}\right)\frac{d^n}{dx^n}\left(\frac{1}{1+x^2}\right)\,dx.$$

The published solution [2] is based on integration by parts and the Fourier transform:

$$\mathcal{F}[\sqrt{\pi/2}\,e^{-|t|}](x) = \frac{1}{1+x^2} \quad \text{and} \quad \int_{-\infty}^{\infty} f(t)\overline{g(t)}\,dt = \int_{-\infty}^{\infty} \mathcal{F}[f(t)](x)\overline{\mathcal{F}[g(t)](x)}\,dx$$

The same argument does not work on the current problem because of the nonconstant weights in A_n. To evaluate the proposed integral we need to find another expression for $\frac{d^n}{dx^n}(1/(1+x^2))$, and so for A_n.

Solution.
The expression for the nth derivative of $1/(1+x^2)$ in our mind is, for $n \geq 0$,

$$\frac{d^n}{dx^n}\left(\frac{1}{1+x^2}\right) = \frac{(-1)^n n! \sin(n+1)\theta}{(1+x^2)^{(n+1)/2}}, \tag{3.30}$$

where $\theta = \arcsin(1/\sqrt{1+x^2})$. Indeed, let $f(x) = \frac{1}{1+x^2}$. Applying the Leibniz rule to $(1+x^2)f(x) = 1$ yields

$$(x^2+1)f^{(n+2)}(x) + 2(n+2)xf^{(n+1)}(x) + (n+2)(n+1)f^{(n)}(x) = 0,$$

where $f^{(k)}$ indicates the kth derivative of $f(x)$. Let $f^{(n)}(x) = (-1)^n n! a_n$. Then

$$(1+x^2)a_{n+2} - 2xa_{n+1} + a_n = 0,$$

subject to

$$a_0 = \frac{1}{1+x^2}, \quad a_1 = \frac{2x}{(1+x^2)^2}.$$

Now let $\theta = \arcsin(1/\sqrt{1+x^2})$. Solving this difference equation gives

$$a_n = \frac{\sin(n+1)\theta}{(1+x^2)^{(n+1)/2}}, \quad n = 0,1,2,\dots. \tag{3.31}$$

Hence,

$$A_n = (-1)^n \sqrt{\frac{2}{\pi}} \frac{\sin(n+1)\theta}{(1+x^2)^{1/2}},$$

and so

$$\int_{-\infty}^{\infty} A_m(x) A_n(x)\, dx = (-1)^{m+n} \frac{2}{\pi} \int_{-\infty}^{\infty} \frac{\sin(m+1)\theta \sin(n+1)\theta}{1+x^2}\, dx.$$

Appealing to $x = \tan\theta$ and $dx = \sec^2\theta\, d\theta$, we have

$$\int_{-\infty}^{\infty} A_m(x) A_n(x)\, dx = (-1)^{m+n} \frac{2}{\pi} \int_{-\pi/2}^{\pi/2} \sin(m+1)\theta \sin(n+1)\theta\, d\theta.$$

Finally, since $\{\sin nx\}$ is orthogonal on $(-\pi/2, \pi/2)$, i.e.,

$$\int_{-\pi/2}^{\pi/2} \sin(m+1)\theta \sin(n+1)\theta\, d\theta = \begin{cases} 0, & \text{if } m \neq n, \\ \frac{\pi}{2}, & \text{if } m = n, \end{cases}$$

this implies that

$$\int_{-\infty}^{\infty} A_m(x) A_n(x)\, dx = \delta(m,n).$$

$\square$

Remark. Notice that the above argument leads to another solution to **Problem 10777**. In fact, let $f(x) = 1/(1+x^2)$. By (3.31), for $n \geq 0$, we have

$$f^{(n)}(0) = n! \cos(n\pi/2), \quad f^{(n)}(\infty) = 0.$$

To evaluate $I(m,n)$, we consider two cases:
Case I. When $m + n$ is even. Without loss of generality, we assume that $m > n$. Integrating by parts yields

$$I(m,n) = \int_0^\infty f^{(m)}(x) f^{(n)}(x)\, dx = -f^{(m)}(0) f^{(n-1)}(0) - I(m+1, n-1).$$

Repeatedly using this reduction formula we find

$$I(m,n) = (-1)^n \int_0^\infty f^{(m+n)}(x) f(x)\, dx = (-1)^n \int_0^\infty f^{(m+n)}(x) f(x)\, dx.$$

Since

$$\frac{1}{1+x^2} = \int_0^\infty e^{-t} \cos(xt)\, dt,$$

and $\cos^{(k)}(x) = \cos(x + k\pi/2)$, we have

$$f^{(m+n)}(x) = \int_0^\infty t^{m+n} e^{-t} \cos(xt + (m+n)\pi/2)\, dt$$

$$= (-1)^{(m+n)/2} \int_0^\infty t^{m+n} e^{-t} \cos(xt)\, dt,$$

and so

$$I(m,n) = (-1)^{(m-n)/2} \int_0^\infty \left(\frac{1}{1+x^2} \int_0^\infty t^{m+n} e^{-t} \cos(xt) \, dt \right) dx$$

$$= (-1)^{(m-n)/2} \int_0^\infty t^{m+n} e^{-t} \left(\int_0^\infty \frac{\cos(xt)}{1+x^2} \, dx \right) dt.$$

The interchange the order of integrations is justified by the dominate convergence theorem. Notice the inner integral is the well-known Laplace integral which is given by

$$\int_0^\infty \frac{\cos(xt)}{1+x^2} \, dx = \frac{\pi}{2} e^{-t}.$$

We finally find that

$$I(m,n) = (-1)^{(m-n)/2} \frac{\pi}{2} \int_0^\infty t^{m+n} e^{-2t} \, dt = (-1)^{(m-n)/2} \frac{(m+n)!\pi}{2^{m+n+2}}.$$

Case II. When $m+n$ is odd. We assume that $m > n$ again. Integrating by parts $(m-n)$ times (with $u = f^{(n)}, v = f^{(m-1)}$ initially) yields

$$I(m,n) = (-1)^m \sum_{k=1}^{m-n} f^{(m-k)}(0) f^{(n+k-1)}(0) - I(m-(m-n), n+(m-n)).$$

In view of $I(m,n) = I(n,m)$, we conclude

$$I(m,n) = \frac{(-1)^m}{4} \sum_{k=1}^{m-n} (\sin((m+n)\pi/2)$$
$$- (-1)^k \sin((m-n)\pi/2))(m-k)!(n+k-1)!.$$

In particular, we have $I(2n+2, 2n+1) = 0, I(2n+1, 2n) = -[(2n)!]^2/2$.

It is also interesting to see that we can establish (3.30) by partial fractions. In fact, since

$$\frac{d^n}{dx^n} \left(\frac{1}{x+a} \right) = (-1)^n \frac{n!}{(x+a)^{n+1}},$$

we have

$$\frac{d^n}{dx^n} \left(\frac{1}{1+x^2} \right) = \frac{1}{2i} \frac{d^n}{dx^n} \left(\frac{1}{x-i} - \frac{1}{1+i} \right)$$

$$= (-1)^n \frac{n!}{2i} \left(\frac{1}{(x-i)^{n+1}} - \frac{1}{(x+i)^{n+1}} \right).$$

Now let $\theta = \arcsin(1/\sqrt{1+x^2})$. Then $x = \cot\theta$ and

$$x - i = \cot\theta - i = \frac{2ie^{-i\theta}}{e^{i\theta} - e^{-i\theta}} = \frac{1}{\sin\theta} e^{-i\theta}.$$

Similarly, we have

$$x + i = \cot\theta + i = \frac{2e^{i\theta}}{e^{i\theta} - e^{-i\theta}} = \frac{1}{\sin\theta} e^{i\theta}.$$

Since $\sin^2\theta = 1/(1 + x^2)$, we conclude that

$$\frac{d^n}{dx^n}\left(\frac{1}{1 + x^2}\right) = (-1)^n n! \frac{\sin^{n+1}\theta}{2i}\left(e^{i(n+1)\theta} - e^{-i(n+1)\theta}\right)$$

$$= \frac{(-1)^n n! \sin(n+1)\theta}{(1 + x^2)^{(n+1)/2}}.$$

3.15   An definite integral Quickie

Problem 4212 (Proposed by H. F. Sandham, 53(4), 1946). Evaluate

$$\int_0^\infty \frac{e^{-x^2}}{(x^2 + 1/2)^2}\, dx.$$

Discussion.
In the MAA journal *Mathematics Magazine*, a problem submitted as a Quickie should have an unexpected, succinct solution. Thus a Quickie provides a double challenge to the reader — to solve the problem and to devise a solution that is clear, concise and surprising. This problem was proposed in 1946. Two published solutions ([38], 1947) were elegant, but not unexpected. Later this problem was solved by using parametric differentiation. Is there any special approach apparently unrelated to the problem provide the quickness sought? A recurrence relation!

Solution.
We consider one more general variation: For nonnegative integer n, let

$$I_n := \int_0^\infty \frac{e^{-x^2}}{(1 + 2x^2)^n}\, dx.$$

Then the proposed integral is $4I_2$. Applying integration by parts, we have

$$
\begin{aligned}
I_n - I_{n+1} &= \int_0^\infty \frac{2x^2 e^{-x^2}}{(1 + 2x^2)^{n+1}}\, dx \\
&= \int_0^\infty \frac{e^{-x^2}}{(1 + 2x^2)^{n+1}}\, dx - (n+1)\int_0^\infty \frac{4x^2 e^{-x^2}}{(1 + 2x^2)^{n+2}}\, dx \\
&= I_{n+1} - 2(n+1)\int_0^\infty \frac{[(1 + 2x^2) - 1]e^{-x^2}}{(1 + 2x^2)^{n+2}}\, dx \\
&= I_{n+1} - 2(n+1)I_{n+1} + 2(n+1)I_{n+2}.
\end{aligned}
$$

This implies that
$$I_n + 2nI_{n+1} - 2(n+1)I_{n+2} = 0.$$

This enables us to evaluate I_2 without knowing I_1. Indeed,

$$I_2 = \frac{1}{2} I_0 = \frac{1}{2} \int_0^\infty e^{-x^2} \, dx = \frac{1}{4}\sqrt{\pi}.$$

Thus the proposed integral equals $\sqrt{\pi}$.   □

Remark. Here we list two problems for your practice.

1. Let $\sinh x \cdot \sinh y = 1$. Show that $\int_0^\infty y \, dx = \frac{\pi^2}{4}$.

2. **Problem 11933** (Proposed by J. M. Pacheco and A. Plaza, 123(8), 2016). For positive integer n, let $H_n = \sum_{k=1}^n 1/k$. Prove

$$\int_0^1 \frac{1}{x+1} \, dx \cdot \int_0^1 \frac{x+1}{x^2+x+1} \, dx \cdots \int_0^1 \frac{x^{n-2}+\cdots+x+1}{x^{n-1}+\cdots+x+1} \, dx \geq \frac{1}{H_n}.$$

4

Inequalities

We select 10 inequality problems that have appeared in the Monthly over the past half century. We will encounter many familiar characters, among them: Cauchy's inequality, L'Hôpital's monotone rule, Majorization and Convexity, Hilbert's identity, Hardy's inequality, and differential equations. What makes our trip worthwhile is the realization that these old chestnuts still have something new to tell us. They all appear here in novel and surprising ways and you should add them to your inequality tool box.

4.1 An inequality from Klamkin

Problem E2483 (Proposed by M. S. Klamkin, 81(6), 1974). Let x be non-negative and let m, n be integers with $m \geq n \geq 1$. Prove that

$$(m + n)(1 + x^m) \geq 2n \frac{1 - x^{m+n}}{1 - x^n}.$$

Discussion.
For $x = 1$, the right-hand side of the inequality is understood as a limit where the inequality actually becomes an equality. Also, for $x = 0$, the inequality becomes $m + n \geq 2n$, which is true since $m \geq n$. Moreover, notice that when $m = n$, the inequality becomes equality again, and replacing x by $1/x$ in the inequality yields the same inequality after multiplying by x^m. Thus, it suffices to prove the inequality for $x \in (0, 1)$ and $m > n$. There are many verifications for this inequality. We single out three. The first proof belongs to the proposer. He worked backward — assume the inequality holds then arrive at an equivalent inequality that can be easily proven. The second proof is due to Allen Stenger, then an undergraduate student at Penn State. He transformed the inequality into an area comparison. The last one is based on the following:

L'Hôpital's Monotone Rule (LMR): Let $f, g : [a, b] \to \mathbb{R}$ be continuous functions that are differentiable on (a, b) with $g'(x) \neq 0$ on (a, b). If f'/g' is increasing (decreasing) on (a, b), then the functions

$$\frac{f(x) - f(b)}{g(x) - g(b)} \quad \text{and} \quad \frac{f(x) - f(a)}{g(x) - g(a)}$$

are correspondingly increasing (decreasing) on (a, b).

Solution I — by M. S. Klamkin.
Based on the above discussion, we can assume that $x \in (0, 1)$ and $m > n$. Let $m + n = r$, $m - n = s$. Then $r > s > 0$. Let $t = \sqrt{x}$ so that $0 < t < 1$. The desired inequality becomes

$$r(1 + t^{r+s}) \geq (r - s)\frac{1 - t^{2r}}{1 - t^{r-s}}.$$

Rearranging this inequality gives

$$\frac{1 - t^{2r}}{rt^r} \geq \frac{1 - t^{2s}}{st^s}.$$

Substituting $t = e^{-y}$ yields

$$\frac{\sinh(ry)}{ry} \geq \frac{\sinh(sy)}{sy}, \tag{4.1}$$

where now $y \in (0, \infty)$. Since

$$\frac{\sinh x}{x} = \frac{1}{x}\sum_{n=0}^{\infty} \frac{1}{(2n+1)!}x^{2n+1} = \sum_{n=0}^{\infty} \frac{1}{(2n+1)!}x^{2n},$$

this implies that $\sinh x / x$ is strictly increasing. Hence (4.1) is established, and so is the proposed inequality.

$\square$

Solution II — by A. Stenger.
Assume that $x \in (0, 1)$ and $m > n$. Let $y = x^n$ and $r = m/n > 1$. Then $0 < y < 1$ and the desired inequality is equivalent to

$$(r + 1)(1 + y^r) \geq \frac{2(1 - y^{r+1})}{1 - y}.$$

Rearranging this inequality gives

$$\frac{1}{2}(1 - y)(1 + y^r) \geq \frac{1 - y^{r+1}}{r + 1}. \tag{4.2}$$

Consider the function $f(x) = x^r$ on $[y, 1]$. Since f is concave up, the region under $y = x^r$ on $[y, 1]$ is contained in the trapezoid with vertices $(y, 0), (y, y^r), (1, 1)$ and $(1, 0)$. The standard area formulas conclude that

$$\text{Area of the trapezoid} = \frac{1}{2}(1 - y)(1 + y^r) \geq \int_y^1 x^r \, dx = \frac{1 - y^{r+1}}{r + 1}.$$

This proves (4.2) and so the proposed inequality.

□

Solution III — by L'Hôpital's monotone rule (LMR).

As above, we assume that $x \in (0,1)$ and $m > n$. The desired inequality is equivalent to

$$\frac{(1-x^n)(1+x^m)}{1-x^{m+n}} \geq \frac{2n}{m+n}. \tag{4.3}$$

Let $f(x) = (1-x^n)(1+x^m), g(x) = 1 - x^{m+n}$. Then

$$\frac{f'(x)}{g'(x)} = \frac{-nx^{n-1}+mx^{m-1}-(m+n)x^{m+n-1}}{-(m+n)x^{m+n-1}} = \frac{1}{m+n}[nx^{-m}-mx^{-n}+(m+n)].$$

To show that $f'(x)/g'(x)$ is decreasing for $0 < x < 1$, it suffices to prove that $nx^{-m} - mx^{-n} + (m+n)$ is decreasing for $0 < x < 1$. This follows from

$$[nx^{-m} - mx^{-n} + (m+n)]' = -mnx^{-m-1} + nmx^{-n-1}$$

$$= -mnx^{-m-1}(1 - x^{m-n}) < 0.$$

Thus, LMR deduces that

$$F(x) := \frac{f(x) - f(1)}{g(x) - g(1)} = \frac{f(x)}{g(x)} = \frac{(1-x^n)(1+x^m)}{1-x^{m+n}}$$

is decreasing for $0 < x < 1$. Moreover, applying L'Hôpital's rule yields

$$F(1) = \lim_{x \to 1^-} F(x) = \lim_{x \to 1} \frac{(1-x^n)(1+x^m)}{1-x^{m+n}} = \frac{2n}{m+n}.$$

Now (4.3) follows from the fact that $F(x) \geq F(1)$ immediately. □

Remark. Consider

$$G(x) := \frac{(1+x^n)(1-x^m)}{1-x^{m+n}}.$$

Along the same lines in the Solution III, applying the LMR, we see that $G(x)$ is increasing for $0 < x < 1$. Thus,

$$\frac{(1+x^n)(1-x^m)}{1-x^{m+n}} \leq \lim_{x \to 1^-} G(x) = \frac{2m}{m+n}.$$

This, together with (4.3), yields the double inequalities

$$2n\frac{1-x^{m+n}}{1-x^n} \leq (m+n)(1+x^m) \leq (m+n)(1+x^n) \leq 2m\frac{1-x^{m+n}}{1-x^m}.$$

Let $m+n = r$, $m-n = s$. As we did in the Solution I, Klamkin's inequality becomes

$$\frac{x^m - 1}{x^n - 1} \geq \frac{m}{n}x^{(m-n)/2}.$$

Replacing x by x/y yields

$$\left(\frac{x^m - y^m}{x^n - y^n} \cdot \frac{n}{m}\right)^{1/(m-n)} \geq \sqrt{xy}.$$

When $n = 1$, the left-hand side is often called the *Stolarsky mean*. It is well-known that

$$\sqrt{xy} \leq \frac{x-y}{\ln x - \ln y} \leq \left(\frac{x^m - y^m}{x-y} \cdot \frac{1}{m} \right)^{1/(m-1)}.$$

It is interesting to see that LMR provides a powerful method for proving the monotonicity of a large class of quotients. The LMR first appeared in Gromov and Taylor's work [51] for volume estimation in differential geometry. Since then, the LMR and its variants have been used in approximation theory, quasi-conformal theory and probability. But, for most analysis readers, the LMR is not as well-known as it should be. To popularize and promote this monotonicity rule, we present a proof of the LMR and two more examples within the realm of elementary analysis. The reader can see the wide applicability of the LMR and find more applications in exercises.

Proof. We may assume that $g'(x) > 0$ and $f'(x)/g'(x)$ is increasing for all $x \in (a, b)$. By the mean value theorem, for a given $x \in (a, b)$ there exists $c \in (a, x)$ such that

$$\frac{f(x) - f(a)}{g(x) - g(a)} = \frac{f'(c)}{g'(c)} \leq \frac{f'(x)}{g'(x)},$$

and so

$$f'(x)(g(x) - g(a)) - g'(x)(f(x) - f(a)) \geq 0.$$

Therefore,

$$\left(\frac{f(x) - f(a)}{g(x) - g(a)} \right)' = \frac{f'(x)(g(x) - g(a)) - g'(x)(f(x) - f(a))}{(g(x) - g(a))^2} \geq 0.$$

This shows that $(f(x) - f(a))/(g(x) - g(a))$ is increasing on (a, b) as desired. Similarly, we can show that $(f(x) - f(b))/(g(x) - g(b))$ is increasing. □

So far we have assumed that a and b are finite. This rule can be extended easily to the case where a or b is infinity. Our first example is to confirm the monotonicity of $(x^p - 1)/(x^q - 1)$ for $p > q > 0$, $x \geq 1$ by using the LMR. To see this, let $f(x) = x^p - 1$ and $g(x) = x^q - 1$. Consider

$$G(x) = \begin{cases} f(x)/g(x), & \text{if } x \neq 1, \\ p/q, & \text{if } x = 1. \end{cases}$$

We have $f'(x)/g'(x) = (p/q)x^{p-q}$, which is increasing on $(1, \infty)$ as long as $p > q > 0$. Hence, by the LMR, $G(x)$ is increasing on $(1, \infty)$. In particular, $G(x) > G(1) = p/q$. This yields the following bonus inequality:

$$\frac{x^p - 1}{p} > \frac{x^q - 1}{q}, \qquad \text{for } x > 1, p > q > 0. \tag{4.4}$$

Our second example comes from Wilker's inequality, which appeared as Monthly Problem E3306. Wilker asked for a proof that

$$\left(\frac{\sin x}{x}\right)^2 + \frac{\tan x}{x} > 2, \quad 0 < x < \frac{\pi}{2}. \tag{4.5}$$

The published solution by Sumner et. al. [90] is elementary but relatively long. Based on the LMR, we give a short alternative solution for (4.5). Our solution is almost algorithmic in nature. Set

$$F(x) = \begin{cases} \left(\frac{\sin x}{x}\right)^2 + \frac{\tan x}{x}, & \text{if } x \neq 0, \\ 2, & \text{if } x = 0. \end{cases}$$

Let $f(x) = \sin^2 x + x \tan x$, $g(x) = x^2$. Now

$$f'(x) = 2\sin x \cos x + x\sec^2 x + \tan x$$

and

$$\frac{f''(x)}{g''(x)} = \cos^2 x - \sin^2 x + \sec^2 x + x\tan x \sec^2 x.$$

Since

$$\left(\frac{f''(x)}{g''(x)}\right)' = 3\tan x \sec^2 x(1 - \cos^4 x) + \sin x \left(\frac{x}{\sin x \cos^4 x} - \frac{1}{\cos x}\right)$$
$$+ 2x\tan^2 x \sec^2 x > 0,$$

$f''(x)/g''(x)$ is increasing on $(0, \pi/2)$. Thus, using the LMR twice and noticing that $f(0) = f'(0) = g(0) = g'(0) = 0$, we deduce that $f'(x)/g'(x)$ and so $F(x) = f(x)/g(x)$ is increasing on $(0, \pi/2)$. Thus (4.5) follows immediately from the fact that $F(x) \geq F(0)$.

We now end this section with some problems for additional practice:

1. Notice that $f(x) = \frac{a^x - 1}{x}$ is increasing for $a > 1, x > 0$ from (4.4). Show that $f(x)$ is log-convex (i.e., $\ln f(x)$ is convex), and then prove that

$$\sqrt{xy} \leq \left(\frac{x^{m-n} - y^{m-n}}{(m-n)(\ln x - \ln y)}\right)^{1/(m-n)} \leq \left(\frac{x^m - y^m}{x^n - y^n} \cdot \frac{n}{m}\right)^{1/(m-n)}.$$

2. Let $x \geq 0$, $m \geq k \geq n \geq 1$ and $2k \geq m + n$. Then

$$(m + k + n)(1 + x^m)(1 + x^k) \geq 3n\frac{1 - x^{m+k+n}}{1 - x^n}.$$

3. Let $x \geq 0$, $\min(1/x, x) \leq y \leq \max(1/x, x)$ and $m \geq k \geq n \geq 1$. Show that

(a) $(m+k+n)(1+x^m)(1+y^k) \geq 3n\frac{1-x^{m+n}y^k}{1-x^n}$, if $2k \geq m+n$;

(b) $(m+k+n)(1+x^k)(1+y^m) \geq 3n\frac{1-x^{k+n}y^m}{1-x^n}$, if $k+n \geq m$.

4. For $0 < x < \pi/2$, prove that

(a) $\left(\frac{\sin x}{x}\right)^3 > \cos x$.

(b) $\frac{2x}{\sin x} + \frac{x}{\tan x} > 3$.

Comment. Using $t^2 \geq 2t-1$ with $t = x/\sin x$ in the second inequality yields

$$\left(\frac{x}{\sin x}\right)^2 + \frac{x}{\tan x} > 2.$$

5. **Problem 11770** (Proposed by S. P. Andriopoulos, 121(3), 2014). Prove, for real numbers a, b, x, y with $a > b > 1$ and $x > y > 1$, that

$$\frac{a^x - b^y}{x - y} > \left(\frac{a+b}{2}\right)^{(x+y)/2} \ln\left(\frac{a+b}{2}\right).$$

6. **Problem 11009** (Proposed by D. Borwein, J. Borwein and J. Rooin, 100(3), 2003). Let $a > b \geq c > d > 0$. Prove that

$$f(x) = \frac{a^x - b^x}{c^x - d^x}$$

is increasing and convex for all x.

Comment. One may show that $f(x)$ is increasing based on the LMR. Without loss of generality, assume that $d = 1$, otherwise divide the ratio by d^x. Now, let $y = c^x$, $\alpha = \ln b/\ln c$, $\beta = \ln a/\ln c$. Then $\beta > \alpha \geq 1$. One can rewrite $f(x)$ as

$$F(y) := \frac{y^\beta - y^\alpha}{y - 1} \qquad y \in (0,1) \cup (1, \infty).$$

7. In Wilker's inequality, prove that

$$\frac{\left(\frac{\sin x}{x}\right)^2 + \frac{\tan x}{x} - 2}{x^3 \tan x} \geq \frac{16}{\pi^4}, \quad (0 < x < \pi/2).$$

The initial published solution by Jean Anglesio is relatively long. Another shorter solution (still about one page long) by David Callan was published half year later.

8. Kober's inequality claims that $\cos x \geq 1 - 2x/\pi$ for $0 < x < \pi/2$. Prove the following refinement:

$$1 - \frac{2}{\pi}x + \frac{\pi - 2}{\pi^2}x(\pi - 2x) \leq \cos x \leq 1 - \frac{2}{\pi}x + \frac{2}{\pi^2}x(\pi - 2x).$$

9. For $0 < x < 1$, prove that

$$\frac{4}{\pi}\frac{x}{1-x^2} < \tan\left(\frac{\pi x}{2}\right) < \frac{\pi}{2}\frac{x}{1-x^2}.$$

10. (**Quarter Circle Lemma**) For all $m \geq 1$, show that

$$\sum_{n=1}^{\infty} \frac{1}{m+n}\left(\frac{m}{n}\right)^{1/2} < \pi.$$

Hint. In the quarter circle, consider the area of a triangle which is similar to the triangle with vertices $(0,0), (\sqrt{m}, \sqrt{n-1})$, and $(\sqrt{m}, \sqrt{n})$.

11. **Problem 11308** (Proposed by O. Furdui, 114(7), 2007). Let n be a positive integer. For $1 \leq i \leq n$, let x_i be a real number in $(0, \pi/2)$ and a_i be a real number in $[1, \infty)$. Prove that

$$\prod_{i=1}^{n}\left(\frac{x_i}{\sin x_i}\right)^{2a_i} + \prod_{i=1}^{n}\left(\frac{x_i}{\tan x_i}\right)^{a_i} > 2.$$

12. **Problem 11869** (Proposed by G. Stoica, 122(9), 2015). Prove that

$$|y \ln y - x \ln x| \leq |y - x|^{1-1/e}$$

for $0 < x < y \leq 1$.
Comment. The exponent $1 - 1/e$ is the best possible.

4.2 Knuth's exponential inequality

Problem 11369 (Proposed by D. Knuth, 115(6), 2008). Prove that for all real t, and all $\alpha \geq 2$,

$$e^{\alpha t} + e^{-\alpha t} - 2 \leq (e^t + e^{-t})^{\alpha} - 2^{\alpha}. \tag{4.6}$$

Discussion.
It is easy to see that (4.6) holds for $t = 0$ or $\alpha = 2$. So we can assume that $t \neq 0$ and $\alpha > 2$. By introducing

$$f(t, \alpha) = [(e^t + e^{-t})^{\alpha} - 2^{\alpha}] - (e^{\alpha t} + e^{-\alpha t} - 2),$$

the published solution by Dale [34] established the positivity of $f(t, \alpha)$ based on

$$f(t, \alpha) = \alpha \int_0^t \left\{ (e^x + e^{-x})^{\alpha} \frac{\sinh x}{\cosh x} - (e^{\alpha x} - e^{-\alpha x}) \right\} dx.$$

We will present two more elementary proofs as follows. First, let $x = e^t$ and $F(x) = (x + x^{-1})^\alpha - x^\alpha - x^{-\alpha}$. Then (4.6) becomes $F(x) \geq F(1)$. Since $F(x) = F(x^{-1})$, it suffices to show that F is decreasing for $0 < x < 1$. The second proof is based on the LMR. We will see that both proofs are almost algorithmic in nature, using only Bernoulli's inequality.

Solution I.
We show that $F'(x) \leq 0$ for $0 < x < 1$, which implies that $F(x)$ is decreasing for $0 < x < 1$. Since

$$F'(x) = \alpha(x + x^{-1})^{\alpha-1}(1 - x^{-2}) - \alpha x^{\alpha-1} + \alpha x^{-\alpha-1},$$

$F'(x) \leq 0$ is evidently equivalent to

$$(1 + x^2)^{\alpha-1} \geq \frac{1 - x^{2\alpha}}{1 - x^2}.$$

Applying Bernoulli's inequality,

$$(1 + t)^p \geq 1 + pt \qquad \text{for } t \geq -1, p \geq 1,$$

we have

$$(1 + x^2)^{\alpha-1} \geq 1 + (\alpha - 1)x^2.$$

Thus,

$$1 + (\alpha - 1)x^2 \geq \frac{1 - x^{2\alpha}}{1 - x^2} = 1 + \left(\frac{1 - x^{2(\alpha-1)}}{1 - x^2}\right) x^2,$$

as long as

$$\alpha - 1 \geq \frac{1 - x^{2(\alpha-1)}}{1 - x^2}$$

or

$$x^{2(\alpha-1)} \geq 1 + (\alpha - 1)(x^2 - 1).$$

This follows from Bernoulli's inequality again:

$$x^{2(\alpha-1)} = (1 + (x^2 - 1))^{\alpha-1} \geq 1 + (\alpha - 1)(x^2 - 1).$$

$\square$

Solution II.
We show that (4.6) holds for all $t > 0$. Let $x = e^t$. Then $x > 1$ and (4.6) is equivalent to

$$\frac{(x^2 + 1)^\alpha - x^{2\alpha} - 1}{x^\alpha} \geq 2^\alpha - 2.$$

To apply the LMR, let $f(x) = (x^2 + 1)^\alpha - x^{2\alpha} - 1$ and $g(x) = x^\alpha$. Then

$$\frac{f'(x)}{g'(x)} = \frac{2x(x^2 + 1)^{\alpha-1} - 2x^{2\alpha-1}}{x^{\alpha-1}}.$$

Define

$$G(x) = \frac{f'(x)}{g'(x)} = 2x \left(x + \frac{1}{x} \right)^{\alpha - 1} - 2x^\alpha.$$

To show that $G(x)$ is increasing for $x > 1$, it suffices to prove that $G'(x) \geq 0$ for $x > 1$. Indeed,

$$G'(x) = 2 \left(x + \frac{1}{x} \right)^{\alpha - 1} \left[\alpha - \frac{2(\alpha - 1)}{x^2 + 1} \right] - 2\alpha x^{\alpha - 1}.$$

Using Bernoulli's inequality, we have

$$\left(x + \frac{1}{x} \right)^{\alpha - 1} = x^{\alpha - 1} \left(1 + \frac{1}{x^2} \right)^{\alpha - 1} \geq x^{\alpha - 1} + (\alpha - 1)x^{\alpha - 3}.$$

Therefore, for $x > 1$ and $\alpha \geq 2$,

$$G'(x) \geq \frac{2(\alpha - 1)(\alpha - 2)x^{\alpha - 3}(x^2 - 1)}{x^2 + 1} \geq 0.$$

Now, the LMR deduces that

$$\frac{f(x) - f(1)}{g(x) - g(1)} = \frac{(x^2 + 1)^\alpha - x^{2\alpha} - 2^\alpha + 1}{x^\alpha - 1}$$

is increasing for $x > 1$. On the other hand, L'Hôpital's rule implies that

$$\lim_{x \to 1} \frac{f(x) - f(1)}{g(x) - g(1)} = \lim_{x \to 1} \frac{(x^2 + 1)^\alpha - x^{2\alpha} - 2^\alpha + 1}{x^\alpha - 1} = 2^\alpha - 2.$$

Thus,

$$\frac{f(x) - f(1)}{g(x) - g(1)} = \frac{(x^2 + 1)^\alpha - x^{2\alpha} - 2^\alpha + 1}{x^\alpha - 1} \geq 2^\alpha - 2.$$

This proves (4.6) as desired. □

Remark. When we search for a general context for (4.6), we find a path connected to two important principles in the theory of inequalities — Majorization and Schur convexity. Since Majorization and Schur convexity are not as well-known as they should be, we will take a few paragraphs to explain them.

We say an n-tuple $\mathbf{x} = (x_1, x_2, \cdots, x_n)$ is *majorized* by another n-tuple $\mathbf{y} = (y_1, y_2, \cdots, y_n)$ (written as $\mathbf{x} \prec \mathbf{y}$) if

M1 $x_1 \geq x_2 \geq \cdots \geq x_n$ and $y_1 \geq y_2 \geq \cdots \geq y_n$,

M2 $x_1 + x_2 + \cdots + x_k \leq y_1 + y_2 + \cdots + y_k$ for all $1 \leq k \leq n - 1$,

M3 $x_1 + x_2 + \cdots + x_n = y_1 + y_2 + \cdots + y_n$.

For example, we have

$$(1,1,1,1) \prec (2,1,1,0) \prec (3,1,0,0) \prec (4,0,0,0).$$

Moreover, for any $\mathbf{x} = (x_1, x_2, \cdots, x_n)$, with the arithmetic mean $\bar{\mathbf{x}} = (x_1 + x_2 + \cdots + x_n)/n$,

$$(\bar{\mathbf{x}}, \bar{\mathbf{x}}, \cdots, \bar{\mathbf{x}}) \prec (x_1, x_2, \cdots, x_n) \prec (x_1 + x_2 + \cdots + x_n, 0, \cdots, 0). \qquad (4.7)$$

Next, if $D \subset \mathbb{R}^n$ and $f : D \to \mathbb{R}$, we say that f is *Schur convex (concave)* on D if

$$f(\mathbf{x}) \le (\ge) f(\mathbf{y}) \qquad \text{for all } \mathbf{x}, \mathbf{y} \in D \text{ and } \mathbf{x} \prec \mathbf{y}.$$

Schur provided the following criterion to check whether a function is Schur convex or concave:

Schur's Criterion: Let $f : (a, b)^n \to \mathbb{R}$ be continuously differentiable and symmetric. Then $f(\mathbf{x})$ is Schur convex (concave) on $(a, b)^n$ if and only if

$$(x_j - x_k) \left(\frac{\partial f}{\partial x_j} - \frac{\partial f}{\partial x_k} \right) \ge (\le) 0$$

for all $1 \le j < k \le n$ and all $\mathbf{x} \in (a, b)^n$.

In particular, for $x_j > 0$ for $1 \le j \le n$, we see that $f(\mathbf{x}) = x_1 x_2 \cdots x_n$ is Schur concave because

$$(x_j - x_k) \left(\frac{\partial f}{\partial x_j} - \frac{\partial f}{\partial x_k} \right) = -(x_j - x_k)^2 \frac{f(\mathbf{x})}{x_j x_k} \le 0.$$

In view of (4.7), the Schur concavity of f yields

$$f(\bar{\mathbf{x}}, \bar{\mathbf{x}}, \cdots, \bar{\mathbf{x}}) = \bar{\mathbf{x}}^n = \left(\frac{x_1 + x_2 + \cdots + x_n}{n} \right)^n \ge f(\mathbf{x}) = x_1 x_2 \cdots x_n.$$

This is the well-known AM-GM inequality, which is a consequence of Jensen's inequality.

In general, Schur provided the following powerful theorem that shows that almost every invocation of Jensen's inequality can be replaced by the Schur convexity.

Schur's Majorization Inequality. Let $\phi : (a, b) \to \mathbb{R}$ be a convex function. Then the function $f : (a, b)^n \to \mathbb{R}$ defined by

$$f(x_1, x_2, \cdots, x_n) = \sum_{j=1}^{n} \phi(x_j)$$

is Schur convex. Thus, for $\mathbf{x}, \mathbf{y} \in (a, b)^n$ with $\mathbf{x} \prec \mathbf{y}$, we have

$$\sum_{j=1}^{n} \phi(x_j) \le \sum_{j=1}^{n} \phi(y_j). \qquad (4.8)$$

We now present a few concrete examples to illustrate how majorization and Schur convexity are used in practice. First, let f be convex on $(0, \infty)$ and $x_j \in (0, \infty)$. Applying (4.8) to (4.7) yields

$$nf\left(\frac{x_1 + x_2 + \cdots + x_n}{n}\right) \le \sum_{j=1}^{n} f(x_j) \le f(x_1 + x_2 + \cdots + x_n) + (n-1)f(0).$$

The left-hand side inequality indicates that Jensen's inequality is a special case of (4.8). Next, we consider

Problem 11139 (Proposed by G. Bennett, 112(3), 2005). Show that if $p > 1$ or $p < 0$, then

$$\frac{1^p}{3^p} < \frac{1^p + 3^p}{5^p + 7^p} < \frac{1^p + 3^p + 5^p}{7^p + 9^p + 11^p} < \cdots < \frac{1^p + 3^p + \cdots + (2n-1)^p}{(2n+1)^p + \cdots + (4n-1)^p} < \cdots .$$

$$(4.9)$$

Here we see that the first inequality in (4.9), rephrased as

$$5^p + 7^p \le 3^p + 9^p,$$

follows at once from the facts that x^p is convex on $(0, \infty)$ for $p > 1$ or $p < 0$ and $(7, 5) \prec (9, 3)$. The second inequality, after some simplifications, becomes

$$9^p + 11^p + 27^p + 33^p \le 5^p + 15^p + 25^p + 35^p,$$

which follows from the majorization

$$(33, 27, 11, 9) \prec (35, 25, 15, 5).$$

Along the same lines, we can establish the entire (4.9).

Finally, we return to Knuth's inequality (4.6). Rewrite (4.6) as

$$(e^t)^\alpha + (e^{-t})^\alpha + 2^\alpha \le (e^t + e^{-t})^\alpha + 1^\alpha + 1^\alpha.$$

Naturally, we have $\phi(x) = x^\alpha$, which is convex for $x > 0$ and $\alpha > 1$, and

$$\{x_1, x_2, x_3\} = \{e^t, e^{-t}, 2\}, \quad \{y_1, y_2, y_3\} = \{e^t + e^{-t}, 1, 1\}. \qquad (4.10)$$

Unfortunately, here $\mathbf{x} \prec \mathbf{y}$ is not valid. Failure occurs at **M2** for $k = 2$. Hardy, Littlewood, and Pólya [54] proved that (4.8) is valid for all convex function $\phi : (0, \infty) \to \mathbb{R}$ if and only if hypotheses **M1–M3** hold. Thus, it is reasonable to consider new forms of majorization by replacing the hypotheses **M1–M3**, wherein the inequality (4.8) is assumed to hold for classes of functions, other than for all convex ones. Recently, Bennett [11] characterized the p-power functions by replacing **M1–M3** with

B1 $x_1 + x_2 + x_3 = y_1 + y_2 + y_3,$

B2 $x_1^2 + x_2^2 + x_3^2 = y_1^2 + y_2^2 + y_3^2,$

B3 $\max\{x_1, x_2, x_3\} \leq \max\{y_1, y_2, y_3\}$,

and proved that (4.8) is valid for $\phi(x) = x^p$ ($p \geq 2$ or $0 \leq p \leq 1$) if and only if the above hypotheses **B1–B3** hold. This establishes (4.6) immediately because (4.10) satisfy the hypotheses **B1–B3**. It is interesting to see that his proof for sufficiency is inventoried in his following earlier proposed problem.

Monthly Problem 11397 (115(10), 2008). Let a, b, c, x, y, z be positive numbers such that $a + b + c = x + y + z$ and $abc = xyz$. Show that if $\max\{x, y, z\} \geq \max\{a, b, c\}$, then $\min\{x, y, z\} \geq \min\{a, b, c\}$.

Without the regular majorization, this fact ensures no loss of generality in assuming that $x > a \geq b > y \geq z > c$ in his proof.

We end this section with a few problems for additional practice.

1. Let $\alpha \geq 1$.

   (a) If $x \in (0, \infty)$, show that $(1 + x)^{\alpha} \geq x^{\alpha} + 2^{\alpha} - 1$.

   (b) If $a, b \in (0, \infty)$, show that $a^{\alpha} + b^{\alpha} \leq (a + b)^{\alpha} - (2^{\alpha} - 2) \min(a^{\alpha}, b^{\alpha})$.

2. **SIAM Problem 68-1, A Network Inequality** (Proposed by J. C. Turner and V. Conway). If $p + q = 1, 0 < p < 1$, and $m, n > 1$ are positive integers, give a direct analytic proof of the inequality

$$(1 - p^m)^n + (1 - q^n)^m > 1.$$

3. (Due to G. Bennett) Let a, b, c, x, y, z be positive numbers. Then the inequality

$$a^p + b^p + c^p \leq x^p + y^p + z^p$$

   is valid whenever $p \geq 1$, and it reverses direction whenever $p < 1$, if and only if the following three conditions are satisfied:

   (a) $a + b + c = x + y + z$,

   (b) $abc = xyz$,

   (c) $\max\{a, b, c\} \leq \max\{x, y, z\}$.

4. (**Ratio Principle**) Let a_n and b_n both be positive sequences. Show that if the sequence a_n/b_n is increasing then same goes for the ratio of the sum:

$$\frac{a_1 + a_2 + \cdots + a_n}{b_1 + b_2 + \cdots + b_n}.$$

5. **Problem 11139 Revisited.** The inequality (4.9), tipped upside down and with 1 added, can be rewritten as

$$\frac{1^p + 3^p}{1^p} \geq \frac{1^p + 3^p + 5^p + 7^p}{1^p + 3^p} \geq \frac{1^p + 3^p + 5^p + 7^p + 9^p + 11^p}{1^p + 3^p + 5^p} \geq \cdots.$$

$$(4.11)$$

Using the majorizations shows

$$\frac{1^p + 3^p}{1^p} \geq \frac{5^p + 7^p}{3^p} \geq \frac{9^p + 11^p}{5^p} \geq \cdots .$$

Use this to prove (4.11).

6. **Problem 12013** (Proposed by D. Stoner, 124(10), 2017). Suppose that a, b, c, d, e, and f are nonnegative real numbers that satisfy $a + b + c = d + e + f$. Let t be a real number greater than 1. Prove that at least one of the inequalities

$$a^t + b^t + c^t > d^t + e^t + f^t,$$

$$(ab)^t + (bc)^t + (ca)^t > (de)^t + (ef)^t + (fd)^t, \text{ and}$$

$$(abc)^t > (def)^t$$

is false.

7. **Problem 12065** (Proposed by H. Lee, 125(8), 2018). Let n be a positive integer, and let $x_1, \ldots, x_n$ be a list of n positive real numbers. For $k \in \{1, \ldots, n\}$, let $y_k = x_k(n+1)/(n+1-k)$ and let

$$z_k = \frac{(k!)^{1/k}}{k+1} \left(\prod_{j=1}^{k} y_j \right)^{1/k}.$$

Prove that the arithmetic mean of $x_1, \ldots, x_n$ is greater than or equal to the arithmetic mean of $z_1, \ldots, z_n$, and determine when equality holds.

4.3 Tight bounds for the normal distribution

Problem 10611 (Proposed by Z. Sasvári, 104(5), 1997). Find the largest value of a and the smallest value of b for which the inequality

$$\frac{1 + \sqrt{1 - e^{-ax^2}}}{2} < \Phi(x) < \frac{1 + \sqrt{1 - e^{-bx^2}}}{2}$$

holds for all $x > 0$, where

$$\Phi(x) = \frac{1}{\sqrt{2\pi}} \int_{-\infty}^{x} e^{-y^2/2} \, dy.$$

Discussion.
Since

$$\int_{-\infty}^{0} e^{-y^2/2}\, dy = \int_{0}^{\infty} e^{-y^2/2}\, dy = \sqrt{\frac{\pi}{2}},$$

the stated inequalities are equivalent to

$$\frac{\sqrt{1 - e^{-ax^2}}}{2} < f(x) < \frac{\sqrt{1 - e^{-bx^2}}}{2}, \tag{4.12}$$

where

$$f(x) = \frac{1}{\sqrt{2\pi}} \int_{0}^{x} e^{-y^2/2}\, dy.$$

Solution.
First, we show that $a = 1/2$ and $b = 2/\pi$ are the best possible constants for which (4.12) holds. Indeed, if the right-hand inequality of (4.12) holds for all $x > 0$, then

$$0 < \frac{\sqrt{1 - e^{-bx^2}}}{2} - f(x) = \left(\frac{\sqrt{b}}{2} - \frac{1}{\sqrt{2\pi}} \right) x + O(x^3)$$

as $x \to 0$, which implies that $b \geq 2/\pi$. Similarly, notice that, for large x,

$$
\begin{aligned}
f(x) &= \frac{1}{\sqrt{2\pi}} \left(\int_{0}^{\infty} e^{-y^2/2}\, dy - \int_{x}^{\infty} e^{-y^2/2}\, dy \right) \\
&= \frac{1}{\sqrt{2\pi}} \left(\int_{0}^{\infty} e^{-y^2/2}\, dy + \int_{x}^{\infty} \frac{1}{y} d\left(e^{-y^2/2} \right) \right) \\
&= \frac{1}{2} - \frac{1}{\sqrt{2\pi}\, x} e^{-x^2/2} + O\left(\frac{e^{-x^2/2}}{x^3} \right).
\end{aligned}
$$

If the left-hand inequality of (4.12) holds for all $x > 0$, then

$$0 < f(x) - \frac{\sqrt{1 - e^{-ax^2}}}{2} = \frac{1}{4} e^{-ax^2} + O(e^{-2ax^2}) - \frac{1}{\sqrt{2\pi}\, x} e^{-x^2/2} + O\left(\frac{e^{-x^2/2}}{x^3} \right)$$

as $x \to \infty$. Dividing each side by $e^{x^2/2}$ yields $a \leq 1/2$.

Next, we show that (4.12) hold for all $x > 0$ when $a = 1/2$ and $b = 2/\pi$. To this end, we transform the single integral into a double integral:

$$f^2(x) = \frac{1}{2\pi} \int_{0}^{x} \int_{0}^{x} e^{-(u^2 + v^2)/2}\, du\, dv.$$

Let $D = [0, x]^2$, $D_1 = \{(u, v) : 0 \leq u, 0 \leq v, u^2 + v^2 \leq x^2\}$, and $D_2 = \{(u, v) : 0 \leq u, 0 \leq v, u^2 + v^2 \leq (4/\pi)x^2\}$. Then we have the inequalities

$$\frac{1}{2\pi} \iint_{D_1} e^{-(u^2 + v^2)/2}\, du\, dv < \frac{1}{2\pi} \iint_{D} e^{-(u^2 + v^2)/2}\, du\, dv$$

$$\leq \frac{1}{2\pi} \iint_{D_2} e^{-(u^2 + v^2)/2}\, du\, dv,$$

where the left-hand inequality holds because $D_1 \subset D$; the right-hand because D and D_2 have the same area and

$$e^{-(u^2+v^2)/2} \leq e^{-(2/\pi)x^2}, \qquad \text{for } (u,v) \in D - D_2;$$

$$e^{-(u^2+v^2)/2} \geq e^{-(2/\pi)x^2}, \qquad \text{for } (u,v) \in D_2 - D.$$

Evaluating the above outer integrals via polar coordinates, we obtain

$$\frac{1 - e^{-x^2/2}}{2} < f^2(x) < \frac{1 - e^{-2x^2/\pi}}{2},$$

which is equivalent to (4.12). □

Remark. Recall the *error function* defined by

$$\mathrm{erf}(x) = \frac{2}{\sqrt{\pi}} \int_0^x e^{-y^2}\, dy.$$

We have

$$f(x) = \frac{1}{\sqrt{2\pi}} \int_0^x e^{-y^2/2}\, dy = \frac{1}{\sqrt{\pi}} \int_0^{x/\sqrt{2}} e^{-t^2}\, dt = \frac{1}{2}\mathrm{erf}(x/\sqrt{2}),$$

and are able to obtain the more detailed asymptotic series of f via

$$
\begin{aligned}
\mathrm{erf}(x) &= \frac{2}{\sqrt{\pi}} e^{-x^2} \left(x + \frac{2x^3}{1 \cdot 3} + \frac{4x^5}{1 \cdot 3 \cdot 5} + \cdots \right) \\
&= \frac{2}{\sqrt{\pi}} \left(x + \frac{1}{3}x^3 + \frac{1}{10}x^5 - \frac{1}{42}x^7 + \cdots \right), \qquad \text{as } x \to 0; \\
\mathrm{erf}(x) &= 1 - \frac{1}{\sqrt{\pi}} e^{-x^2} \left(\frac{1}{x} - \frac{1}{2x^3} + \frac{3}{4x^5} - \frac{15}{8x^7} + \cdots \right), \qquad \text{as } x \to \infty.
\end{aligned}
$$

An interesting survey on inequalities involving the *complementary error function*

$$\mathrm{erfc}(x) = \frac{2}{\sqrt{\pi}} \int_x^\infty e^{-y^2}\, dy = 1 - \mathrm{erf}(x)$$

is given in Mitrinovic's [67]. In particular, one can find several inequalities for *Mills' ratio*

$$R(x) := e^{x^2/2} \int_x^\infty e^{-y^2/2}\, dy. \tag{4.13}$$

Motivated by the proposed inequality and Mills's ratio, we consider a more general expression

$$F(x) = \frac{1}{\Gamma(1 + 1/p)} \int_0^x e^{-y^p}\, dy, \qquad (\text{for } p > 0, p \neq 1).$$

Here the coefficient $\Gamma(1+1/p)$ comes from the substitution $t = y^p$ and

$$\int_0^\infty e^{-y^p}\, dy = \frac{1}{p}\int_0^\infty t^{1/p-1}e^{-t}\, dt = \frac{1}{p}\Gamma(1/p) = \Gamma(1+1/p).$$

We now collect some inequalities related to Mills's ratio and $F(x)$, and leave the proofs to the reader.

1. **(Komatu's inequality)** For $x > 0$ and $R(x)$ defined by (4.13), show that
$$\frac{2}{\sqrt{x^2+4}+x} < R(x) < \frac{2}{\sqrt{x^2+2}+x}.$$
   Also show that the upper bound can be replaced by $4/(\sqrt{x^2+8}+3x)$.

2. For $x > 0$, can one find the best possible constants a and b such that
$$\frac{2}{\sqrt{x^2+a}+x} < R(x) < \frac{2}{\sqrt{x^2+b}+x}?$$

3. Show that if $x > \sqrt{(\sqrt{5}-1)/2}$, then $R(x) < \frac{1}{\sqrt{1+x^2}}$; and $R(x) > \frac{1+x^2}{x(2+x^2)}$ if $x > \sqrt{2}$. This indicates that
$$R(x) \simeq \frac{1}{\sqrt{1+x^2}}, \qquad \text{as } x \to \infty.$$

4. (Approximation of Mills' ratio by rational functions, due to O. Kouba). Define two sequences $P_n(x), Q_n(x)$ as
$$(P_0(x), P_1(x)) = (1, x), \quad P_{n+1}(x) = xP_n(x) + nP_{n-1}(x);$$
$$(Q_0(x), Q_1(x)) = (0, 1), \quad Q_{n+1}(x) = xQ_n(x) + nQ_{n-1}(x).$$

   Show that

   (a) For all $n \geq 0, P_n(x) = e^{-x^2/2}(e^{x^2/2})^{(n)}$, where $f^{(n)}$ indicates the nth derivative of f.

   (b) For all $n \geq 0$,
$$P_n(x) = \frac{1}{\sqrt{2\pi}}\int_{-\infty}^\infty t^n \exp\left(-\frac{(t-x)^2}{2}\right)\, dt.$$

   (c) For all $n \geq 0, x > 0$,
$$\frac{Q_{2n}(x)}{P_{2n}(x)} < R(x) < \frac{Q_{2n+1}(x)}{P_{2n+1}(x)}.$$

   Comment. Let $H_n(x)$ be the nth *Hermite polynomial*. Then $P_n(x) = \frac{-i}{\sqrt{2}}H_n(ix/\sqrt{2})$ with $i = \sqrt{-1}$.

5. (An extension of (4.12), due to H. Alzer [5]). If $p > 1$, for all $x > 0$, prove that

$$\left(1 - e^{-ax^p}\right)^{1/p} < F(x) < \left(1 - e^{-bx^p}\right)^{1/p},$$

where $a = 1, b = [\Gamma(1+1/p)]^{-p}$. If $0 < p < 1$, the above inequalities hold with the values of a and b switched.

Comment. Alzer's proof involves a very clever analysis of the function

$$I(p, x) = \int_0^x e^{-y^p} \, dy - \Gamma(1+1/p)[1 - e^{-x^p}]^{1/p}.$$

While reading through his proof, one may find that the monotone arguments acquire new layers of meaning.

6. Show that

$$\ln\left(\frac{1}{1 - e^{-ax}}\right) \le \int_x^\infty \frac{e^{-y}}{y} \, dy \le \ln\left(\frac{1}{1 - e^{-bx}}\right)$$

are valid for all $x > 0$ if and only if $a \ge e^\gamma$ and $0 < b \le 1$, where γ is Euler's constant.

7. **Problem 11219** (Proposed by R. A. Strubel, 113(4), 2006). Prove that when n is a positive integer and s is a real number greater than 1

$$1 + n(\zeta(s) - 1) < \sum_{k=0}^\infty \left(\frac{n}{n+k}\right)^s \le n\zeta(s).$$

Hint: Use

$$\left(\frac{n}{n+k}\right)^s = \frac{1}{\Gamma(s)} \int_0^\infty t^{s-1} e^{-t} \exp(-kt/n) \, dt.$$

4.4 An inequality due to Knopp

Problem 11145 (Proposed by J. Zinn, 112(4), 2005). Find the least c such that if $n \ge 1$ and $a_1, \ldots, a_n > 0$ then

$$\sum_{k=1}^n \frac{k}{\sum_{j=1}^k 1/a_j} \le c \sum_{k=1}^n a_k. \tag{4.14}$$

Discussion.

We notice that the summand in the left-hand side of (4.14) is the harmonic mean of $a_1, \ldots, a_k$. Here (4.14) indicates that the sum of the harmonic means is bounded by the sum of the original sequence. Naturally, we apply the AM-HM inequality to the summands on the left-hand side of (4.14) and find that

$$\sum_{k=1}^{n} \frac{k}{\sum_{j=1}^{k} 1/a_j} \leq \sum_{k=1}^{n} \frac{1}{k} \sum_{j=1}^{k} a_j = \sum_{j=1}^{n} a_j \sum_{k=j}^{n} \frac{1}{k}.$$

Here we fall short of our goal because the upper bound diverges. To overcome this failure, we rewrite the Cauchy-Schwarz inequality as

$$(b_1 + b_2 + \cdots + b_k)^2 \leq \left(\sum_{j=1}^{k} \frac{1}{a_j} \right) \left(\sum_{j=1}^{k} b_j^2 a_j \right), \qquad (4.15)$$

and deduce that

$$\sum_{k=1}^{n} \frac{k}{\sum_{j=1}^{k} 1/a_j} \leq \sum_{k=1}^{n} \left(\frac{k}{\left(\sum_{j=1}^{k} b_j \right)^2} \left(\sum_{j=1}^{k} b_j^2 a_j \right) \right)$$

$$= \sum_{j=1}^{n} a_j b_j^2 \left(\sum_{k=j}^{n} \frac{k}{\left(\sum_{j=1}^{k} b_j \right)^2} \right),$$

where b_j, for $j = 1, 2, \ldots, k$, can be viewed as *parameters*. Thus, to prove (4.14), we just need to choose $b_j, j = 1, 2, \ldots, n$ such that

$$b_j^2 \left(\sum_{k=j}^{n} \frac{k}{\left(\sum_{j=1}^{k} b_j \right)^2} \right)$$

is bounded.

Next, we test the bound c by letting $a_k = 1/k$. When we substitute this sequence into (4.14), we see that it implies

$$\sum_{k=1}^{n} \frac{2}{k+1} \leq c \sum_{k=1}^{n} \frac{1}{k} \qquad \text{for all } n \geq 1.$$

Since the harmonic series diverges, for the bound in (4.14) to hold in general, we must have $c \geq 2$.

Solution.
We show that $c = 2$ is the best possible bound in (4.14). Repeating our foregoing calculation in the discussion, applying (4.15) with $b_k = k$, we have

$$(1 + 2 + \cdots + k)^2 \leq \left(\frac{1}{a_1} + \frac{1}{a_2} + \cdots + \frac{1}{a_k} \right) (1^2 a_1 + 2^2 a_2 + \cdots + k^2 a_k).$$

Since $1 + 2 + \cdots k = k(k+1)/2$, the above inequality is equivalent to

$$\frac{k}{\frac{1}{a_1} + \frac{1}{a_2} + \cdots + \frac{1}{a_k}} \leq \frac{4k}{k^2(k+1)^2} \sum_{j=1}^{k} j^2 a_j.$$

Summing over k gives

$$\sum_{k=1}^{n} \frac{k}{\frac{1}{a_1} + \frac{1}{a_2} + \cdots + \frac{1}{a_k}} \leq \sum_{k=1}^{n} \frac{4k}{k^2(k+1)^2} \sum_{j=1}^{k} j^2 a_j = 2 \sum_{j=1}^{n} j^2 a_j \sum_{k=j}^{n} \frac{2k}{k^2(k+1)^2}.$$

Using partial fractions,

$$\sum_{k=j}^{n} \frac{2k+1}{k^2(k+1)^2} = \sum_{k=j}^{n} \left(\frac{1}{k^2} - \frac{1}{(k+1)^2} \right) = \frac{1}{j^2} - \frac{1}{(n+1)^2},$$

so we find that

$$\sum_{k=1}^{n} \frac{k}{\sum_{i=1}^{k} \frac{1}{a_i}} \leq 2 \sum_{j=1}^{n} j^2 a_j \left(\frac{1}{j^2} - \frac{1}{(n+1)^2} - \sum_{k=j}^{n} \frac{1}{k^2(k+1)^2} \right) \leq 2 \sum_{j=1}^{n} a_j.$$

This completes the proof of (4.14). $\qquad\qquad\qquad\qquad\qquad\qquad\square$

Remark. We see that (4.15) has played a key role in the above proof. Indeed, the flexibility of the parameters in (4.15) turns out to have a remarkable number of variations. Here, we single out one particularly charming case. With $a_j = j^2$ for $1 \leq j \leq n$, we get

$$(b_1 + b_2 + \cdots + b_n)^2 \leq \left(\sum_{j=1}^{n} \frac{1}{j^2} \right) \left(\sum_{j=1}^{n} j^2 b_j^2 \right) \leq \frac{\pi^2}{6} \left(\sum_{j=1}^{n} j^2 b_j^2 \right).$$

Furthermore, with $a_j = t + j^2/t$ for $1 \leq j \leq n$, let

$$S := \sum_{j=1}^{n} b_j^2, \text{ and } T := \sum_{j=1}^{n} j^2 b_j^2.$$

(4.15) leads to

$$(b_1 + b_2 + \cdots + b_n)^2 \leq C_n \left(tS + \frac{T}{t} \right),$$

where

$$C_n = \sum_{j=1}^{n} \frac{1}{t + j^2/t} = \sum_{j=1}^{n} \frac{t}{t^2 + j^2}.$$

Since $t/(t^2 + x^2)$ is decreasing in $x \in (0, \infty)$ for all $t > 0$, invoking the right-end Riemann sum yields

$$C_n \leq \int_0^n \frac{t}{t^2 + x^2} \, dx \leq \int_0^\infty \frac{t}{t^2 + x^2} \, dx = \frac{\pi}{2}.$$

Thus,
$$(b_1 + b_2 + \cdots + b_n)^2 \le \frac{\pi}{2}\left(tS + \frac{T}{t}\right), \quad \text{for all } t > 0.$$

In particular, setting $t = \sqrt{T/S}$ yields
$$(b_1 + b_2 + \cdots + b_n)^2 \le \pi\sqrt{ST},$$

or

$$(b_1 + b_2 + \cdots + b_n)^4 \le \pi^2(b_1^2 + b_2^2 + \cdots + b_n^2)(b_1^2 + 2^2 b_2^2 + \cdots + n b_n^2). \quad (4.16)$$

This is known as *Carlson's inequality*.

Based on the Monthly Editorial comment, the inequality (4.14) with $c = 2$ is actually a special case of the following inequality: For $p > 0$,

$$\sum_{k=1}^{\infty}\left(\frac{k}{\sum_{j=1}^{k} 1/a_j}\right)^p \le \left(\frac{p+1}{p}\right)^p \sum_{k=1}^{\infty} a_k^p.$$

This inequality was first established by Knopp [59]. With $c = 4$, (4.14) appeared as the **Putnam Problem 1964-A5**.

It is interesting to see the historic evolution of several famous mean inequalities during the early twentieth century. The tale begins with David Hilbert in 1906. In his research on integral equations, Hilbert was led to determine the convergence of some double series. For this purpose, he established *Hilbert's inequality*. If $\sum_{m=1}^{\infty} a_m^2 < \infty$ and $\sum_{n=1}^{\infty} b_n^2 < \infty$, then

$$\sum_{n=1}^{\infty}\sum_{m=1}^{\infty} \frac{a_m b_n}{m+n} \le C\left(\sum_{m=1}^{\infty} a_m^2\right)^{1/2}\left(\sum_{n=1}^{\infty} b_n^2\right)^{1/2} \quad (4.17)$$

with the bound $C = 2\pi$. Hilbert's original proof was based on trigonometric integrals including the following remarkable formula:

$$\sum_{n=1}^{N}\sum_{m=1}^{N}\left(\frac{1}{n+m} + \frac{1}{n-m}\right) a_m b_n$$
$$= \frac{1}{2\pi}\int_{-\pi}^{\pi} t\left(\sum_{k=1}^{N}(-1)^k(a_k \sin kt - b_k \cos kt)\right)^2 dt. \quad (4.18)$$

Here, when $n = m, 1/(n-m)$ is treated to be zero. Five years later, Schur provided a new proof that exploited the theory of analytic functions, and showed that (4.17), as well as its integral analogue, actually holds with the best possible bound $C = \pi$. But none of these proofs satisfied with Hardy's desire: Simple and elementary. He aimed to derive (4.17) from the Cauchy-Schwarz inequality only. Indeed, Hardy (1915) proved that the convergence of any of the three series

$$(1) \sum_{n=1}^{\infty} \frac{a_n A_n}{n} \quad (2) \sum_{n=1}^{\infty}\left(\frac{A_n}{n}\right)^2 \quad (3) \sum_{n=1}^{\infty}\sum_{m=1}^{\infty} \frac{a_m a_n}{m+n}$$

implies that of the others, where $A_n = \sum_{k=1}^{n} a_k$. In 1920, he finally established *Hardy's inequality*. If $a_1, a_2, \ldots$ is a sequence of nonnegative real numbers, then for $p > 1$,

$$\sum_{k=1}^{\infty} \left(\frac{\sum_{j=1}^{k} a_j}{k} \right)^p \leq \left(\frac{p}{p-1} \right)^p \sum_{k=1}^{\infty} a_k^p. \tag{4.19}$$

Two years later, Carleman revealed the geometric mean case:

Carleman's inequality. If $a_1, a_2, \ldots$ is a sequence of positive real numbers, then

$$\sum_{k=1}^{\infty} \sqrt[k]{a_1 a_2 \cdots a_k} \leq e \sum_{k=1}^{\infty} a_k. \tag{4.20}$$

The proofs of (4.17), (4.19), and (4.20) perfectly illustrate how to introduce parameters to squeeze the bounds, and have served as a benchmark for new ideas and methods. For example, Pólya's elegant proof of (4.20) used little more than the AM-GM inequality, but demonstrated where to use it most effectively. These inequalities have been generalized and applied in analysis and differential equations. For the reader interested in learning more about these inequalities, please refer to Michael Steele's fascinating book [88]. With the Cauchy-Schwarz inequality as the initial guide, this lively, problem-oriented book will coach one toward mastery of more classical inequalities, including those of Hölder, and (4.14)–(4.20).

We now end this section with some problems for additional practice.

1. Let $a_k \geq 0$ for $1 \leq k \leq n$. Show that

$$\sqrt[n]{a_1 a_2 \cdots a_n} \leq \frac{2}{n(n-1)} \sum_{1 \leq i < j < \leq n} \sqrt{a_i a_j}.$$

2. (Due to R. M. Redheffer) Let $a_1, a_2, \ldots$ be a positive real number sequence. Show that

$$\frac{(k+1)^2}{1/a_1 + 1/a_2 + \cdots + 1/a_k} + \sum_{j=1}^{k-1} \frac{2j+1}{1/a_1 + 1/a_2 + \cdots + 1/a_j} \leq 4 \sum_{j=1}^{k} a_j. \tag{4.21}$$

Use (4.21) to prove (4.14).

3. Let $i = \sqrt{-1}$. Show that

$$\sum_{n=1}^{N} \sum_{m=1}^{N} \frac{a_m b_n}{m+n} = \frac{i}{2\pi} \int_0^{2\pi} (t - \pi) \left(\sum_{n=1}^{N} a_n e^{int} \right) \left(\sum_{m=1}^{N} b_m e^{imt} \right) dt.$$

Use this identity to prove the finite version of (4.17).

4. Let a_n and b_n be real vectors. Show that

$$\left(\sum_{i=1}^{n} (a_i + b_i) \right) \left(\sum_{i=1}^{n} \frac{a_i b_i}{a_i + b_i} \right) \leq \left(\sum_{i=1}^{n} a_i \right) \left(\sum_{i=1}^{n} b_i \right).$$

5. Let a_n and b_n be real sequences. Show that

$$\sum_{i=1}^{\infty}\sum_{j=1}^{\infty} \frac{a_i b_j}{\max(i,j)} \le 4 \left(\sum_{i=1}^{\infty} a_i^2\right)^{1/2} \left(\sum_{j=1}^{\infty} b_j^2\right)^{1/2},$$

where the 4 can not be replaced by a smaller constant.

6. Let $A_k = \sum_{j=1}^{k} a_j$. Show that

$$\sum_{k=1}^{n} \left(\frac{A_k}{k}\right)^2 - 2\sum_{k=1}^{n} \frac{a_k A_k}{k} = -\frac{A_n^2}{n} - \sum_{k=2}^{n} (k-1)\left(\frac{A_k}{k} - \frac{A_{k-1}}{k-1}\right)^2.$$

Use this identity to derive (4.19) with $p = 2$.

Hint: Show that, for $k \ge 2$,

$$\frac{2a_k A_k}{k} - \left(\frac{A_k}{k}\right)^2 = \frac{A_k^2}{k} - \frac{A_{k-1}^2}{k-1} + (k-1)\left(\frac{A_k}{k} - \frac{A_{k-1}}{k-1}\right)^2.$$

7. Use $(1/x)\int_0^x f(t)\,dt = \int_0^1 f(tx)\,dt$ and Minkowski's integral inequality to prove the integral version of (4.19): If $p > 1$ and $f : [0,\infty) \to [0,\infty)$ with $\int_0^{\infty} f^p\,dx < \infty$, then

$$\int_0^{\infty} \left(\frac{1}{x}\int_0^x f(t)\,dt\right)^p dx \le \left(\frac{p}{p-1}\right)^p \int_0^{\infty} f^p(x)\,dx.$$

8. (*Weighted Hardy's inequality*). Suppose that $a_k \ge 0$ and $\alpha_k > 0$ for all $k \ge 1$. If $\sum_{k=1}^{\infty} \alpha_k a_k^p < \infty$, prove that

$$\sum_{k=1}^{\infty} \alpha_k \left(\frac{\sum_{j=1}^{k} \alpha_j a_j}{\sum_{j=1}^{k} \alpha_j}\right) \le \left(\frac{p}{p-1}\right)^p \sum_{k=1}^{\infty} \alpha_k a_k^p.$$

Comment. Replacing a_k by $a_k^{1/p}$ and letting $p \to \infty$ in the above inequality, since

$$\lim_{p \to \infty} \left(\frac{\sum_{j=1}^{k} \alpha_j a_j^{1/p}}{\sum_{j=1}^{k} \alpha_j}\right)^p = (a_1^{\alpha_1} a_2^{\alpha_2} \cdots a_k^{\alpha_k})^{1/\sum_{j=1}^{k} \alpha_j},$$

this yields the *Weighted Carleman's inequality:* If $\sum_{k=1}^{\infty} \alpha_k a_k < \infty$, then

$$\sum_{k=1}^{\infty} \alpha_k (a_1^{\alpha_1} a_2^{\alpha_2} \cdots a_k^{\alpha_k})^{1/\sum_{j=1}^{k} \alpha_j} \le e \sum_{k=1}^{\infty} \alpha_k a_k.$$

9. Let $a_1, a_2, \ldots$ be a positive real number sequence. Show that

$$\sqrt[k]{a_1 a_2 \cdots a_k} \le \frac{e}{2k^2} \sum_{j=1}^{k} (2j-1)a_j.$$

Use this term-wise bound to prove (4.20).

10. (A refinement of Carleman's inequality, due to Chen [31]) Let $a_1, a_2, \ldots$ be a positive real number sequence. Show that

$$\sum_{k=1}^{\infty} \sqrt[k]{a_1 a_2 \cdots a_k} \le e \sum_{k=1}^{\infty} \left(1 - \sum_{j=1}^{n} \frac{b_j}{(1+k)^j}\right) a_k,$$

where n is any positive integer and b_j is given by

$$b_1 = \frac{1}{2}, \quad b_{n+1} = \frac{1}{(n+1)(n+2)} - \frac{1}{n+1} \sum_{i=1}^{n} \frac{b_i}{n-i+2}.$$

11. **Problem 6663** (Proposed by W. Janous, 98(6), 1991). Show that

$$\sum_{j=1}^{N} \left(\frac{1 + x + x^2 + \cdots + x^{j-1}}{j}\right)^2 < (4\ln 2)(1 + x^2 + x^4 + \cdots + x^{2N-2})$$

for $0 < x < 1$ and all positive integers N; also show that the constant $4 \ln 2$ is the least possible.

Comment. Notice that if we drop the factor $\ln 2$ in the bound, the stated inequality is a direct application of (4.19) with $p = 2$. Here the bound 4 in (4.19) is the best possible in general, since $\ln 2 = 0.693\ldots$, this problem provides us a special case in which the bound 4 can be improved.

12. **Problem 11202** (Proposed by G. Bennett, 113(2), 2006). Prove that if a_n is a sequence of positive numbers with $\sum_{n=1}^{\infty} a_n < \infty$, then for all $p \in (0, 1)$

$$\lim_{n \to \infty} n^{1 - 1/p} (a_1^p + a_2^p + \cdots + a_n^p)^{1/p} = 0.$$

13. **Problem 12004** (Proposed by M. Omarjee, 124(8), 2017). Let $a_1, a_2, \ldots$ be a strictly increasing sequence of real numbers satisfying $a_n \le n^2 \ln n$ for all $n \ge 1$. Prove that the series $\sum_{n=1}^{\infty} 1/(a_{n+1} - a_n)$ diverges.

14. (**Independent Study**) Replacing a_j by $a_j^{1/p}$ and p by $1/p$ in (4.19), we get

$$\sum_{k=1}^{\infty} \left(\frac{\sum_{j=1}^{k} a_j^p}{k}\right)^{1/p} \le \left(\frac{1}{1-p}\right)^{1/p} \sum_{k=1}^{\infty} a_k$$

for $0 < p < 1$. The summand on the left-hand side is often called pth power mean or Hölder mean of $a_1, a_2, \ldots, a_k$. Thus, this inequality, Hardy's inequality, and Carleman's inequality are unified as

$$\sum_{k=1}^{\infty} M(a_1, a_2, \ldots, a_k) \le C \sum_{k=1}^{\infty} a_k, \qquad (4.22)$$

where M is a mean and C is some finite positive constant. Can you offer a characterization of mean M (for example, in the class of symmetric, increasing, Jensen concave) and also a formula for the best possible constant C satisfying (4.22)?

4.5 A discrete inequality by integral

Problem 11680 (Proposed by B. Bogosel and C. Lupu, 119(10), 2012). Let $x_1, \ldots, x_n$ be nonnegative real numbers. Show that

$$\left(\sum_{i=1}^{n} \frac{x_i}{i} \right)^4 \le 2\pi^2 \sum_{i,j=1}^{n} \frac{x_i x_j}{i+j} \sum_{i,j=1}^{n} \frac{x_i x_j}{(i+j)^3}.$$

Discussion.
Inspired by Hilbert's identity (4.18), we represent the terms in the proposed inequality as integrals and expect that they can be estimated efficiently. To illustrate how this approach is used in practice, we consider **Problem E1682** (Proposed by V. R. Rao, 71(3), 1964), which asked to prove the following reverse Hilbert-type inequality:

$$\left(\sum_{i=1}^{n} \frac{a_i}{i} \right)^2 \le \sum_{i,j=1}^{n} \frac{a_i a_j}{i+j-1}.$$

Rewrite $a_i/i = \int_0^1 a_i x^{i-1}\, dx$. The Cauchy-Schwarz inequality enables us to prove this inequality in one line:

$$\left(\sum_{i=1}^{n} \frac{a_i}{i} \right)^2 = \left(\int_0^1 \sum_{i=1}^{n} a_i x^{i-1}\, dx \right)^2 \le \int_0^1 \left(\sum_{i=1}^{n} a_i x^{i-1} \right)^2 dx = \sum_{i,j=1}^{n} \frac{a_i a_j}{i+j-1}.$$

To apply this approach to the proposed inequality, we need to choose a function f that represents the three sums in the inequality as integrals. The search for a suitable function of f can take various paths, but (4.18) offers the insight and suggests that $f(x) = \sum_{i=1}^{n} x_i e^{-ix}$. Finally, we need to estimate the corresponding integral inequality.

Solution.

Following the foregoing discussion, let $f(x) = \sum_{i=1}^{n} x_i e^{-ix}$. We compute

$$\int_0^\infty f(x)\,dx = \sum_{i=1}^n x_i \int_0^\infty e^{-ix}\,dx = \sum_{i=1}^n \frac{x_i}{i},$$

$$\int_0^\infty f^2(x)\,dx = \sum_{i,j=1}^n x_i x_j \int_0^\infty e^{-(i+j)x}\,dx = \sum_{i,j=1}^n \frac{x_i x_j}{i+j},$$

$$\int_0^\infty x^2 f^2(x)\,dx = \sum_{i,j=1}^n x_i x_j \int_0^\infty x^2 e^{-(i+j)x}\,dx = 2\sum_{i,j=1}^n \frac{x_i x_j}{(i+j)^3}.$$

Apply the integral version of Carlson's inequality: If $f \geq 0$ on $[0,\infty)$ such that $f, xf \in L^2([0,\infty))$, then

$$\left(\int_0^\infty f(x)\,dx\right)^4 \leq \pi^2 \int_0^\infty f^2(x)\,dx \int_0^\infty x^2 f^2(x)\,dx,$$

which matches the proposed inequality exactly. $\qquad\square$

Remark. The above solution offered a powerful example: Once a quantity is represented as an integral, it may be reshaped into alternative forms, and so it can be estimated more efficiently. We elaborate on this idea via two more examples.

Example 1. We want to find a simple bound for the nth derivative of $\sin x / x$. Observe that

$$\frac{d^4}{dx^4}\left(\frac{\sin x}{x}\right) = \frac{\sin x}{x} + \frac{2\cos x}{x^2} - \frac{12\sin x}{x^3} - \frac{24\cos x}{x^4} + \frac{25\sin x}{x^5}.$$

It is hard to see that this expression will be bounded by $1/5$. However, using

$$\frac{\sin x}{x} = \int_0^1 \cos(xt)\,dt,$$

for all $n \in \mathbb{N}$, we have

$$\left|\frac{d^n}{dx^n}\left(\frac{\sin x}{x}\right)\right| \leq \int_0^1 \left|t^n \cos\left(xt + \frac{n\pi}{2}\right)\right|\,dt \leq \int_0^1 t^n\,dt = \frac{1}{n+1}.$$

Example 2. Problem 11769 (Proposed by P. P. Dályay, 121(4), 2014). Let a_i, b_i $(1 \leq i \leq n)$ be positive real numbers. Show that

$$\left(\sum_{j=1}^n \frac{a_i}{b_j}\right)^2 - 2\sum_{j,k=1}^n \frac{a_j a_k}{(b_j + b_k)^2} \leq 2\sqrt{2}\left(\sum_{j,k=1}^n \frac{a_j a_k}{(b_j + b_k)} \sum_{j,k=1}^n \frac{a_j a_k}{(b_j + b_k)^3}\right)^{1/2}.$$

In the original statement of this problem, it was missing the $\sqrt{2}$. Here we quote a brilliant solution due to Roberto Tauraso (`http://www.mat.uniroma2.it/~tauraso/AMM/AMM11769.pdf`), who used the same idea of proving Carlson's equality (4.16). Define $f : (0, \infty) \to (0, \infty)$ by

$$f(x) = \sum_{j=1}^{n} a_j e^{-b_j x}.$$

Then, for $m \in \{0, 1, 2\}$,

$$C := \int_0^\infty f(x)\, dx = \sum_{j=1}^{n} \frac{a_i}{b_j},$$

$$D_m := \frac{1}{m!} \int_0^\infty x^m f^2(x)\, dx = \frac{1}{m!} \sum_{j,k=1}^{n} \frac{a_j a_k}{(b_j + b_k)^{m+1}}.$$

By the Cauchy-Schwarz inequality, for all $t > 0$

$$\begin{aligned}
\left(\int_0^\infty f(x)\, dx \right)^2 &= \left(\int_0^\infty \frac{1}{x+t}(x+t)f(x)\, dx \right)^2 \\
&\leq \left(\int_0^\infty \frac{dx}{(x+t)^2} \right) \left(\int_0^\infty (x+t)^2 f^2(x)\, dx \right) \\
&= t \int_0^\infty f^2(x)\, dx + 2 \int_0^\infty x f^2(x)\, dx + \frac{1}{t} \int_0^\infty x^2 f^2(x)\, dx.
\end{aligned}$$

It follows that for any $t > 0$,

$$C^2 - 2D_1 \leq t D_0 + \frac{2}{t} D_2.$$

In particular, letting $t = (2D_2/D_0)^{1/2}$, which gives the minimum value over all t of $tD_0 + 2D_2/t$, yields

$$C^2 - 2D_1 \leq 2\sqrt{2}(D_0 D_2)^{1/2},$$

which is equivalent to the proposed inequality.   □

We now end this section with a few problems for additional practice.

1. Let n be a positive integer. Use induction to prove that, for $x > 0$,

$$\frac{d^n}{dx^n} \cos(\sqrt{x}) = \frac{(-1)^n}{2^{2n}(n-1)!} \int_0^1 (1-t)^{n-1} t^{-1/2} \cos(\sqrt{xt})\, dt.$$

    Then show that

$$\left| \frac{d^n}{dx^n} \cos(\sqrt{x}) \right| \leq \frac{n!}{(2n)!}.$$

2. **Problem 11746** (Proposed by P. P. Dalyay, 120(10), 2013). Let f be a continuous function from $[0, \infty)$ to $\mathbb{R}$ such that the following integrals converge: $S = \int_0^\infty f^2(x)\, dx$, $T = \int_0^\infty x^2 f^2(x)\, dx$, and $U =$

$\int_0^\infty x^4 f^2(x)\,dx$. Let $V = T + \sqrt{T^2 + 3SU}$. Given that f is not identically 0, show that

$$\left(\int_0^\infty |f(x)|\,dx \right)^4 \leq \frac{\pi^2 S(T+V)^2}{9V}.$$

3. **Problem 11819** (Proposed by C. Lupu, 122(2), 2015). Let f be a continuous, nonnegative function on $[0,1]$. Show that

$$\int_0^1 f^3(x)\,dx \geq 4 \left(\int_0^1 x^2 f(x)\,dx \right) \left(\int_0^1 x f^2(x)\,dx \right).$$

Comment. This problem can be generalized in the following manner: Let f, g be continuous and nonnegative functions on $[0,1]$. If $\alpha, \beta \geq 0$, then

$$\left(\int_0^1 f^{\alpha+\beta}(x)\,dx \right) \left(\int_0^1 g^{\alpha+\beta}(x)\,dx \right)$$
$$\geq \left(\int_0^1 f^\alpha(x) g^\beta(x)\,dx \right) \left(\int_0^1 f^\beta(x) g^\alpha(x)\,dx \right).$$

4. **Problem 11840** (Proposed by G. Stoica, 122(5), 2015). Let $z_1, z_2, \ldots, z_n$ be complex numbers. Prove that

$$\left(\sum_{k=1}^n |z_k| \right)^2 - \left| \sum_{k=1}^n z_k \right|^2 \geq \left(\sum_{k=1}^n |\mathrm{Re} z_k| - \left| \sum_{k=1}^n \mathrm{Re} z_k \right| \right)^2.$$

5. **Problem 11925** (Proposed by L. Giugiuc, 123(7), 2016). Let n be an integer with $n \geq 4$. Find the largest k such that for any list a of n real numbers that sum to 0,

$$\left(\sum_{j=1}^n a_j^2 \right)^3 \geq k \left(\sum_{j=1}^n a_j^3 \right)^2.$$

4.6 An inequality by power series

Problem 11989 (Proposed by S. P. Andriopoulos, 124(6), 2016). Let x be a number between 0 and 1. Prove

$$\prod_{n=1}^\infty (1 - x^n) \geq \exp\left(\frac{1}{2} - \frac{1}{2(1-x)^2} \right).$$

Discussion.
Taking the logarithm of each side of the proposed inequality yields

$$\sum_{n=1}^\infty \ln(1 - x^n) \geq \frac{1}{2} - \frac{1}{2(1-x)^2}.$$

As usual, we let

$$f(x) = \sum_{n=1}^{\infty} \ln(1 - x^n) - \frac{1}{2} + \frac{1}{2(1 - x)^2}.$$

Since $f(0) = 0$, the proposed inequality can be proved by showing that

$$f'(x) = -\sum_{n=1}^{\infty} \frac{nx^{n-1}}{1 - x^n} + \frac{1}{(1 - x)^3} \geq 0 \quad \text{for all } x \in (0, 1).$$

Unfortunately, the preceding expression of $f'(x)$ makes it hard to determine whether or not $f'(x) \geq 0$. Now we will carry the analysis through the power series: If $f(x) = \sum_{n=1}^{\infty} a_n x^n$ on $(-1, 1)$ with $a_n \geq 0$ for all n, then $f(x) \geq 0$ for all $x \in (0, 1)$. It is interesting to see that an *arithmetic function* $\sigma(n) = \sum_{d|n} d$, which denotes the sum of all positive divisors of n, comes to the rescue.

Solution.
For $0 < x < 1$, since $1/(1 - x) = \sum_{n=0}^{\infty} x^n$, differentiating gives

$$\frac{1}{(1 - x)^2} = \sum_{n=1}^{\infty} nx^{n-1} = \sum_{n=0}^{\infty} (n + 1)x^n.$$

Invoking $\sum_{k=1}^{n} k = n(n + 1)/2$ yields

$$-\frac{1}{2} + \frac{1}{2(1 - x)^2} = \sum_{n=1}^{\infty} \frac{n + 1}{2} x^n = \sum_{n=1}^{\infty} \frac{x^n}{n} \sum_{k=1}^{n} k.$$

On the other hand, since $\sigma(n) = \sum_{d|n} d$, we have

$$\sum_{n=1}^{\infty} \ln(1 - x^n) = -\sum_{n=1}^{\infty} \sum_{k=1}^{\infty} \frac{(x^n)^k}{k} = -\sum_{n=1}^{\infty} \sum_{k=1}^{\infty} \frac{nx^{nk}}{nk}$$

$$= -\sum_{m=1}^{\infty} x^m \sum_{k|m} \frac{1}{k} = -\sum_{m=1}^{\infty} \frac{\sigma(m)}{m} x^m,$$

where the interchange of the order of summations is justified by the positivity of all summands. Since $\sum_{k=1}^{n} k \geq \sigma(n)$, it follows that

$$\sum_{n=1}^{\infty} \ln(1 - x^n) - \frac{1}{2} + \frac{1}{2(1 - x)^2} = \sum_{n=1}^{\infty} \frac{1}{n} \left(\sum_{k=1}^{n} k - \sigma(n) \right) x^n \geq 0$$

for all $0 < x < 1$. This proves the proposed inequality as desired. □

Remark. Along the same lines, we can establish a similar reversed inequality:

$$\prod_{n=1}^{\infty} (1 - x^n) \leq \exp\left(\frac{1}{2} - \frac{1}{2(1 - x^2)} \right).$$

Indeed, for $0 < x < 1$, this immediately follows from

$$-\sum_{n=1}^{\infty} \ln(1 - x^n) = \sum_{n=1}^{\infty}\sum_{k=1}^{\infty} \frac{(x^n)^k}{k} = \sum_{k=1}^{\infty} \frac{1}{k} \sum_{n=1}^{\infty} x^{nk}$$

$$= \sum_{k=1}^{\infty} \frac{x^k}{k(1 - x^k)} \geq \frac{x^2}{2(1 - x^2)}$$

$$= -\frac{1}{2}\left(1 - \frac{1}{1 - x^2}\right).$$

The procedure used in the preceding solution is very useful. The next two examples will recapitulate and elaborate on this idea.

Example 1. (Nesbitt's inequality) For $a, b, c > 0$, Show that

$$\frac{a}{b + c} + \frac{b}{c + a} + \frac{c}{a + b} \geq \frac{3}{2}.$$

Because the left-hand side of the stated inequality is homogenous, without loss of generality, we assume that $a + b + c = 1$. So it suffices to prove that, for $a, b, c \in (0, 1)$,

$$\frac{a}{1 - a} + \frac{b}{1 - b} + \frac{c}{1 - c} \geq \frac{3}{2}.$$

Using geometric series and Jensen's inequality, we conclude that

$$\frac{a}{1 - a} + \frac{b}{1 - b} + \frac{c}{1 - c} = \sum_{n=1}^{\infty} (a^n + b^n + c^n)$$

$$= 3 \cdot \sum_{n=1}^{\infty} \frac{a^n + b^n + c^n}{3}$$

$$\geq 3 \cdot \sum_{n=1}^{\infty} \left(\frac{a + b + c}{3}\right)^n$$

$$= 3 \cdot \sum_{n=1}^{\infty} \left(\frac{1}{3}\right)^3 = \frac{3}{2}.$$

Example 2. (Wilker's inequality revisited, see (4.5) and Problem 7 in Section 4.1) Here we present a power series proof of the following improved Wilker's inequality: If $x \in (0, \pi/2)$, then

$$\frac{16}{\pi^4} x^3 \tan x < \left(\frac{\sin x}{x}\right)^2 + \frac{\tan x}{x} - 2 < \frac{8}{45} x^3 \tan x, \qquad (4.23)$$

where both $16/\pi^4$ and $8/45$ are the best possible constants in (4.23).

Let

$$F(x) := \frac{\left(\frac{\sin x}{x}\right)^2 + \frac{\tan x}{x} - 2}{x^3 \tan x} = \frac{\sin 2x}{2x^5} + \frac{1}{x^4} - \frac{2\cot x}{x^3}.$$

Notice that

$$F(0) = \lim_{x \to 0^+} F(x) = \frac{8}{45} \quad \text{and} \quad F(\pi/2) = 16/\pi^4.$$

Thus, to prove (4.23), it suffices to show that $F(x)$ is strictly decreasing on $(0, \pi/2)$. But

$$F'(x) = -\frac{5 \sin 2x}{2x^6} + \frac{\cos 2x}{x^5} - \frac{4}{x^5} + \frac{6 \cos x}{x^4 \sin x} + \frac{2}{x^3 \sin^2 x},$$

which does not render $F'(x) < 0$ immediately clear. Instead we consider the power series of $F(x)$. Recall that

$$\sin 2x = 2x - \frac{4}{3}x^3 + \sum_{n=0}^{\infty} \frac{(-1)^n 2^{2n+5}}{(2n+5)!} x^{2n+5};$$

$$\cot x = \frac{1}{x} - \frac{1}{3}x - \sum_{n=0}^{\infty} \frac{2^{2n+4} B_{n+2}}{(2n+4)!} x^{2n+3},$$

where B_n is the nth Bernoulli number. We obtain

$$\begin{aligned} F(x) &= \sum_{n=0}^{\infty} \frac{2^{2n+4}}{(2n+5)!} \left[2(2n+5)B_{n+2} + (-1)^n \right] x^{2n} \\ &= \frac{8}{45} - \frac{8}{945}x^2 + \frac{16}{14175}x^4 + \frac{8}{467775}x^6 + \cdots. \end{aligned}$$

Invoking that

$$\zeta(2n) = \frac{2^{2n-1} \cdot \pi^{2n}}{(2n)!} B_n,$$

by induction, for $n > 2$, we find that

$$2(2n+5)B_{n+2} = \frac{4 \cdot (2n+5)!}{(2\pi)^{2n+4}} \zeta(2n+4) > \frac{4 \cdot (2n+5)!}{(2\pi)^{2n+4}} > 1.$$

Let

$$G(x) := F(\sqrt{x}) = \frac{8}{45} - \frac{8}{945}x + \frac{16}{14175}x^2 + \frac{8}{467775}x^3 + \cdots.$$

Then $G''(x) \geq 0$, and so $G'(x)$ is increasing on $(0, \pi^2/4)$. Since $G'(x) = F'(\sqrt{x})/(2\sqrt{x})$ and

$$G'(\pi^2/4) = \frac{1}{\pi} F'(\pi/2) = \frac{16}{\pi^4} \left(1 - \frac{10}{\pi^2} \right) < 0,$$

it follows that $G'(x) < 0$ on $(0, \pi^2/4)$, and so $F'(x) < 0$ on $(0, \pi/2)$. Thus, for $x \in (0, \pi/2)$,

$$\frac{8}{45} = F(0) > F(x) > F(\pi/2) = \frac{16}{\pi^4},$$

which proves (4.23) as desired.

Now we compile a few problems for additional practice.

1. For $a, b \in (0,1)$ and $p, q > 0$ with $1/p + 1/q = 1$, show that

$$\frac{q}{1-a^p} + \frac{p}{1-b^q} \geq \frac{pq}{1-ab} \quad \text{and} \quad \frac{a^p}{p(1-a^p)^2} + \frac{b^p}{q(1-b^q)^2} \geq \frac{ab}{(1-ab)^2}.$$

2. **Problem 11692** (Proposed by C. Lupu and S. Spataru, 120(2), 2013). Let a_1, a_2, a_3, a_4 be real number in $(0,1)$ with $a_4 = a_1$. Show that

$$\frac{3}{1-a_1 a_2 a_3} + \sum_{k=1}^{3} \frac{1}{1-a_k^3} \geq \sum_{k=1}^{3} \left(\frac{1}{1-a_k^2 a_{k+1}} + \frac{1}{1-a_k a_{k+1}^2} \right).$$

3. **CMJ Problem 1119** (Proposed by Spiros P. Andriopoulos, 49(1), 2018). For any number x with $0 < x < 1$, prove that

$$\sum_{n=1}^{\infty} \frac{x^n}{1+x+x^2+\cdots+x^n} < \ln\left(\frac{1}{1-x}\right).$$

4. **Problem 11504** (Proposed by F. Holland, 117(5), 2010). Let N be a positive integer and x a positive real number. Prove that

$$\sum_{m=0}^{N} \frac{1}{m!} \left(\sum_{k=1}^{N-m+1} \frac{x^k}{k} \right)^m \geq 1 + x + \cdots + x^N.$$

5. *Refinement of Shafer-Fink's inequality.* For $0 \leq x \leq 1$, show that

$$\frac{1}{180} x^5 + \frac{1}{189} x^7 \leq \sin^{-1} x - \frac{3x}{2+\sqrt{1-x^2}} \leq \frac{\pi-3}{2}.$$

6. *Hyperbolic analogue of Wilker's inequality.* For $x > 0$, show that

$$\left(\frac{\sinh x}{x} \right)^2 + \frac{\tanh x}{x} - 2 > \frac{8}{45} x^3 \tanh x.$$

7. *A Wilker-type inequality.* For $x \in (0, \pi/2)$, show that

$$\frac{2}{45} x^3 \sin x < \left(\frac{x}{\sin x} \right)^2 + \frac{x}{\tan x} - 2 < \left(\frac{2}{\pi} - \frac{16}{\pi^3} \right) x^3 \sin x, \quad (4.24)$$

where both $2/45$ and $2/\pi - 16/\pi^3$ are the best possible constants in (4.24).

8. Let the power series $f(x) = \sum_{n=0}^{\infty} a_n x^n$ and $g(x) = \sum_{n=0}^{\infty} b_n x^n$ converge for $|x| < R$. If $b_n > 0$ and a_n/b_n is strictly increasing for all $n = 0, 1, \ldots$, show that $f(x)/g(x)$ is also strictly increasing on $(0, R)$.

9. Let $n \in \mathbb{N}$. Assume that f is a real function defined on (a, b) such that $f^{(k)}(a^+), f^{(k)}(b^-)$ exist for $k = 0, 1, \ldots, n$. If $f^{(n)}(x)$ is increasing on (a, b), prove that

$$\sum_{k=0}^{n-1} \frac{f^{(k)}(a^a)}{k!}(x-a)^k + \frac{1}{(b-a)^n}\left(f(b^-) - \sum_{k=0}^{n-1} \frac{(b-a)^k f^{(k)}(a^+)}{k!}\right)(x-a)^n$$

$$> f(x) > \sum_{k=0}^{n} \frac{f^{(k)}(a^+)}{k!}(x-a)^k.$$

Comment. Applying this to $F(x)$ in Example 2 above yields a refinement of (4.23):

$$\left(\frac{8}{45} - \frac{8}{945}x^2\right) x^3 \tan x < \left(\frac{\sin x}{x}\right)^2 + \frac{\tan x}{x} - 2$$

$$< \left(\frac{8}{45} - \frac{8}{945}x^2 + \frac{\alpha}{14174}\right) x^3 \tan x,$$

where $\alpha = (480\pi^6 - 40320\pi^4 + 3628800)/\pi^8 = 17.15041\ldots$.

10. Let $x \in (0, \pi/2)$ and $p \geq 1$. Show that

$$\left(\frac{\sin x}{x}\right)^{2p} + \left(\frac{\tan x}{x}\right)^p > \left(\frac{x}{\sin x}\right)^{2p} + \left(\frac{x}{\tan x}\right)^p > 2.$$

What is the smallest p such that this inequality holds?

4.7   An inequality by differential equation

Problem 11923 (Proposed by O. Kouba, 123(7), 2016). Let f_p be the function on $(0, \pi/2)$ given by

$$f_p(x) = (1 + \sin x)^p - (1 - \sin x)^p - 2\sin(px).$$

Prove $f_p > 0$ for $0 < p < 1/2$ and $f_p < 0$ for $1/2 < p < 1$.

Discussion.
Since $f_p(0) = 0$, we naturally proceed to show that $f_p'(x) > 0$ for $0 < p < 1/2$ and $f_p'(x) < 0$ for $1/2 < p < 1$ on $(0, \pi/2)$. However, notice that

$$f_p'(x) = p(1 + \sin x)^{p-1} \cos x + p(1 - \sin x)^{p-1} \cos x - 2p\cos(px).$$

We see that the analysis of positivity and negativity of $f_p'(x)$ is even harder than $f_p(x)$ itself, although the following graph offers an affirmative answer.

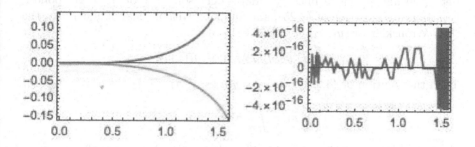

FIGURE 4.1
The graph of $f_p(x)$ with $p = 1/3, 2/3$, and $1/2$, respectively

The graphs with different p-values are shown in Figure 4.1. We see that $f_p(x)$ changes sign when $p = 1/2$.

The published solution by Mumbai brought to mind Hadamard's famous dictum ([52], p. 123): "The shortest and best way between two truths of the real domain often passes through the imaginary one." Using the binomial series and de Moivre's formula, Mumbai established

$$f_p(x) = 4\cos^{2p}(x/2) \sum_{k=0}^{\infty} \binom{2p}{4k+3} \tan^{4k+1}(x/2).$$

Since

$$a_k := \binom{2p}{4k+3} = \frac{1}{(4k+3)!} (2p) \underbrace{(2p-1)(2p-2)\cdots(2p-4k-2)}_{4k-2 \text{ terms}},$$

the desired inequality follows from the fact that $a_k > 0$ if $0 < 2p < 1$ and $a_k < 0$ if $1 < 2p < 2$.[1]

Here we present a real-variable proof via the method of variation of constants in differential equations. Consider the second-order inhomogeneous ordinary differential equation

$$\mathcal{L}[y] = y'' + p(x)y' + q(x)y = g(x), \qquad x \in (a,b). \tag{4.25}$$

---

[1] Several solutions in this book have illustrated how complex-variables can be used to provide quick proofs of a wide variety of problems, especially for the series and integration. Lax and Zalcman's beautiful slim book [64] offered us an extended meditation on Hadamard's dictum. Their discussion begins with the shortest proof of the *fundamental theorem of algebra* and concludes with Newman's proof of the *prime number theorem*.

Let $y_1(x)$ and $y_2(x)$ be linearly independent solutions of the corresponding homogeneous equation of (4.25), and suppose, without loss of generality, that their Wronskian $W(y_1, y_2) = 1$. Define

$$K(x, t) := y_2(x)y_1(t) - y_1(x)y_2(t).$$

Then the solution of (4.25) satisfying $y(a) = y'(a) = 0$ is given by

$$y(x) = \int_a^x K(x, t)g(t)\, dt.$$

In the following solution we will show that $f_p(x)$ displays such an integral representation.

Solution.
We show that indeed $f_p, f_p' > 0$ for $0 < p < 1/2$ and $f_p, f_p' < 0$ for $1/2 < p < 1$. First, we prove that $f_p(x)$ satisfies a second-order ordinary differential equation with constant coefficients. To this end, we compute

$$\begin{aligned}
f_p'(x) &= p(1 + \sin x)^{p-1} \cos x + p(1 - \sin x)^{p-1} \cos x - 2p\cos(px),\\
f_p''(x) &= p(p-1)(1 + \sin x)^{p-2} \cos^2 x - p(1 + \sin x)^{p-1} \sin x\\
&\quad + p(p-1)(1 - \sin x)^{p-2} \cos^2 x - p(1 - \sin x)^{p-1} \sin x + 2p^2 \sin(px)\\
&= -p^2(1 + \sin x)^p + p(2p - 1)(1 + \sin x)^{p-1}\\
&\quad + p^2(1 - \sin x)^p - p(2p - 1)(1 - \sin x)^{p-1} + 2p^2 \sin(px),
\end{aligned}$$

where we have used the identity: $\cos^2 x = 2(1 \pm \sin x) - (1 \pm \sin x)^2$. Thus, we find that

$$f_p''(x) + p^2 f_p(x) = g(x), \tag{4.26}$$

where

$$g(x) = p(2p - 1)\left[(1 + \sin x)^{p-1} - (1 - \sin x)^{p-1}\right].$$

For $x \in (0, \pi/2)$ and $0 < p < 1$, x^{p-1} is a decreasing function because its derivative is negative. This implies that

$$(1 + \sin x)^{p-1} - (1 - \sin x)^{p-1} < 0,$$

and so $g(x) > 0$ if $0 < p < 1/2$ and $g(x) < 0$ if $1/2 < p < 1$.

It is well-known that $y'' + p^2 y = 0$ has linearly independent solutions $\cos(px)$ and $\sin(px)$. Let

$$y_1(x) = \frac{1}{\sqrt{p}} \cos(px), \quad y_2(x) = \frac{1}{\sqrt{p}} \sin(px).$$

Then $W(y_1, y_2) = 1$ and

$$\begin{aligned}
K(x, t) &= y_2(x)y_1(t) - y_1(x)y_2(t) = \frac{1}{p}(\sin(px)\cos(pt) - \cos(px)\sin(pt))\\
&= \frac{1}{p}\sin(p(x - t)),
\end{aligned}$$

which is positive for $0 < t < x < \pi/2$ and $0 < p < 1$. In view of $f_p(0) = f'_p(0) = 0$, we find the unique solution of (4.26) as

$$f_p(x) = \int_0^x K(x,t)g(t)\,dt \qquad (0 < x < \pi/2). \tag{4.27}$$

Now the proposed inequality follows from the positivity and negativity properties of g.

From (4.27) and $K(x,x) = 0$ we have

$$f'_p(x) = \int_0^x \frac{\partial}{\partial x}K(x,t)g(t)\,dt = \int_0^x \cos(p(x-t))g(t)\,dt.$$

Since $\cos(p(x - t)) > 0$ for $0 < t < x < \pi/2$ and $0 < p < 1$, from the property of $g(x)$, we conclude that $f'_p(x) > 0$ if $0 < p < 1/2$ and $f'_p(x) < 0$ if $1/2 < p < 1$ as well. $\qquad\square$

Remark. Another interesting approach for the analysis of $f_p(x)$ is to apply the following *Maximum Principle*: Let $f : [a, b] \to \mathbb{R}$ be a function satisfying

$$f''(x) + p(x)f'(x) \geq 0, \qquad \text{for all } x \in (a,b),$$

where $p(x)$ is bounded on $[a,b]$. Then $f(x)$ achieves its maximum and minimum at the end points.

In this case, we have $f_p(0) = 0, f_p(\pi/2) = 2^p - 2\sin(p\pi/2)$. Since $f_{1/2}(\pi/2) = f_1(\pi/2) = 0$ and

$$\frac{d}{dp}f_p(\pi/2) = 2^p \ln 2 - \pi \cos(p\pi/2), \quad \frac{d^2}{dp^2}f_p(\pi/2) = 2^p(\ln 2)^2 + \frac{\pi^2}{2}\sin(p\pi/2) > 0$$

we conclude that

$$\min_{x\in[0,\pi/2]}\{f_p(x)\} = 0 \quad \text{for } 0 < p < 1/2; \qquad \max_{x\in[0,\pi/2]}\{f_p(x)\} = 0 \quad \text{for } 1/2 < p < 1.$$

See Figure 4.2. Now the proposed problem follows from the maximum principle.

The above solution illustrates the power of the maximum principle in the qualitative analysis of some differential equations or differential inequalities. For more details and examples on this subject, we refer the reader to Protter and Weinberger's excellent book [76].

We now conclude this section with a few problems for additional practice.

1. **SIAM Problem 86-17** (Proposed by C. L. Frenzen). The function

$$f(t) = \left(\frac{1 - e^{-t}}{t}\right)^{-1/2}$$

is analytic for $|t| < 2\pi$ and has a Taylor series of the form $f(t) = \sum_{n=0}^{\infty} b_n t^n$. Prove that $b_0 = 1, (-1)^{n+1}b_{2n} > 0, (-1)^n b_{2n+1} > 0$ for all $n \in \mathbb{N}$.
Hint: Let $t = 4xi$. Then $f(t) = (x \cot x)^{1/2} + ix(x \cot x)^{-1/2}$.

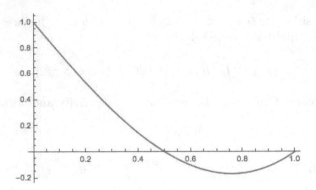

FIGURE 4.2
The graph of $f_p(\pi/2)$ for $0 < p < 1$

2. **Problem 11024** (Proposed by V. Rădulescu, 110(5), 2003). Consider a continuous function
$g : (0, \infty) \to (0, \infty)$ such that for some $\alpha > 0$,

$$\lim_{x \to \infty} \frac{g(x)}{x^{1+\alpha}} = \infty.$$

Let $f : \mathbb{R} \to (0, \infty)$ be a twice-differentiable function for which there exist $a > 0$ and $x_0 \in \mathbb{R}$ such that for all $x \geq x_0$,

$$f''(x) + f'(x) > ag(f(x)).$$

Prove that $\lim_{x \to \infty} f(x)$ exists and is finite, and evaluate the limit.
Comment: It is interesting to see what happens when the superlinear growth assumption on g is replaced by linear growth.

3. **Problem 11137** (Proposed by V. Rădulescu, 112(2), 2005). Let ϕ be a continuous positive function on the open interval (A, ∞), and assume that f is a C^2-function on (A, ∞) satisfying the differential equation
$$f''(t) = (1 + \phi(t))(f^2(t) - 1)f(t).$$

(a) Given that there exists $a \in (A, \infty)$ such that $f(a) \geq 1$ and $f'(a) \geq 0$, prove that there is a positive constant K such that $f(x) \geq Ke^x$ whenever $x \geq a$.

(b) Given instead that there exists $a \in (A, \infty)$ such that $f'(a) < 0$ and $f(x) \geq 1$ for $x > a$, prove that there is a positive constant K such that $f(x) \geq Ke^x$ for $x \geq a$.

(c) Given that f is bounded on (A, ∞) and that there exists $\alpha > 0$ such that $\phi(x) = O(e^{-(1+\alpha)x})$, prove that $\lim_{x \to \infty} e^x f(x)$ exists and is finite.

Comment: The nonlinearity $f(f^2 - 1)$ goes back to the *Ginzburg-Landau theory* arising in superconductivity.

4. **Problem 11135** (Proposed by R. Redheffer, 112(2), 2005). Let A and b be continuous nondecreasing functions from $[0, \infty)$ into $[0, \infty)$. Let $\lambda, \mu \in [0, 1]$ with $\mu\lambda < 1$. Let

$$\rho = 1/(1 - \mu\lambda) \quad \text{and} \quad B(t) = \int_0^t b(s)ds.$$

Show that if $v : [0, \infty) \to \mathbb{R}$ is continuous, $v(0) \leq 0$, and

$$v(t) \leq tA(t) + \int_0^t b(s)v(s)\, ds + \lambda v(\mu t)$$

on $[0, \infty)$, then for $t \geq 0$,

$$v(t) \leq \rho t A(t) e^{\rho B(t)}.$$

5. Let p be a continuous real-valued function on $\mathbb{R}$ and let f satisfy the differential equation

$$f''(x) + p(x)f'(x) - f(x) = 0.$$

If f has more than one zero, show that $f(x) \equiv 0$.

4.8   Bounds for a reciprocal of log sum

Problem 11847 (Proposed by M. Bencze, 122(6), 2015). Prove that for $n \geq 1$,

$$\frac{n(n + 1)(n + 2)}{3} < \sum_{k=1}^n \frac{1}{\ln^2(1 + 1/k)} < \frac{n}{4} + \frac{n(n + 1)(n + 2)}{3}.$$

Discussion.
Since we cannot find the exact sum that appears in the proposed inequality, it is natural to estimate the sum by applying the well-known inequality

$$\frac{x}{1 + x} < \ln(1 + x) < x, \qquad \text{for } x > 0. \tag{4.28}$$

This leads us to

$$\sum_{k=1}^n k^2 < \sum_{k=1}^n \frac{1}{\ln^2(1 + 1/k)} < \sum_{k=1}^n (k + 1)^2.$$

Using $\sum_{k=1}^{n} k^2 = n(n+1)(2n+1)/6$, we find

$$\frac{n(n+1)(2n+1)}{6} < \sum_{k=1}^{n} \frac{1}{\ln^2(1+1/k)} < \frac{(n+1)(n+2)(2n+3)}{6} - 1.$$

But

$$\frac{n(n+1)(2n+1)}{6} < \frac{n(n+1)(n+2)}{3} \quad \text{and}$$

$$\frac{(n+1)(n+2)(2n+3)}{6} - 1 > \frac{n}{4} + \frac{n(n+1)(n+2)}{3},$$

so (4.28) is too vague to achieve the desired bounds.

Notice that the proposed inequality depends on a positive integer n, hence induction may lead to a proof. More than that, working backward, we find an insight to tighten the loose bounds in (4.28) from the proof. Indeed, the inductive step would require

$$\frac{n(n+1)(n+2)}{3} + \frac{1}{\ln^2(1+1/(k+1))} > \frac{(n+1)(n+2)(n+3)}{3}.$$

This is equivalent to

$$\ln\left(1 + \frac{1}{k+1}\right) < \frac{1}{\sqrt{(k+1)(k+2)}} = \frac{\frac{1}{k+1}}{\sqrt{1 + \frac{1}{k+1}}},$$

so it suffices to prove that for $x > 0$,

$$\ln(1+x) < \frac{x}{\sqrt{1+x}}.$$

Similarly, to prove the right inequality we need

$$\frac{x}{1+x/2} < \ln(1+x).$$

Thus, the sharpened inequality

$$\frac{x}{1+x/2} < \ln(1+x) < \frac{x}{\sqrt{1+x}} \tag{4.29}$$

will lead us quickly to the proof of the proposed inequality.

Solution.

To prove (4.29), let

$$f(x) = \ln(1+x) - \frac{x}{1+x/2}.$$

For $x > 0$, we have

$$f'(x) = \frac{1}{1+x} - \frac{4}{(2+x)^2} = \frac{x^2}{(1+x)(2+x)^2} > 0.$$

Thus $f(x)$ is strictly increasing on $(0, \infty)$ and so $f(x) > f(0) = 0$. This proves the left-hand side inequality of (4.29). Similarly, let

$$g(x) = \frac{x}{\sqrt{1+x}} - \ln(1+x).$$

For $x > 0$, we have

$$g'(x) = \frac{x+2}{2(1+x)\sqrt{1+x}} - \frac{1}{1+x} = \frac{x+2-2\sqrt{1+x}}{2(1+x)\sqrt{1+x}}.$$

Thus $g'(x) > 0$ as long as $x + 2 - 2\sqrt{1+x} > 0$. This can be asserted by

$$\left((x+2) - 2\sqrt{1+x}\right)' = 1 - \frac{1}{\sqrt{1+x}} > 0, \quad (x > 0).$$

Therefore, $g(x)$ is also strictly increasing on $(0, \infty)$ and so $g(x) > g(0) = 0$, which proves the right-hand side inequality of (4.29).

Now we square (4.29), and then take the reciprocal to get

$$\frac{1+x}{x^2} < \frac{1}{\ln^2(1+x)} < \frac{(x+2)^2}{4x^2}.$$

Taking $x = 1/k$ yields

$$k(k+1) < \frac{1}{\ln^2(1+1/k)} < \frac{(2k+1)^2}{4} = k(k+1) + \frac{1}{4}.$$

Summing over k gives

$$\sum_{k=1}^{n} k(k+1) < \sum_{k=1}^{n} \frac{1}{\ln^2(1+1/k)} < \sum_{k=1}^{n} \left(k(k+1) + \frac{1}{4}\right).$$

This yields the desired inequality since $\sum_{k=1}^{n} k(k+1) = n(n+1)(n+2)/3$. $\square$

Remark. There is a much shorter proof of (4.29): By the elementary inequality $2\sqrt{1+t} < t+2$ for $t > 0$, we have

$$\frac{4}{(t+2)^2} < \frac{1}{1+t} < \frac{t+2}{2\sqrt{1+t}(1+t)}.$$

Now (4.29) follows from integrating the above inequality from 0 to x.

Taking a careful look at the search for optimal bounds for $\ln(1+x)$, we see that the process evolves in the following basic pattern: We start with a familiar inequality, but the consequences disappoint us; then we appeal to either deeper mathematical tools or a new idea to reframe the problem so that the applied inequality is at its best. This proposed problem provided us an excellent instructive example.[2]

---

[2]Chapter 7 in Chen's book [29] should serve another example of using this kind of methodology.

We end this section by selecting some problems for additional practice.

1. Determine the best possible constants α and β such that inequalities

$$\frac{e}{2n+\alpha} \le e - \left(1 + \frac{1}{n}\right)^n \le \frac{e}{2n+\beta}$$

hold for every $n \ge 1$.

2. **Problem E3432** (Proposed by L. Toth, 98(3), 1991). Prove that for every positive integer n we have

$$\frac{1}{2n+2/5} < H_n - \ln n - \gamma < \frac{1}{2n+1/3}.$$

Show that $2/5$ can be replaced by a slightly smaller number, but $1/3$ cannot be replaced by a slightly larger number.
Comment. The best possible constant to replace $2/5$ is $1/(1-\gamma)-2$.

3. **Problem 11954** (Proposed by P. Bracken, 124(1), 2017). Determine the largest constant c and the smallest constant d such that, for all positive integers n

$$\frac{1}{n-c} \le \sum_{k=n}^{\infty} \frac{1}{k^2} \le \frac{1}{n-d}.$$

4. Let $I_n = \sum_{k=1}^{n-1} \csc(k\pi/n)$. Show that

$$I_n = \frac{2n}{\pi}(\ln n + \gamma - \ln(\pi/2)) + O(1)$$

Hint. Apply the trapezoidal rule to the integral

$$\int_{1/2n}^{1-1/2n} \left(\frac{1}{\sin(\pi x)} - \frac{1}{\pi x(1-x)}\right) dx.$$

5. Prove the complete asymptotic expansion

$$I_n \sim \frac{2n}{\pi}(\ln n + \gamma - \ln(\pi/2)) + \frac{2n}{\pi}\sum_{k=1}^{\infty} \frac{2^{2k-1}-1}{k(2k)!} B_{2k}^2 \left(-\frac{\pi^2}{n^2}\right)^k,$$

where B_{2k} are Bernoulli numbers.

6. **(Independent Study)** In 1890 Mathieu defined $S(x)$, now called *Mathieu's series*, by

$$S(x) := \sum_{k=1}^{\infty} \frac{2k}{(k^2+x^2)^2}.$$

Various inequalities have been established on $S(x)$. For example, Alzer et al. [6] proved that

$$\frac{1}{x^2 + 1/(2\zeta(3))} < S(x) < \frac{1}{x^2+1/6}, \qquad \text{for all } x \in \mathbb{R}.$$

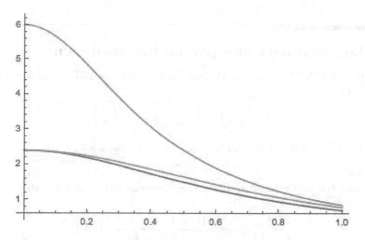

FIGURE 4.3
The graph of three functions

Although this estimate is asymptotically sharp, it leaves a big gap for the upper bound when $x \approx 0$, See Figure 4.3. Can one find a best possible estimate of $S(x)$ applicable on the entire real line? The following interesting observation,

$$S(x) = \sum_{k=1}^{\infty} \frac{1}{x} \int_0^{\infty} e^{-kt} t \sin(xt)\, dt = \frac{1}{x} \int_0^{\infty} \frac{t}{e^t - 1} \sin(xt)\, dt,$$

may be useful. Furthermore, define a parametric Mathieu's series

$$S_p(x) := \sum_{k=1}^{\infty} \frac{2k}{(k^2 + x^2)^{1+p}}.$$

Show that

(a) For $p > 0$, $S_p(x) == \frac{\sqrt{\pi}}{(2x)^{p-1/2}\Gamma(p+1)} \int_0^{\infty} \frac{t^{p+1/2}}{e^t - 1} J_{p-1/2}(xt)\, dt$,
where J_α denotes the ordinary Bessel function of order α.

(b) For $p > 0$,

$$\int_0^{\infty} S_p(x)\, dx = \frac{\sqrt{\pi}\,\Gamma(p+1/2)}{\Gamma(p+1)} \zeta(2p).$$

(c) A function $f : (0, \infty) \to \mathbb{R}$ is said to be *completely monotonic* if $(-1)^n f^{(n)}(x) \geq 0$ for all $x \geq 0$ and $n \in \mathbb{N}$. Show that $S_p(x)$ is completely monotonic and log-convex for each $x > 0$.

4.9 Log-concavity of a partial binomial sum

Problem 11985 (Proposed by Donald Knuth, 124(6), 2017). For fixed $s, t \in \mathbb{N}$ with $s \leq t$, let

$$a_n = \binom{n}{s} + \binom{n}{s+1} + \cdots \binom{n}{t}.$$

Prove that this sequence is log-concave, namely that $a_n^2 \geq a_{n-1}a_{n+1}$ for $n \geq 1$.

Discussion.
Notice that the binomial sequence $\binom{n}{k}$ is log-concave for fixed n and $0 \leq k \leq n$ since

$$\frac{\binom{n}{k}^2}{\binom{n}{k-1}\binom{n}{k+1}} = \frac{(k+1)(n-k+1)}{k(n-k)} > 1.$$

It is also known that the linear transformation

$$y_n = \sum_{k=0}^{n} \binom{n}{k} x_k, \quad n = 0, 1, 2, \ldots$$

preserves log-concavity, i.e., the log-concavity of $\{x_n\}$ implies that of $\{y_n\}$. In particular, with $x_n \equiv 1$, which is log-concave, we find that $z_n = \sum_{k=0}^{n} \binom{n}{k} = 2^n$, the sum of nth row of Pascal's triangle, is log-concave. Notice here a_n is only the partial sum of the entries in the nth row of Pascal's triangle. To show that $a_n^2 \geq a_{n-1}a_{n+1}$, we use the q-log-concavity technique, which was first suggested by Stanley ([87], also available at http://math.mit.edu/~rstan/pubs/pubfiles/72.pdf). Define

$$f_n(q) = \sum_{k=s}^{t} \binom{n}{k} q^k. \tag{4.30}$$

We show that $f_n^2(q) - f_{n-1}(q)f_{n+1}(q)$ has nonnegative coefficients. Thus $a_n^2 \geq a_{n-1}a_{n+1}$ immediately follows from setting $q = 1$.

Solution.
Let $f_n(q)$ be defined by (4.30). Applying the identity $\binom{n}{k} = \binom{n-1}{k-1} + \binom{n-1}{k}$ yields

$$
\begin{aligned}
f_n(q) &= \sum_{k=s}^{t} \left(\binom{n-1}{k-1} + \binom{n-1}{k} \right) q^k \quad \text{(shifting the index in the first sum)} \\
&= \binom{n-1}{s-1} q^s - \binom{n-1}{t} q^{t+1} + (q+1) \sum_{k=s}^{t} \binom{n-1}{k} q^k \\
&= \binom{n-1}{s-1} q^s - \binom{n-1}{t} q^{t+1} + (q+1) f_{n-1}(q).
\end{aligned}
$$

Next, let $\Delta_n(q) = f_n^2(q) - f_{n-1}(q)f_{n+1}(q)$. We show that all coefficients in $\Delta_n(q)$ are nonnegative. Indeed,

$$
\begin{aligned}
\Delta_n(q) \;=\;& f_n(q)\left(\binom{n-1}{s-1}q^s - \binom{n-1}{t}q^{t+1} + (q+1)f_{n-1}(q)\right) \\
& - f_{n-1}(q)\left(\binom{n}{s-1}q^s - \binom{n}{t}q^{t+1} + (q+1)f_n(q)\right) \\
=\;& \left(\binom{n-1}{s-1}f_n(q) - \binom{n}{s-1}f_{n-1}(q)\right)q^s \\
& + \left(\binom{n}{t}f_{n-1}(q) - \binom{n-1}{t}f_n(q)\right)q^{t+1} \\
=\;& \sum_{k=s}^{t}\left(\binom{n-1}{s-1}\binom{n}{k} - \binom{n}{s-1}\binom{n-1}{k}\right)q^{k+s} \\
& + \sum_{k=s}^{t}\left(\binom{n}{t}\binom{n-1}{k} - \binom{n-1}{t}\binom{n}{k}\right)q^{k+t+1}.
\end{aligned}
$$

Since, for $i \le j \le n$,

$$
\binom{n-1}{i}\binom{n}{j} = \frac{n-i}{n}\binom{n}{i}\binom{n}{j} \ge \frac{n-j}{n}\binom{n}{i}\binom{n}{j} = \binom{n}{i}\binom{n-1}{j},
$$

applying this inequality with $(i, j) = (s-1, k)$ and $(i, j) = (k, t)$ yields that all coefficients in $\Delta_n(q)$ are nonnegative, which proves the desired inequality. $\square$

Remark. Log-concave sequences occur naturally in combinatorics, algebra, analysis, geometry, computer science, probability, and statistics. There has been a considerable amount of research devoted to this topic in recent years. It provides rich materials which are suitable for undergraduate research. We refer the reader to Stanley [87] and Brenti [24]. Motivated by this proposed problem, let $\{a(n,k)\}_{0 \le k \le n}$ be a triangular array of nonnegative numbers. We consider the linear transformation

$$
y_n = \sum_{k=0}^{n} a(n,k)x_k, \qquad n = 0, 1, 2, \ldots. \tag{4.31}
$$

It is interesting to determine the conditions on $\{a(n,k)\}_{0 \le k \le n}$ under which (4.31) preserves the log-concavity property. By taking the special log-concavity sequences, we find the following necessary conditions. If (4.31) preserves log-concavity, then for $m \in \mathbb{N}, q > 0$,

1. the column sequence $\{a(n,m)\}_{n \ge m}$ is log-concave;
2. the row-sum sequence $a_n = \sum_{k=0}^{n} a(n,k)$ is log-concave; and
3. the sequence $f_m(n,q) = \sum_{k=m}^{n} a(n,k)q^k$ is log-concave for $n \ge m$;

4. the diagonal sequence $\{a(n, n)\}_{n \geq 0}$ is log-concave.

The proofs are left to the reader. Here we compile some problems for addition practice.

1. Let $c(n, k)$ and $S(n, k)$ be *Stirling numbers of the first kind and the second kind*, respectively. Show that both $c(n, k)$ and $S(n, k)$ are log-concave.

2. Let $f : [a, b] \to \mathbb{R}$ be a nonnegative and continuous function. Define

$$I_n = \int_a^b f^n(x) \, dx, \quad n \geq 1.$$

Show that $\{I_n\}_{n \geq 2}$ is log-concave.
Comment. Recall that *Legendre polynomials* $P_n(x)$ in integral form: for $x \geq 1$,

$$P_n(x) = \frac{1}{\pi} \int_0^\pi (x + \sqrt{x^2 - 1} \cos \theta)^n \, d\theta.$$

Thus, for fixed $x \geq 1$, $\{P_n(x)\}_{n \geq 0}$ is log-concave.

3. Let $1, a_1, a_2, \ldots$ be a positive log-concave sequence. Define

$$\sum_{n=0}^\infty b_n \frac{x^n}{n!} = \exp\left(\sum_{k=1}^\infty a_k \frac{x^k}{k!} \right).$$

Show that $\{b_n\}$ is log-convex (i.e., $b_n^2 \leq b_{n-1} b_{n+1}$ for all n) and $\{b_n/n!\}$ is log-concave.

4. Let D_n be the number of derangements on n objects. Show that $\{D_n\}_{n \geq 3}$ is log-convex.

5. A *Motzkin path* of length n is a lattice path in the (x, y)-plane from $(0, 0)$ to $(n, 0)$ with steps $(1, 1)$ (up), $(1, -1)$ (down) and $(1, 0)$ (level), never falling below the x-axis. Define

$$M_n = |\{\text{the set of all Motzkin paths of length } n\}|.$$

It is known that

$$M_n = \sum_{k \geq 0} \binom{n}{2k} C_k,$$

where $C_n = \binom{2n}{n}/(n + 1)$ is the nth Catalan number. Show that $\{M_n\}$ is log-convex.

6. (Newton's Real Roots Theorem). If the polynomial $\sum_{k=0}^n a_k x^k$ has positive coefficients and all of its roots are real, then $\{a_n\}_{0 \leq k \leq n}$ is log-concave.
Comment. The q-log-concavity technique is based on this theorem, which provides an approach for proving the log-concavity of a sequence.

7. Let i, j be nonnegative integers. Show that the linear transformation

$$y_n = \sum_{k=0}^{n} \binom{n+i}{k+j} x_k, \quad n = 0, 1, 2, \dots$$

preserves log-concavity.

8. Define

$$f_m(n, q) = \sum_{k=m}^{n} \binom{n}{k} q^k = (q+1) f_m(n-1, q) + \binom{n-1}{m-1} q^m.$$

Show that

$$f_m^2(n, q) - f_m(n-1, q) f_m(n+1, q)$$

$$= \sum_{k=m}^{n} \left[\binom{n-1}{m-1} \binom{n-1}{k-1} - \binom{n-1}{m-2} \binom{n-1}{k} \right] q^{k+m}.$$

9. (A test for (4.31) to preserve log-concavity). Let $\{a(n, k)\}_{0 \le k \le n}$ be a triangular array of nonnegative numbers. Define

$$a_k(n, t) = \begin{cases} 2a(n, k) a(n, t-k) \\ \quad -a(n-1, k) a(n+1, t-k) \\ \quad -a(n+1, k) a(n-1, t-k), & \text{for } k < t/2; \\ a^2(n, k) \\ \quad -a(n-1, k) a(n+1, t-k) \\ \quad -a(n+1, k) a(n-1, t-k), & \text{for } t \text{ even and } k = t/2 \end{cases}$$

and

$$A_m(n, t) = \sum_{k=m}^{\lfloor t/2 \rfloor} a_k(n, t).$$

Show that (4.31) preserves log-concavity if and only if $A_m(n, t) \ge 0$ for all $2m \le t \le 2n$.

Comment. A explicit example other than Pascal's triangle is $\{\binom{n}{k} \binom{i-n}{j-k}\}$, where $i, j \in \mathbb{N}, i \ge j$.

4.10 A bound of divisor sums related to the Riemann hypothesis

Problem 10949 (Proposed by J. Lagarias, 109(6), 2002). Let $H_n = \sum_{j=1}^{n} 1/j$. Show that for each positive integer n,

$$\sum_{d|n} d \le H_n + 2 \exp(H_n) \ln(H_n), \tag{4.32}$$

with equality only for $n = 1$.

Discussion.
Before Monthly published this problem, the proposer's article [62] appeared at https://arxiv.org/abs/math/0008177. In this paper, he proved that the Riemann hypothesis is equivalent to

$$\sigma(n) = \sum_{d|n} d \le H_n + \exp(H_n)\ln(H_n), \qquad (4.33)$$

which is derived from Robin's criterion [79]: The Riemann hypothesis is true if and only if

$$\sigma(n) = \sum_{d|n} d \le e^\gamma n \ln(\ln n) \qquad \text{for all } n \ge 5041,$$

where γ is Euler's constant. The current proposed problem is a weaker version of (4.33).

Clearly, $\sigma(n) \le \sum_{k=1}^n k = n(n+1)/2$. But, the proposed upper bound is only on the scale of $n\ln(\ln n)$ since $H_n \sim \ln n$. To get a feel for achieving the required upper bound, we derive a similar upper bound by an elementary argument. In addition to $\sigma(n)$, we presume three more arithmetic functions that are defined using number-theoretic properties in some way.

$\omega(n)$ = the number of distinct prime divisors of n;

$\tau(n)$ = the number of positive divisors of n, including 1 and n;

$\phi(n)$ = the number of positive integers not exceeding n that are relatively prime to n.

These functions can explicitly be evaluated in terms of the prime factorization of n. Let

$$n = p_1^{a_1} \cdots p_k^{a_k},$$

where p_i are distinct primes and $a_i \in \mathbb{N}$ for $1 \le i \le k$. Then $\omega(n) = k$ and

$$\tau(n) = \sum_{d|n} 1 = \prod_{i=1}^k (a_i + 1);$$

$$\phi(n) = n \prod_{i=1}^k \left(1 - \frac{1}{p_i}\right);$$

$$\sigma(n) = \sum_{d|n} d = \prod_{i=1}^k \frac{p_i^{a_i+1} - 1}{p_i - 1}.$$

Note that

$$\frac{\sigma(n)}{n} = \prod_{i=1}^k \frac{p_i^{a_i+1} - 1}{p_i^{a_i}(p_i - 1)}.$$

Since

$$\frac{p_i^{a_i+1} - 1}{p_i^{a_i}(p_i - 1)} = \frac{p_i - 1/p_i^{a_i}}{p_i - 1} < \frac{p_i}{p_i - 1},$$

we find that

$$\frac{\sigma(n)}{n} \leq \left(\prod_{i=1}^{k} \frac{p_i}{p_i - 1} \right). \tag{4.34}$$

In particular, if $\omega(n) = 5$, in view of the fact that $x/(x - 1)$ is decreasing, we obtain

$$\frac{\sigma(n)}{n} \leq \frac{2}{1} \cdot \frac{3}{2} \cdot \frac{5}{4} \cdot \frac{7}{6} \cdot \frac{11}{10} = \frac{77}{16}.$$

On the other hand, we have

$$\sigma(n) = \prod_{i=1}^{k} (p_i^{a_i} + p_i^{a_i-1} + \cdots + 1) \geq \prod_{i=1}^{k} (p_i^{a_i} + p_i^{a_i-1})$$

$$= \prod_{i=1}^{k} p_i^{a_i} \left(1 + \frac{1}{p_i}\right) = n \prod_{i=1}^{k} \left(1 + \frac{1}{p_i}\right).$$

Using the well-known inequality

$$\prod_{i=1}^{k} (1 + \alpha_i) \geq 1 + \sum_{i=1}^{k} \alpha_i$$

for all $\alpha_i \geq 0$, we have

$$\frac{\sigma(n)}{n} \geq \prod_{i=1}^{k} \left(1 + \frac{1}{p_i}\right) \geq 1 + \sum_{i=1}^{k} \frac{1}{p_i}.$$

Since the sum of reciprocals of all primes is divergent, this implies that $\sigma(n)/n$ has no finite upper bound. For proceeding with (4.32), the above line of reasoning suggests that we establish that

$$\prod_{p|n} \frac{p_i}{p_i - 1} \leq C \ln(\ln n)$$

for some positive constant C. Along the way, we presume some approximation formulas from [80], and use *Mathematica* extensively to do numerical verifications.

Solution.

We prove (4.32) by two steps. First, we show that, for $n \geq 7$,

$$\sigma(n) < 2.59n \ln(\ln n). \tag{4.35}$$

The proof of (4.35) consists of the following three facts:

F1 Let p_n be the nth prime. If $n \geq 39$, then $\ln p_n < \frac{7}{5} \ln n$.

F2 If $n > 31$, then

$$\prod_{p \mid n} \left(1 + \frac{1}{p}\right) < \frac{28}{15} \ln(\ln n).$$

F3 If $n \geq 13, n \neq 42$ and 210, then

$$\prod_{p \mid n} \frac{p}{p-1} < 2.59 \ln(\ln n).$$

To prove **F1**, we invoke a sharp explicit estimate ([80], Corollary 1, (3.5)):

$$\pi(x) > \frac{x}{\ln x} \qquad \text{for } x \geq 17, \tag{4.36}$$

where $\pi(x)$ denotes the number of primes not exceeding x. Since $p_7 = 17$, letting $x = p_n$ in (4.36) yields

$$p_n < n \ln p_n \qquad \text{for } n \geq 7.$$

Notice that $p_n < n\sqrt{p_n}$ for $n \geq 2$. Taking the logarithm of both sides gives $\ln p_n < 2 \ln n$. Thus,

$$p_n < 2n \ln n \qquad \text{for } n \geq 7.$$

Since $2n \ln n \leq n^{3/2}$ for $n \geq 75$, the above inequality leads to $p_n < n^{3/2}$ for $n \geq 75$. Using the *Mathematica* command

```
Negative[Table[n^3/2 - Prime[n], {n, 7, 74}]]
```

we can easily verify that $p_n < n^{3/2}$ for $n \geq 7$. Thus,

$$p_n < \frac{3}{2} n \ln n \qquad \text{for } n \geq 7.$$

Similarly, since $\frac{3}{2} n \ln n \leq n^{7/5}$ for $n \geq 160$, the above inequality leads to $p_n < n^{7/5}$ for $n \geq 160$, and so

$$\ln p_n < \frac{7}{5} \ln n \qquad \text{for } n \geq 160.$$

With

```
Negative[Table[7. Log[n]/5 - Log[Prime[n]], {n, 39, 159}]]
```

we can verify that $\ln p_n < \frac{7}{5} \ln n$ indeed holds for all $n \geq 39$. This proves **F1**. Notice that for $n = 38, p_n = 163$, the inequality is reversed. Figure 4.4 explains the values of $39, 75$ and 160 in the proof of **F1**.

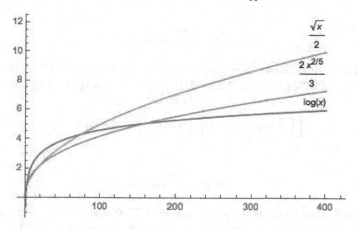

FIGURE 4.4
Find the determined numbers c, p, n such that $\ln x < c\,n^p$ for $x \geq n$

Recall that ([80], Theorem 5, (3.18))

$$\sum_{p \leq x} \frac{1}{p} < \ln(\ln x) + B + \frac{1}{2\ln^2 x} \qquad \text{for } x \geq 286, B = 0.261497\ldots.$$

When $x \geq 286$, since

$$\frac{1}{2\ln^2 x} \leq \frac{1}{2\ln^2 286} = 0.01562\ldots < 0.016; \quad \ln(4/3) = 0.287682\ldots > B + 0.016$$

we have

$$\sum_{p \leq x} \frac{1}{p} < \ln(\ln x) + \ln(4/3) = \ln\left(\frac{4}{3}\ln x\right).$$

Therefore, for $p_k \geq 286$,

$$\ln\left(\prod_{p|n}\left(1+\frac{1}{p}\right)\right) = \sum_{p|n}\ln\left(1+\frac{1}{p}\right) \leq \sum_{p|n}\frac{1}{p} \leq \sum_{p \leq p_k}\frac{1}{p} < \ln\left(\frac{4}{3}\ln p_k\right),$$

which is valid for $k \geq 62$ because $p_{61} = 283, p_{62} = 293$.
Notice that $\omega(n) = k$. From **F1** we have

$$\prod_{p|n}\left(1+\frac{1}{p}\right) < \frac{4}{3}\left(\frac{7}{5}\ln k\right) = \frac{28}{15}\ln(\ln n) \qquad \text{for } \omega(n) \geq 62,$$

where we have used the fact that $\omega(n) \leq \ln n$ for $n \geq 7$. We now deal with the case $\omega(n) \leq 61$. Using *Mathematica*, we have

$$\prod_{p|n}\left(1+\frac{1}{p}\right) \leq \prod_{i=1}^{61}\left(1+\frac{1}{p_i}\right) = 6.21964\ldots,$$

and
$$6.22 < \frac{28}{15} \ln(\ln n)$$

if $n \geq 1.5 \times 10^{12}$. So we just need to check the case $n < 1.5 \times 10^{12}$. Since $\prod_{i=1}^{12} p_i = 7420738134810 > 7.42 \times 10^{12}$, this implies that $\omega(n) \leq 11$. Notice that

$$\prod_{p|n} \left(1 + \frac{1}{p}\right) \leq \prod_{i=1}^{11} \left(1 + \frac{1}{p_i}\right) = 4.00195\ldots,$$

and
$$4.002 < \frac{28}{15} \ln(\ln n)$$

if $n \geq 5500$. This leave us with the case $n < 5500$ and $\omega(n) \leq 5$. By applying the same line of reasoning, we can prove **F2** all the way to $n \geq 31$.

Let γ be Euler's constant. Notice that $e^\gamma < 1.7810725$ and $1/\ln^2 x < 0.03836$ for $x > 165$. By the known inequality

$$\prod_{p \leq x} \frac{p}{p-1} < e^\gamma \ln x \left(1 + \frac{1}{\ln^2 x}\right),$$

we have

$$\prod_{p \leq x} \frac{p}{p-1} < 1.8497 \ln x \quad \text{for } x \geq 165. \tag{4.37}$$

By **F1** and $k = \omega(n) < \ln n$ for $n \geq 7$, if $k \geq 39$, since $p_k \geq p_{39} = 167$, (4.37) implies

$$\prod_{p|n} \frac{p}{p-1} < \prod_{p \leq p_k} \frac{p}{p-1} < 1.8497 \ln p_k < 2.5896 \ln k < 2.5896 \ln(\ln n).$$

Thus, it suffices to prove **F3** for $k \leq 38$. *Mathematica* confirms that

$$\prod_{p|n} \frac{p}{p-1} < \prod_{p \leq p_{38}} \frac{p}{p-1} < 9.5 < 2.59 \ln(\ln n)$$

for $n \geq 5 \cdot 10^{17}$. On the other hand, since $\prod_{p \leq p_{17}} > 10^{18}$, it suffices to prove **F3** for $k \leq 16$ and $n < 5 \cdot 10^{17}$. Notice that

$$\prod_{p|n} \frac{p}{p-1} < \prod_{p \leq p_{16}} \frac{p}{p-1} < 7.5 < 2.59 \ln(\ln n)$$

for $n \geq 10^8$. Since $\prod_{p \leq p_9} > 2 \cdot 10^8$, if $n < 10^8$, then $k \leq 8$. Moreover, we have

$$\prod_{p|n} \frac{p}{p-1} < \prod_{p \leq p_8} \frac{p}{p-1} < 5.9 < 2.59 \ln(\ln n)$$

for $n \geq 3 \cdot 10^4$. Since $\prod_{p \leq p_6} = 30030 > 3 \cdot 10^3$, if $n < 3 \cdot 10^4$, then $k \leq 5$. As we did in the discussion, we have

$$\prod_{p|n} \frac{p}{p-1} < \prod_{p \leq p_5} \frac{p}{p-1} \leq \frac{77}{16} < 2.59 \ln(\ln n)$$

for $n \geq 1000$. If $n < 1000, k \leq 4$, and we have

$$\prod_{p|n} \frac{p}{p-1} \leq \frac{35}{8} < 2.59 \ln(\ln n)$$

if $n \geq 300$. If $n < 300$, then $k \leq 3$. For $n < 300, n \neq 210 = 2 \cdot 3 \cdot 5 \cdot 7$, we have

$$\prod_{p|n} \frac{p}{p-1} \leq 3.75 < 2.59 \ln(\ln n)$$

if $n \geq 71$. Finally, we use *Mathematica* to verify **F3** all the way from 70 to 31 ($n = 42$ is an exception). This proves **F3**. Now (4.35) follows from (4.34) and **F3** immediately.

Next we show that, for $n \geq 3$,

$$\exp(H_n) \ln(H_n) \geq e^\gamma n \ln(\ln n). \tag{4.38}$$

To this end, notice that

$$H_n - 1 = \sum_{k=1}^{n} \left(\frac{1}{k} - \frac{1}{n} \right) = \sum_{k=1}^{n} \int_{k}^{n} \frac{dx}{x^2} = \int_{1}^{n} \frac{1}{x^2} \left(\sum_{1 \leq k \leq x} 1 \right) dx = \int_{1}^{n} \frac{\lfloor x \rfloor}{x^2} \, dx.$$

Hence

$$H_n = 1 + \int_{1}^{n} \frac{x - \{x\}}{x^2} \, dx = \ln n + 1 - \int_{1}^{n} \frac{\{x\}}{x^2} \, dx,$$

where $\{x\}$ denotes the fractional part of x. Recall that

$$\gamma = 1 - \int_{1}^{\infty} \frac{\{x\}}{x^2} \, dx.$$

We obtain that

$$H_n = \ln n + \gamma + \int_{n}^{\infty} \frac{\{x\}}{x^2} \, dx.$$

This implies

$$H_n > \ln n + \gamma,$$

which upon exponentiating yields

$$\exp(H_n) \geq e^\gamma n.$$

Combining this with $\ln(H_n) \geq \ln(\ln n)$ for $n \geq 3$ proves (4.38).

Since $2.59 < 2e^\gamma$, we see that (4.32) is valid for $n \geq 7$. The remaining cases $1 \leq n \leq 6$ now follow by numerical verification via *Mathematica*. □

Remark. There are lots of unsolved problems in mathematics, but none more important and intriguing than the *Riemann hypothesis*. To herald the new century, Hilbert in 1900 challenged mathematicians with a list of 23 problems (the Riemann hypothesis is the eighth on the list) that he believed should set the course for the mathematical explorers of the twentieth century. To date, 17 of the 23 questions have been answered. Those who discovered the solutions make up "the honors class" and their solutions have transformed the mathematical landscape and indeed have become fundamental to the development of twentieth century mathematics. But the Riemann hypothesis remains unsolved.

Exactly 100 years later, the Clay Mathematics Institute published a list of the seven unsolved problems that they predicted would be the most important questions of the twenty-first century. The official problem description on Riemann hypothesis was written by Enrico Bombieri, one of the greatest mathematicians of the age (see "Problems of the Millennium: The Riemann Hypothesis," available at `https://www.claymath.org/sites/default/files/official_problem_description.pdf`). Of the unsolved problems on Hilbert's list, the Riemann hypothesis was the only one to be included on the new list. Moreover, the Riemann hypothesis and the other six problems have a reward of one million dollars each. Now, finding a proof will make you both famous and rich!

Various mathematicians have made some headway toward a proof. Terence Tao arranged the existing facts together in "The Riemann hypothesis in various settings" (see `https://terrytao.wordpress.com/2013/07/19/the-riemann-hypothesis-in-various-settings/`). This proposed problem provided an elementary gateway to the Riemann hypothesis. Its major appeal is that anyone with rudimentary exposure to number theory can play with it. Recently, Veen and Craats's book [92] offered a nice introduction of the Riemann hypothesis for mathematically talented secondary school students. The resolution of the Riemann hypothesis is most likely to require a different level of mathematics.

Now we collect some problems for additional practice. They may reveal new perspectives on the resolution of the proposer's equivalent theorem.

1. Show that

$$\left| \sum_{\text{prime } p \leq x} \frac{1}{p} - \ln(\ln x) \right| < 6 \quad \text{for } x > 4.$$

2. For $n \geq 2$, show that

$$\sigma(n) \leq \frac{n^2}{\phi(n)} \leq \phi(n)\tau^2(n); \quad \sigma(n) \leq n\tau(n) - \phi(n);$$

$$\sigma(n) \leq \phi(n) + \tau(n)(n - \phi(n)); \quad \sigma(n) \leq \frac{n+1}{2}\tau(n).$$

3. When n is odd, show that $\sigma(n) \leq \phi(n)\tau^2(n)$ can be improved to $\sigma(n) \leq \phi(n)\tau(n)$. Using this result prove that

$$\sigma(n) \leq 2\phi(n)\tau(n) \text{ for } n \geq 2 \text{ even.}$$

4. (J. Sandor and L. Kovacs) Let $P(n)$ be the largest prime factor of n. Show that

$$\sigma(n) \leq \frac{3}{4}\phi(n)P(n) \text{ for } n \geq 3 \text{ odd, and } \sigma(n) \leq 3\phi(n)P(n) \text{ for } n \text{ even.}$$

Comment. Since $\tau(n)$ and $P(n)$ are not comparable, the inequalities in Problems 3 and 4 are independent of each other.

5. Let $n = p_1^{a_1} \cdots p_k^{a_k}$, where p_i are distinct primes and $a_i \in \mathbb{N}$ for $1 \leq i \leq k$. Define

$$\sigma^*(n) = \sum_{i=1}^{k} p_i^{a_i}.$$

For example, $90 = 2 \cdot 3^2 \cdot 5, \sigma^*(90) = 2 + 3^2 + 5 = 16$. Show that

(a) $\sigma^*(n) \leq n$ and the equality holds only if $k = 1$, i.e., $n = p^a$.

(b) Let n be an odd integer with $k \geq 2$. If $n > 15$, then

$$\sigma^*(n) \leq \frac{n-1}{2}.$$

(c) If $k \geq 2$ and $p_i^{a_i} \geq k$ for all $1 \leq i \leq k$, then

$$\sigma^*(n) \leq \frac{n}{k^{k-2}}.$$

6. Recall that Dedekind's totient function $\psi(n)$ is defined by

$$\psi(n) = n \prod_{i=1}^{k} \left(1 + \frac{1}{p_i}\right).$$

Show that, for $n \geq 1$, $\sigma(n) < \zeta(2)\psi(n)$.

7. (A. Grytczuk) If n is odd and $n > 3^9/2$, show that

$$\sigma(2n) < \frac{39}{40}e^{\gamma}(2n)\ln\ln(2n).$$

8. For $n \geq 3$, prove

$$H_n + \exp(H_n)\ln(H_n) \leq e^\gamma n \ln(\ln n) + \frac{4n}{\ln n}.$$

9. (**Independent Study**) Prove (4.32) by replacing 2 with $c \in (1,2)$. Here $c = 1$ is equivalent to the Riemann hypothesis.

5

Monthly Miniatures

This chapter introduces 10 Monthly problems related to the mean value theorem, eigenvalues and determinants of some special matrices, weighted trigonometric sums and Dirichlet series, infinite sum-product identity, and polynomial zero identities. In addition to developing valuable techniques for solving these problems, it also includes some interesting historical accounts of ideas and contexts. Among them, we will encounter various extensions of the mean value theorem, Chebyshev polynomials, $(0,1)$-matrices, the Cauchy matrix, Dodgson condensation, Dirichlet convolution, the Woods-Robbbins identity and the alternating sign matrix conjecture. It is intended to facilitate a natural transition that bridges problem-solving to independent exploration of new results.

5.1 Value defined by an integral

Problem 11555 (Proposed by D. V. Thong, 118(2), 2011). Let f be a continuous real-valued function on $[0,1]$ such that $\int_0^1 f(x)dx = 0$. Prove that there exists $c \in (0,1)$ such that

$$c^2 f(c) = \int_0^c (x + x^2)f(x)\,dx.$$

Discussion.
The equality to be proved suggests we consider the auxiliary function

$$F(x) = x^2 f(x) - \int_0^x (t + t^2)f(t)dt,$$

which is continuous on $[0,1]$. Thus, by the intermediate value theorem, there is a point $c \in (0,1)$ such that $F(c) = 0$ provided we can find distinct points $a, b \in [0,1]$ such that $F(a) < 0$ and $F(b) > 0$.

Another path is to apply *Flett's mean value theorem* [42]: Let $f : [a, b] \to \mathbb{R}$ be continuous on $[a, b]$ and differentiable on (a, b). If $f'(a) = f'(b)$, then there exists a number $c \in (a, b)$ such that

$$f'(c) = \frac{f(c) - f(a)}{c - a}.$$

This enables us to show that if $f : [0, b] \to \mathbb{R}$ is continuous and $\int_0^b f(x)\, dx = 0$, then there exists $a \in (0, b)$ such that $\int_0^a xf(x)\, dx = 0$. Thus, the auxiliary function

$$H(x) = \int_0^x (t^2 + (1 - x)t)f(t)\, dt$$

fulfills the assumptions of Flett's mean value theorem, so that $H(c) - H(0) = cH'(c)$ leads to the desired equality.

Based on the discussion above, we give two solutions.

Solution I.

If f is identically zero, there is nothing to prove, so assume that $f(x)$ is not identically zero. Since f is continuous on $[0, 1]$ such that $\int_0^1 f(x)dx = 0$, there are distinct $a, b \in [0, 1]$ such that

$$f(a) = \max_{x \in [0,1]} f(x) > 0, \quad f(b) = \min_{x \in [0,1]} f(x) < 0. \tag{5.1}$$

Define

$$F(x) = x^2 f(x) - \int_0^x (t + t^2)f(t)dt.$$

Note that $F(x)$ is continues on $[0, 1]$. Since $(t + t^2)f(a) \geq (t + t^2)f(t)$ for all $t \in [0, 1]$,

$$F(a) \geq a^2 f(a) - \int_0^a (t + t^2)f(a)dt = a^2 \left(\frac{1}{2} - \frac{a}{3}\right) f(a) > 0.$$

Similarly, $(t + t^2)f(b) \leq (t + t^2)f(t)$ for all $t \in [0, 1]$, so

$$F(b) \leq b^2 f(b) - \int_0^b (t + t^2)f(b)dt = b^2 \left(\frac{1}{2} - \frac{b}{3}\right) f(b) < 0.$$

The intermediate value theorem implies that there is a number c between a and b such that $F(c) = 0$, so that $c^2 f(c) = \int_0^c (x + x^2)f(x)dx$.   □

Solution II.

First, for $x \in [0, 1]$, define

$$F(x) = x \int_0^x f(t)\, dt - \int_0^x tf(t)\, dt.$$

Note that $F(x)$ is differentiable on $(0, 1)$ and $F'(x) = \int_0^x f(t)\, dt$. Since $F'(0) = 0$ and $F'(1) = \int_0^1 f(t)\, dt = 0$, by Flett's mean value theorem, there exists $a \in (0, 1)$ such that

$$F'(a) = \frac{F(a) - F(0)}{a - 0}.$$

This implies that

$$\int_0^a t f(t)\, dt = 0. \tag{5.2}$$

Next, let $G(x) = e^{-x} \int_0^x t f(t)\, dt$. Then $G(0) = G(a) = 0$. By Rolle's theorem, there exists $b \in (0, a)$ such that

$$G'(b) = -e^{-b} \int_0^b t f(t)\, dt + e^{-b} b f(b) = 0.$$

Thus,

$$\int_0^b t f(t)\, dt = b f(b). \tag{5.3}$$

Finally, let $H(x) = \int_0^x (t^2 + (1-x)t) f(t)\, dt$. Then

$$H'(x) = x f(x) - \int_0^x t f(t)\, dt.$$

Thus, $H'(0) = 0$ and from (5.3) $H'(b) = 0$. Once again by Flett's mean value theorem, there exists $c \in (0, b)$ such that

$$H'(c) = \frac{H(c) - H(0)}{c - 0}.$$

This is equivalent to $c^2 f(c) = \int_0^c (x + x^2) f(x)\, dx$ and the desired equality is confirmed. $\qquad\square$

Remark. Based on Solution I, the hypothesis $\int_0^1 f(x)\, dx = 0$ can be replaced by the hypothesis that f changes sign in $(0, 1)$. Similarly, by changing the hypothesis $\int_0^1 f(x)\, dx = 0$ to $\int_0^1 f(x)\, dx = \int_0^1 x f(x)\, dx$, the argument in Solution II yields the same result (see Problem 5 below).

It is interesting to see that we can prove $H(x)$ fulfills the assumptions of Flett's mean value theorem along the same lines as in Solution I. In fact, since clearly $H'(0) = 0$, it suffices to show that there is $c \in (0, 1)$ such that $H'(c) = 0$. Note that $H'(x) = x f(x) - \int_0^x t f(t)\, dt$ is continuous on $[0, 1]$. Using (5.1), we have

$$H'(a) = a f(a) - \int_0^a t f(t)\, dt \geq a f(a) - f(a) \int_0^a t\, dt = a f(a)(1 - a/2) > 0$$

and

$$H'(b) = b f(b) - \int_0^b t f(t)\, dt \leq b f(b) - f(b) \int_0^b t\, dt = b f(b)(1 - b/2) < 0.$$

The intermediate value theorem implies that there is a number c between a and b such that $H'(c) = 0$.

   Mean value theorems play an essential role in analysis. Rolle's and Lagrange's mean value theorems are two common forms. Flett's mean value theorem provides a variant of Lagrange's mean value theorem with a Rolle's type condition. In general, Flett's mean value theorem can be extended as: Let f be n times differentiable on (a, b) and $f^{(n)}(a) = f^{(n)}(b)$. Then there exists $\xi \in (a, b)$ such that

$$f(\xi) - f(a) = \sum_{k=1}^{n} \frac{(-1)^{k+1}}{k!} (\xi - a)^k f^{(k)}(\xi). \tag{5.4}$$

Recall the nth Taylor polynomial of f which is given by

$$T_n(f, x_0)(x) = f(x_0) + \frac{f'(x_0)}{1!}(x - x_0) + \cdots + \frac{f^{(n)}(x_0)}{1!}(x - x_0)^n.$$

(5.4) has the following nice compact form: $f(a) = T_n(f, \xi)(a)$.
   Now we end this section with some problems for additional practice.

1. Let f be continuous on $[a, b]$ and differentiable on (a, b). Prove that there exists $c \in (a, b)$ such that

$$f'(c) = \frac{f(c) - f(a)}{c - a} + \frac{1}{2} \frac{f'(b) - f'(a)}{b - a}(c - a).$$

2. (**A variant of Cauchy's mean value theorem**) Let $f, g : [a, b] \to \mathbb{R}$ be continuous on $[a, b]$ and differentiable on (a, b). If $g'(x) \neq 0$ on $[a, b]$ and

$$\frac{f'(a)}{g'(a)} = \frac{f'(b)}{g'(b)},$$

   then there exists a number $c \in (a, b)$ such that

$$\frac{f'(c)}{g'(c)} = \frac{f(c) - f(a)}{g(c) - g(a)}.$$

3. Let f be n-times differentiable on (a, b). Prove that there exists $c \in (a, b)$ such that

$$f(a) = \sum_{k=0}^{n} \frac{f^{(k)}(c)}{k!}(a - c)^k + \frac{(a - c)^{n+1}}{(n + 1)!} \frac{f^{(n)}(b) - f^{(n)}(a)}{b - a}.$$

4. Let f be continuous on $[0, 1]$ such that $\int_0^1 f(x)dx = \int_0^1 xf(x)dx$. Prove that there exists $c \in (0, 1)$ such that

$$\int_0^c f(x)\,dx = 0.$$

5. Let f be continuous on $[0,1]$ such that $\int_0^1 f(x)dx = 0$. Prove that there exists $c \in (0,1)$ such that

   (a) $f(c) = f'(c) \int_0^c f(x)\,dx$, if f is differentiable on $(0,1)$.
   (b) $(1-c)f(c) = c \int_0^c f(x)\,dx$.
   (c) $c^2 f(c) = 2 \int_0^c x f(x)\,dx$, if in addition $f(0) = 0$.

6. **Problem 11581** (Proposed by D. V. Thong, 118(6), 2011). Let f be a continuous, nonconstant function from $[0,1]$ to $\mathbb{R}$ such that $\int_0^1 f(x)\,dx = 0$. Also, let $m = \min_{0 \le x \le 1} f(x)$ and $M = \max_{0 \le x \le 1} f(x)$. Prove that

$$\left| \int_0^1 x f(x)\,dx \right| \le \frac{-1}{2} \frac{mM}{M-m}.$$

Comment. This offers a nice example of problems leading to publications. Let $F(x) = \int_0^x f(t)\,dt$. Integration by parts yields

$$\int_0^1 x f(x)\,dx = x F(x)|_0^1 - \int_0^1 F(x)\,dx = -\int_0^1 F(x)\,dx.$$

Kouba generalized this problem and obtained the following optimal bound:

$$\left(\int_0^1 |F(x)|^p\,dx \right)^{1/p} \le \frac{-1}{\sqrt[p]{1+p}} \frac{mM}{M-m}, \qquad \text{for any } p > 0,$$

which is published in [60].

7. **Problem 11814** (Proposed by C. Lupu, 122(1), 2015). Let ϕ be a continuously differentiable function from $[0,1]$ into $\mathbb{R}$, with $\phi(0) = 0$ and $\phi(1) = 1$, and suppose that $\phi'(x) \neq 0$ for $0 \le x \le 1$. Let f be a continuous function from $[0,1]$ into $\mathbb{R}$ such that $\int_0^1 f(x)\,dx = \int_0^1 \phi(x)f(x)\,dx$. Show that there exists c with $0 < c < 1$ such that $\int_0^c \phi(x)f(x)\,dx = 0$.

5.2   Another mean value theorem

Problem 11872 (Proposed by P. C. Le Van, 122(9), 2015). Let f be a continuous function from $[0,1]$ into $\mathbb{R}$ such that $\int_0^1 f(x)dx = 0$. Prove that for all positive integers n there exists $c \in (0,1)$ such that

$$c^{n+1} f(c) = n \int_0^c x^n f(x)\,dx.$$

Discussion.
We can proceed with this problem in two different ways. First, we apply
(5.2) to $x^k f(x)$ recursively to get $\int_0^{c_n} x^n f(x)\, dx = 0$ for some $c_n \in (0,1)$.
Now, introduce the auxiliary function $F(x) = \frac{1}{x^n} \int_0^x t^n f(t)\, dt$, the equality
to be proved follows from Rolle's theorem. Second, we establish the proposed
equality by the intermediate value theorem as we did in Solution I of **Problem
11555**.

Solution I.
Applying (5.2) to $xf(x)$ yields

$$\int_0^{c_2} x^2 f(x)\, dx = 0, \quad \text{for } c_2 \in (0,1).$$

Repeating this process yields

$$\int_0^{c_n} x^n f(x)\, dx = 0, \quad \text{for } c_n \in (0,1).$$

Define

$$F(x) = \begin{cases} \frac{1}{x^n} \int_0^x t^n f(t)\, dt, & \text{if } x \neq 0, \\ \\ 0, & \text{if } x = 0. \end{cases}$$

Then $F(x)$ is continuous on $[0, c_n]$ and differentiable on $(0, c_n)$ and $F(0) = F(c_n) = 0$. By Rolle's theorem, these exists $c \in (0, c_n)$ such that

$$F'(c) = f(c) - \frac{n}{c^{n+1}} \int_0^c x^n f(x)\, dx = 0.$$

This implies $c^{n+1} f(c) = n \int_0^c x^n f(x)\, dx$.

$\square$

Solution II.
If f is identically zero, then there is nothing to prove. Assume that $f(x)$ is
not identically zero. Since $\int_0^1 f(x) dx = 0$, there exist distinct $a, b \in [0,1]$ such
that

$$f(a) = \max_{x \in [0,1]} f(x) > 0, \quad f(b) = \min_{x \in [0,1]} f(x) < 0.$$

Let

$$F(x) = \begin{cases} f(x) - \frac{n}{x^{n+1}} \int_0^x t^n f(t)\, dt, & \text{if } x > 0, \\ \\ f(0)\left(1 - \frac{n}{n+1}\right), & \text{if } x = 0. \end{cases}$$

Since $\lim_{x \to 0^+} F(x) = F(0)$, the function $F(x)$ is continuous on $[0,1]$. We
claim $F(a) > 0$. This is clear when $a = 0$. For $a > 0$,

$$F(a) \geq f(a) - \frac{n}{a^{n+1}} \int_0^a t^n f(a) dt = \left(1 - \frac{n}{n+1}\right) f(a) > 0.$$

Similarly, we have $F(b) < 0$. The intermediate value theorem implies that there is a number c between a and b such that $F(c) = 0$. That is, $c^{n+1} f(c) = n \int_0^c x^n f(x)\, dx$. $\qquad\square$

Remark. What if we replace x^n by some function $w(x)$? Here we give one generalization: Let $w(x) \in C^1[0, 1]$ with $w(0) = 0$ and $w'(x) > 0$ for $x \in (0, 1)$. Then there exists $c \in (0, 1)$ such that

$$f(c) = \frac{w'(c)}{w^2(c)} \int_0^c w(x) f(x)\, dx.$$

An editorial comment following our featured solution noted that Omajee and Tauraso also obtained the above extension. To this end, define

$$G(x) = \frac{1}{w(x)} \int_0^x w(t) f(t)\, dt.$$

Notice that $w(x)$ is increasing. By the mean value theorem for integrals, we have

$$|G(x)| = \frac{1}{w(x)} \left| \int_0^x w(t) f(t)\, dt \right| = \frac{w(\xi)|f(\xi)|x}{w(x)} \le |f(\xi)|x, \quad \text{for some } \xi \in (0, x).$$

This implies that $\lim_{x \to 0^+} G(x) = 0$. Since

$$G'(x) = -\frac{w'(x)}{w^2(x)} \int_0^x w(t) f(t)\, dt + f(x),$$

by Rolle's theorem, it suffices to show that $G(x_0) = 0$ for some $x_0 \in (0, 1)$. We prove this by contradiction. Without loss of generality, we assume that $G(x) > 0$ for all $x \in (0, 1)$. Let $F_1(x) = \int_0^x f(t)\, dt$. Then integration by parts yields

$$G(t) = \frac{1}{w(x)} (w(t) F_1(t))\Big|_0^x - \frac{1}{w(x)} \int_0^x w'(t) F_1(t)\, dt = F_1(x) - F_2(x),$$

where $F_2(x) := \frac{1}{w(x)} \int_0^x w'(t) F_1(t)\, dt$. The function $F_2(x)$ is differentiable in $(0, 1)$ and

$$F_2'(x) = \frac{w(x) w'(x) F_1(x) - w'(x) \int_0^x w'(t) F_1(t)\, dt}{w^2(x)}$$

$$= \frac{w'(x)(F_1(x) - F_2(x))}{w(x)} = \frac{w'(x) G(x)}{w(x)} > 0.$$

Since $\lim_{x \to 0^+} F_2(x) = \lim_{x \to 0^+} (G(x) - F_1(x)) = 0$, it follows that $\lim_{x \to 1^-} F_2(x) > 0$. But, on the other hand, the assumption $F_1(1) = 0$ and $G(x) > 0$ imply that

$$\lim_{x \to 1^-} F_2(x) = \lim_{x \to 1^-} (F_1(x) - G(x)) \le 0,$$

which is a contradiction. Therefore, there exists $x_0 \in (0,1)$ such that $G(x_0) = 0$.

Here are more problems for your practice.

1. **Problem 10739** (Proposed by O. Ciaurri, 106(6), 1999). Suppose that $f : [0,1] \to \mathbb{R}$ has a continuous second derivative with $f''(x) > 0$ on $(0,1)$, and suppose that $f(0) = 0$. Choose $a \in (0,1)$ such that $f'(a) < f(1)$. Show that there is a unique $b \in (a,1)$ such that $f'(a) = f(b)/b$.

2. **Monthly Problem 11892** (Proposed by F. Perdomo and A. Plaza, 123(2), 2016). Let f be a real-valued continuous and differentiable function on $[a,b]$ with positive derivative on (a,b). Prove that, for all pairs (x_1, x_2) with $a \le x_1 < x_2 \le b$ and $f(x_1)f(x_2) > 0$, there exists $t \in (x_1, x_2)$ such that

$$\frac{x_1 f(x_2) - x_2 f(x_1)}{f(x_2) - f(x_1)} = t - \frac{f(t)}{f'(t)}.$$

3. **Problem 11290** (Proposed by C. Lupu and T. Lupu, 114(4), 2007). Let f and g be continuous real functions on $[0,1]$. Prove that there exists $c \in (0,1)$ such that

$$\int_0^1 f(x)\,dx \int_0^c xg(x)\,dx = \int_0^1 g(x)\,dx \int_0^c xf(x)\,dx.$$

Comment: The x in the above equality can be replaced by any function $w(x) \in C^1[0,1]$ with $w'(x) \ge 0$ on $[0,1]$.

4. **Problem 11313** (Proposed by C. Lupu and T. Lupu, 114(8), 2007). Let f be a four-times differentiable function on $\mathbb{R}$ with $f^{(4)}$ continuous on $[0,1]$ such that

$$\int_0^1 f(x)\,dx + 3f(1/2) = 8\int_{1/4}^{3/4} f(x)\,dx.$$

Prove that there is some c between 0 and 1 such that $f^{(4)}(c) = 0$.

5. **Problem 11429** (Proposed by C. Lupu and T. Lupu, 116(4), 2009). For a continuous real-valued function ϕ on $[0,1]$, let $T\phi$ be the function mapping $C[0,1] \to \mathbb{R}$ given by $T\phi(t) = \phi(t) - \int_0^t \phi(u)\,du$, and similarly define S by $S\phi(t) = t\phi(t) - \int_0^t u\phi(u)\,du$. Show that if f and g are continuous real-valued functions on $[0,1]$, then there exist numbers a, b, and $c \in (0,1)$ such that each of the following is true:

$$Tf(a) = Sf(a).$$

$$Tg(b)\int_0^1 f(u)\,du = Tf(b)\int_0^1 g(u)\,du.$$

$$Sg(c)\int_0^1 f(u)\,du = Sf(c)\int_0^1 g(u)\,du.$$

Comment: These equalities can be viewed as some intermediate value variants.

6. **Problem 11517** (Proposed by C. Lupu and T. Lupu, 117(6), 2010). Let f be a three-times differentiable real-valued function on $[a,b]$ with $f(a) = f(b)$. Prove that

$$\left| \int_a^{(a+b)/2} f(x)\,dx - \int_{(a+b)/2}^b f(x)\,dx \right| \le \frac{(b-a)^4}{192} \sup_{x\in[a,b]} |f'''(x)|.$$

7. **Problem 11981** (Proposed by C. Lupu, 124(5), 2017). Suppose $f : [0,1] \to \mathbb{R}$ is a differentiable function with continuous derivative and with $\int_0^1 f(x)\,dx = \int_0^1 xf(x)\,dx = 1$. Prove that

$$\int_0^1 |f'(x)|^3\,dx \ge \left(\frac{128}{3\pi}\right)^2.$$

8. **Problem 12046** (Proposed by M. Omarjee, 125(5), 2018). Suppose that $f : [0,1] \to \mathbb{R}$ has a continuous and nonnegative third derivative, and suppose that $\int_0^1 f(x)\,dx = 0$. Prove that

$$10\int_0^1 x^3 f(x)\,dx + 6\int_0^1 xf(x)\,dx \ge 15\int_0^1 x^2 f(x)\,dx.$$

5.3 The product of derivatives by Darboux's theorem

Problem 11753 (Proposed by P. Pongsriiam, 121(1), 2014). Let f be a continuous map from $[0,1]$ to $\mathbb{R}$ that is differentiable on $(0,1)$, with $f(0) = 0$ and $f(1) = 1$. Show that for each positive integer n there exist distinct numbers $c_1, \ldots, c_n \in (0,1)$ such that

$$\prod_{k=1}^n f'(c_k) = 1.$$

Discussion.
For $n = 1$, we wish to find $c_1 \in (0,1)$ such that $f'(c_1) = 1$. This follows from the mean value theorem immediately. For $n = 2$, consider the subintervals

$[0, x]$ and $[x, 1]$ where $x \in (0, 1)$ is to be determined. By the mean value theorem, there is a $c_1 \in (0, x)$ and $c_2 \in (x, 1)$ such that

$$f'(c_1) = \frac{f(x) - f(0)}{x - 0} = \frac{f(x)}{x} \quad \text{and} \quad f'(c_2) = \frac{f(1) - f(x)}{1 - x} = \frac{1 - f(x)}{1 - x}.$$

Thus, $f'(c_1)f'(c_2) = 1$ if and only if

$$f(x)(1 - f(x)) = x(1 - x).$$

This does not hold unless $f(x) = x$ or $f(x) = 1 - x$. However, we see that if $f(x) > x$, then $f'(c_1) > 1$ and $f'(c_2) < 1$. Similarly, if $f(x) < x$, then $f'(c_1) < 1$ and $f'(c_2) > 1$.

Let $D = \{f'(t) : t \in (0, 1)\}$. Then there is a $\delta > 0$ such that $(1 - \delta, 1 + \delta) \subset D$. By Darboux's theorem, $f'(x)$ has the intermediate value property. Thus, we can select distinct c_1 and c_2 such that

$$f'(c_1) = y \in (1, 1 + \delta) \quad \text{and} \quad f'(c_2) = \frac{1}{y} \in (1/(1 + \delta), 1) \subset (1 - \delta, 1).$$

For example, let $f(x) = x^2$. We can choose $\delta = 1/2, y = 4/3, c_1 = 2/3, c_2 = 3/8$.

With the above analysis we can carry out the case for an arbitrary positive integer n.

Solution.
If $f(x) \equiv x$, the required equality holds trivially, so assume that $f(x)$ is not identically equal to x. Based on the analysis in the discussion above, we see that the required equality holds for $n = 2$. For any $n > 2$, let $m = \lfloor n/2 \rfloor$. Choose m distinct numbers $y_k \in (1, 1 + \delta)$ $(1 \leq k \leq m)$. Darboux's theorem assures that there are distinct c_k $(1 \leq i \leq 2m)$ such that

$$f'(c_k) = y_k, \quad f'(c_{m+k}) = \frac{1}{y_k}, \quad (1 \leq k \leq m).$$

If n is odd, just take $y_n = 1$. Thus,

$$\prod_{k=1}^{n} f'(c_k) = \prod_{k=1}^{m} f'(c_k) \cdot f'(c_{m+k}) = \prod_{k=1}^{m} \left(y_k \cdot \frac{1}{y_k} \right) = 1.$$

$\square$

Remark. The featured solution by John Hagood [53] used induction. By contradiction, he proved that if there exist distinct $c_1, c_2, \ldots, c_{n-1}$ with $\prod_{i=1}^{n-1} f'(c_i) = 1$, then there is a number $c_n \notin \{c_1, c_2, \ldots, c_{n-1}\}$ such that $f'(c_n) = 1$.

This problem pushes the mean value theorem in another direction: Once there exists c such that $f'(c) = 1$, so does the geometric mean of distinct derivatives. These kinds of extensions have appeared in recent journals and books. Here we collect a list of problems related to arithmetic, geometric and harmonic means for additional practice.

1. Let $f(x)$ be differentiable on $[0,1]$ with $f(0) = 0$ and $f(1) = 1$. For each positive integer n and arbitrary weights w_k, $(w_k > 0, \sum_{k=1}^{n} w_k = 1)$, show that there exist distinct points $c_1, c_2, \ldots, c_n \in [0,1]$ such that

$$\sum_{k=1}^{n} \frac{w_k}{f'(c_k)} = 1.$$

2. Let $f(x)$ be continuous on $[a,b]$ and $c \in (a,b)$ which is not an extreme point of f. For each positive integer n, show that

    (a) there exist distinct points $c_1, c_2, \ldots, c_n \in [a,b]$ such that

$$f(c) = \frac{1}{n} \sum_{k=1}^{n} f(c_k);$$

    (b) there exist distinct points $c_1, c_2, \ldots, c_n \in [a,b]$ such that

$$f(c) = \sqrt[n]{\prod_{k=1}^{n} f(c_k)};$$

    (c) if $f(c) \neq 0$, then there exist distinct points $c_1, c_2, \ldots, c_n \in [a,b]$ such that
$$f(c) = \frac{n}{\sum_{k=1}^{n} 1/f(c_k)}.$$

3. Let $f(x)$ and $g(x)$ be differentiable on $[a,b]$ and $g'(x) \neq 0$ for every $x \in (a,b)$. For each positive integer n, show that

    (a) there exist distinct points $c_1, c_2, \ldots, c_n \in (a,b)$ such that

$$\frac{f(b) - f(a)}{g(b) - g(a)} = \frac{1}{n} \sum_{k=1}^{n} \frac{f'(c_k)}{g'(c_k)};$$

    (b) there exist distinct points $c_1, c_2, \ldots, c_n \in (a,b)$ such that

$$\frac{f(b) - f(a)}{g(b) - g(a)} = \sqrt[n]{\prod_{k=1}^{n} \frac{f'(c_k)}{g'(c_k)}};$$

    (c) if $f(b) \neq f(a)$, then there exist distinct points $c_1, c_2, \ldots, c_n \in (a,b)$ such that

$$\frac{f(b) - f(a)}{g(b) - g(a)} = \frac{n}{\sum_{k=1}^{n} \frac{g'(c_k)}{f'(c_k)}}.$$

4. Let $f(x)$ and $g(x)$ be continuous on $[a, b]$ and $g(x) \neq 0$ for every $x \in (a, b)$. For each positive integer n, show that

    (a) there exist distinct points $c_1, c_2, \ldots, c_n \in (a, b)$ such that

    $$\frac{\int_a^b f(x)\, dx}{\int_a^b g(x)\, dx} = \frac{1}{n} \sum_{k=1}^n \frac{f(c_k)}{g(c_k)};$$

    (b) there exist distinct points $c_1, c_2, \ldots, c_n \in (a, b)$ such that

    $$\frac{\int_a^b f(x)\, dx}{\int_a^b g(x)\, dx} = \sqrt[n]{\prod_{k=1}^n \frac{f(c_k)}{g(c_k)}};$$

    (c) if $\int_a^b f(x)\, dx \neq 0$, then there exist distinct points $c_1, c_2, \ldots, c_n \in (a, b)$ such that

    $$\frac{\int_a^b f(x)\, dx}{\int_a^b g(x)\, dx} = \frac{n}{\sum_{k=1}^n \frac{g(c_k)}{f(c_k)}}.$$

Comments. Math. Magazine Problem 1867, proposed by A. Plaza and C. Rodriguez, is a special case of parts (a) and (c) with $[a, b] = [0, 1]$, $\int_0^1 f(x)\, dx = 1, g(x) \equiv 1$. CMJ Problem 956, proposed by D. V. Thong, is a special case of (b) with $n = 3, [a, b] = [0, 1]$, $\int_0^1 f(x)\, dx = 1, g(x) \equiv 1$.

5.4   An integral-derivative inequality

Problem 11417 (Proposed by C. Lupu and T. Lupu, 116(2), 2009). Let f be a continuous differentiable real-valued function on $[0, 1]$ such that $\int_{1/3}^{2/3} f(x)dx = 0$. Show that

$$\int_0^1 (f'(x))^2\, dx \geq 27 \left(\int_0^1 f(x)\, dx \right)^2. \tag{5.5}$$

Discussion.
A natural approach is to try to apply the Cauchy-Schwarz inequality. In view of (5.5), the Cauchy-Schwarz inequality yields

$$\int_0^1 (f'(x))^2\, dx \int_0^1 g^2(x)\, dx \geq \left(\int_0^1 f'(x)g(x)\, dx \right)^2.$$

Thus, it suffices to find a function $g(x)$ for which

$$\int_0^1 g^2(x)\,dx = \frac{1}{27} \quad \text{and} \quad \int_0^1 f'(x)g(x)\,dx = \int_0^1 f(x)\,dx.$$

Solution.

Integration by parts yields

$$\begin{aligned}
\int_0^1 f'(x)g(x)\,dx &= f(x)g(x)|_0^1 - \int_0^1 f(x)g'(x)\,dx \\
&= f(x)g(x)|_0^1 - \left(\int_0^{1/3} f(x)g'(x)\,dx + \int_{1/3}^{2/3} f(x)g'(x)\,dx \right. \\
&\quad \left. + \int_{2/3}^1 f(x)g'(x)\,dx \right).
\end{aligned}$$

The expected equality

$$\int_0^1 f'(x)g(x)\,dx = \int_0^1 f(x)\,dx$$

suggests that we choose $g(x)$ such that $g(0) = g(1) = 0$ and

$$g'(x) = \begin{cases} -1, & x \in [0, 1/3]; \\ \text{constant}, & x \in [1/3, 2/3]; \\ -1, & \in [2/3, 1]. \end{cases}$$

In the simplest form, we have

$$g(x) = \begin{cases} -x, & x \in [0, 1/3]; \\ 2x - 1, & x \in [1/3, 2/3]; \\ 1 - x, & x \in [2/3, 1]. \end{cases} \tag{5.6}$$

Notice that g is continuous and its derivative is piecewise continuous on $[0, 1]$. Moreover,

$$\int_0^1 g^2(x)\,dx = \int_0^{1/3} x^2\,dx + \int_{1/3}^{2/3} (2x - 1)^2\,dx + \int_{2/3}^1 (1 - x)^2\,dx = \frac{1}{27}.$$

The inequality to be proved then follows from the Cauchy-Schwarz inequality

$$\int_0^1 (f'(x))^2\,dx \int_0^1 g^2(x)\,dx \geq \left(\int_0^1 f'(x)g(x)\,dx \right)^2 = \left(\int_0^1 f(x)\,dx \right)^2.$$

$\square$

Remark. The selection of $g(x)$ is not unique. In fact, for any continuous function $\phi(x)$ with $\int_0^1 \phi(x)\,dx = 1$ and $\int_0^1 f(x)\phi(x)\,dx = 0$, define

$$g(x) = \int_0^x \phi(t)\,dt - x.$$

With the essentially same argument above, regardless of the condition that $\int_{1/3}^{2/3} f(x)\,dx = 0$, we have

$$\int_0^1 (f'(x))^2\,dx \geq \frac{1}{C}\left(\int_0^1 f(x)\,dx\right)^2,$$

where $C = \int_0^1 g^2(x)\,dx$.

The equality

$$\int_0^1 f'(x)g(x)\,dx = \int_0^1 f(x)\,dx$$

has played a key role to establish (5.5). It is interesting to see what happens if we replace $f'(x)$ by $f''(x)$: Let $g(x)$ be given by (5.6). Define $G(x) = \int_0^x g(t)\,dt$. Then

$$G(x) = \begin{cases} -x^2/2, & x \in [0, 1/3]; \\ x^2 - x + 1/6, & x \in [1/3, 2/3]; \\ -x^2/2 + x - 1/2, & x \in [2/3, 1]. \end{cases}$$

Notice that G is continuously differentiable with $G(0) = G(1) = G'(0) = G'(1) = 0$ and its second derivative is piecewise continuous. Integrating by parts twice gives

$$\int_0^1 f''(x)G(x)\,dx = \int_0^1 f(x)\,dx.$$

Since

$$\int_0^1 G^2(x)\,dx = \frac{11}{4860},$$

the Cauchy-Schwarz inequality then yields a solution to the following

Problem 11946 (Proposed by M. Omarjee, 123(10), 2016). Let f be a twice differentiable function from $[0, 1]$ to $\mathbb{R}$ with f'' continuous on $[0, 1]$ and $\int_{1/3}^{2/3} f(x)\,dx = 0$. Prove

$$4860\left(\int_0^1 f(x)\,dx\right)^2 \leq 11\int_0^1 (f''(x))^2\,dx.$$

To end this section, we compile a few more problems for your practice.

1. Let f be twice continuously differentiable on $[0, \pi]$ with $f(0) = f(\pi) = 0$. Let $f(x) = \sum_{k=1}^{\infty} a_k \sin kx$ and $S_n(x) = \sum_{k=1}^{n} a_k \sin kx$. Show that, for every $n \in \mathbb{N}$,

$$\int_0^\pi (f(x) - S_n(x))^2\,dx \leq \frac{1}{3n^2}\int_0^\pi |f''(x)|^2\,dx.$$

2. **Problem 11133** (Proposed by P. Bracken, 112(2), 2005). Let f be a nonnegative, continuous, concave function on $[0, 1]$ with $f(0) = 1$. Prove that

$$2 \int_0^1 x^2 f(x) \, dx + \frac{1}{12} \leq \left(\int_0^1 f(x) \, dx \right)^2.$$

Comment. Seiffert offered the following extension: For any $p > 0$,

$$(p+1) \int_0^1 x^{2p} f(x) \, dx + \frac{2p-1}{8p+4} \leq \left(\int_0^1 f(x) \, dx \right)^2.$$

3. **Problem 11548** (Proposed by C. Lupu and T. Lupu, 118(1), 2011). Let f be a twice-differentiable real-valued function with continuous second derivative, and suppose that $f(0) = 0$. Show that

$$\int_{-1}^1 (f''(x))^2 \, dx \geq 10 \left(\int_{-1}^1 f(x) \, dx \right)^2.$$

4. **Problem 11756** (Proposed by P. Perfetti, 121(2), 2014). Let f be a function from $[-1, 1]$ to $\mathbb{R}$ with continuous derivatives of all orders up to $2n + 2$. Given $f(0) = f''(0) = \cdots = f^{(2n+2)}(0) = 0$, prove

$$\frac{(4n+5)((2n+2)!)^2}{2} \left(\int_{-1}^1 f(x) \, dx \right)^2 \leq \int_{-1}^1 (f^{(2n+2)}(x))^2 \, dx.$$

5. **Problem 11812** (Proposed by C. Chiser, 122(1), 2015). Let f be a twice continuously differentiable function from $[0, 1]$ to $\mathbb{R}$. Let p be an integer greater that 1. Given that

$$\sum_{k=1}^{p-1} f(k/p) = -\frac{1}{2}(f(0) + f(1)),$$

prove that

$$\left(\int_0^1 f(x) \, dx \right)^2 \leq \frac{1}{5! \, p^4} \int_0^1 (f''(x))^2 \, dx.$$

6. **Problem 11861** (Proposed by P. C. Le Van, 122(8), 2015). Let n be a natural number and let f be a continuous function from $[0, 1]$ to $\mathbb{R}$ such that $\int_0^1 f^{2n+1}(x) \, dx = 0$. Prove that

$$\frac{(2n+1)^{2n+1}}{(2n)^{2n}} \left(\int_0^1 f(x) \, dx \right)^{4n} \leq \int_0^1 (f(x))^{4n} \, dx.$$

7. **Problem 11884** (Proposed by C. Lupu and T. Lupu, 123(1), 2016).
   Let f be a real-valued function on $[0,1]$ such that f and its first two
   derivatives are continuous. Prove that if $f(1/2) = 0$, then

$$\int_0^1 (f''(x))^2 \, dx \geq 320 \left(\int_0^1 f(x) \, dx \right)^2.$$

8. **Problem 11918** (Proposed by P. C. Le Van, 123(6), 2016). Let f be
   continuously differentiable on $[0,1]$, with $f(1/2) = 0$ and $f^{(i)}(0) = 0$
   when i is even and less than n. Prove

$$\left(\int_0^1 f(x) \, dx \right)^2 \leq \frac{1}{(2n+1)4^n(n!)^2} \int_0^1 (f^{(n)}(x))^2 \, dx.$$

9. **Monthly Problem 11981** (Proposed by C. Lupu, 124(5), 2017).
   Suppose $f : [0,1] \to \mathbb{R}$ is a differentiable function with continuous
   derivative and with $\int_0^1 f(x) \, dx = \int_0^1 x f(x) \, dx = 1$. Prove that

$$\int_0^1 |f'(x)|^3 \, dx \geq \left(\frac{128}{3\pi} \right)^2.$$

10. **Problem 12088** (Proposed by F. Stanescu, 126(1), 2019). Let k
    be a positive integer with $k \geq 2$, and let $f : [0,1] \to \mathbb{R}$ be a function
    with continuous k-th derivative. Suppose $f^{(k)}(x) \geq 0$ for all $x \in$
    $[0,1]$, and suppose $f^{(i)}(0) = 0$ for all $i \in \{0, 1, \ldots, k-2\}$. Prove

$$\int_0^1 x^{k-1} f(1-x) \, dx \leq \frac{(k-1)!k!}{(2k-1)!} \int_0^1 f(x) \, dx.$$

5.5   Eigenvalues of a $(0,1)$-matrix

Problem 10958 (Proposed by R. Chapman, 124(2), 2017). Let A_n be the
$n \times n$ $(0,1)$-matrix with 1s in exactly those positions (j,k) such that $n \leq$
$j + k \leq n + 1$. Find the eigenvalues of A_n.

Discussion.
The published solution is in a verification manner. Unfortunately, it is hard
to see where the claimed eigenvalues and corresponding eigenfunctions come
from. Let the characteristic polynomial of A_n be $P_n(\lambda)$. Since the zeros of
$P_n(\lambda)$ are the eigenvalues of A_n, to get a feel for $P_n(\lambda) = \det(\lambda I_n - A_n)$,

we compute $P_n(\lambda)$ directly by cofactor expansions along the last row of the matrix for $n = 1, 2, 3, 4$:

$$P_1(\lambda) = \lambda - 1;$$

$$P_2(\lambda) = \begin{vmatrix} \lambda - 1 & -1 \\ -1 & \lambda \end{vmatrix} = \lambda(\lambda - 1) - 1 = \lambda P_1(\lambda) - 1;$$

$$P_3(\lambda) = \begin{vmatrix} \lambda & -1 & -1 \\ -1 & \lambda - 1 & 0 \\ -1 & 0 & \lambda \end{vmatrix} = \lambda \begin{vmatrix} \lambda & -1 \\ -1 & \lambda - 1 \end{vmatrix} - \begin{vmatrix} -1 & -1 \\ \lambda - 1 & 0 \end{vmatrix}$$

$$= \lambda P_2(\lambda) - P_1(\lambda);$$

$$P_4(\lambda) = \begin{vmatrix} \lambda & 0 & -1 & -1 \\ 0 & \lambda - 1 & -1 & 0 \\ -1 & -1 & \lambda & 0 \\ -1 & 0 & 0 & \lambda \end{vmatrix} = \lambda P_3(\lambda) - P_2(\lambda).$$

In general, these facts suggest that $P_n(\lambda)$ satisfies the recurrence relation:

$$P_n(\lambda) = \lambda P_{n-1}(\lambda) - P_{n-2}(\lambda) \quad \text{for } n \geq 2 \tag{5.7}$$

with $P_0(\lambda) = 1, P_1(\lambda) = \lambda - 1$. Thus, we can find P_n by solving the recurrence (5.7), which can be carried out either by a special function or by a generating function.

Solution I.

By induction, for $n \geq 2$, we can verify that $P_n(\lambda) = \det(\lambda I_n - A_n)$ satisfies (5.7) with $P_0(\lambda) = 1, P_1(\lambda) = \lambda - 1$. Now, we determine $P_n(\lambda)$ in terms of the well-known *Chebyshev polynomials of the second kind* $U_n(\lambda)$.

Recall that $U_n(\lambda)$ is defined by

$$U_n(\lambda) = 2\lambda U_{n-1}(\lambda) - U_{n-2}(\lambda), \quad \text{with} \quad U_0(\lambda) = 1, U_1(\lambda) = 2\lambda. \tag{5.8}$$

We claim that $P_n(\lambda) := U_n(\lambda/2) - U_{n-1}(\lambda/2)$ satisfies (5.7) for $n = 2, 3, \ldots$. In fact, this is verified directly for $n = 2$. For $n \geq 3$, using (5.8) twice yields

$$\begin{aligned} P_n(\lambda) &= U_n(\lambda/2) - U_{n-1}(\lambda/2) \\ &= \lambda U_{n-1}(\lambda/2) - U_{n-2}(\lambda/2) - (\lambda U_{n-2}(\lambda/2) - U_{n-3}(\lambda/2)) \\ &= \lambda(U_{n-1}(\lambda/2) - U_{n-2}(\lambda/2)) - (U_{n-2}(\lambda/2) - U_{n-3}(\lambda/2)) \\ &= \lambda P_{n-1}(\lambda) - P_{n-2}(\lambda). \end{aligned}$$

Since

$$U_n(\cos \theta) = \frac{\sin(n + 1)\theta}{\sin \theta},$$

by the sum to product and double angle formulas, we find that

$$\begin{aligned} P_n(2 \cos \theta) &= U_n(\cos \theta) - U_{n-1}(\cos \theta) \\ &= \frac{\sin(n + 1)\theta - \sin n\theta}{\sin \theta} = \frac{\cos[(2n + 1)\theta/2]}{\cos(\theta/2)}. \end{aligned}$$

Hence, the zeros of P_n (i.e., the eigenvalues of A_n) are

$$\lambda_k = 2\cos\theta_k = 2\cos\left(\frac{2k-1}{2n+1}\pi\right),$$

where $k = 1, 2, \ldots, n$.   □

Solution II.
We provide another approach to solve (5.7) by using the generating function method. Let

$$G(x) = \sum_{n=0}^{\infty} P_n(\lambda)x^n.$$

Using (5.7) and the initial conditions, we obtain

$$
\begin{aligned}
G(x) &= P_0(\lambda) + P_1(\lambda)x + \sum_{n=2}^{\infty} P_n(\lambda)x^n \\
&= 1 + (\lambda - 1)x + \lambda\sum_{n=2}^{\infty} P_{n-1}(\lambda)x^n - \sum_{n=2}^{\infty} P_{n-2}(\lambda)x^n \\
&= 1 + (\lambda - 1)x + \lambda x(G(x) - 1) - x^2 G(x).
\end{aligned}
$$

Therefore,

$$G(x) = \frac{1-x}{1-\lambda x + x^2}.$$

Let $\lambda = 2\cos\theta$. Then $1 - \lambda x + x^2 = (1 - xe^{i\theta})(1 - xe^{-i\theta})$. Using partial fractions, we have

$$G(x) = \frac{A}{1 - xe^{i\theta}} + \frac{B}{1 - xe^{-i\theta}},$$

where

$$A = \frac{e^{i\theta}}{1 + e^{i\theta}}, \quad B = \frac{1}{1 + e^{i\theta}}.$$

Invoking the geometric series, we find that

$$P_n(\lambda) = Ae^{in\theta} + Be^{-in\theta}.$$

To find the zeros of $P_n(\lambda)$, rewrite

$$P_n(\lambda) = Be^{-in\theta}(e^{i(2n+1)\theta} + 1).$$

For $P_n(\lambda) = 0$, we must have

$$e^{i(2n+1)\theta} + 1 = 0.$$

This implies that $\theta = (2k-1)\pi/(2n+1)$ for $k = 1, 2\ldots, n$. Hence, the eigenvalues of A_n are

$$\lambda_k = 2\cos\theta_k = 2\cos\left(\frac{2k-1}{2n+1}\pi\right),$$

for $k = 1, 2 \ldots, n$. $\qquad\qquad\qquad\qquad\qquad\qquad\qquad\qquad\qquad$ □

Remark. The key insight of using the substitution $\lambda = 2 \cos \theta$ can be revealed by the structure of P_n itself. Let $z = e^{i\theta}$. Then $\lambda = 2 \cos \theta = z + z^{-1}$. It is interesting to see that

$$
\begin{aligned}
P_1(z + z^{-1}) &= z - 1 + z^{-1}; \\
P_2(z + z^{-1}) &= z^2 - z + 1 - z^{-1} + z^{-2}; \\
&\cdots\cdots \\
P_n(z + z^{-1}) &= z^n - z^{n-1} + \cdots + z^{-n}.
\end{aligned}
$$

Hence,

$$
P_n(z + z^{-1}) = z^{-n} \left(z^{2n} - z^{2n-1} + \cdots + 1 \right) = z^{-n} \frac{1 + z^{2n+1}}{1 + z}.
$$

In this form, we can easily find all zeros of P_n.

The $(0, 1)$-matrices have played a fundamental role in a wide variety of combinatorial problems. For example, let $X = \{x_1, x_2, \ldots, x_n\}$ and let $X_1, X_2, \ldots, X_m$ be subsets of X. We can define a $(0, 1)$-matrix $A = (a_{ij})_{m \times n}$ as follows: $a_{ij} = 1$ if $x_j \in X_i$, otherwise, $a_{ij} = 0$. Here the 1's in the ith row of A indicate the elements that belong to X_i and the 1's in the jth column of A specify the sets that contain x_j. Thus, the matrix A characterizes the m subsets $X_1, \ldots, X_m$ of X. Moreover, let r_i be the sum of the ith row and s_j be the sum of the jth column. Define vectors

$$
R = (r_1, r_2, \ldots, r_m) \in \mathbb{R}^m \quad \text{and} \quad S = (s_1, s_2, \ldots, s_n) \in \mathbb{R}^n.
$$

Let $\mathcal{U}(\mathcal{R}, \mathcal{S})$ be all $m \times n$ $(0, 1)$-matrices with $\sum_{j=1}^{n} a_{ij} = r_i$ and $\sum_{i=1}^{m} a_{ij} = s_j$. As an illustration, let $R = (4, 3, 3, 3, 3)$ and $S = (5, 3, 3, 3, 2)$. Then

$$
A = \begin{pmatrix} 1 & 1 & 1 & 0 & 1 \\ 1 & 1 & 0 & 1 & 0 \\ 1 & 1 & 0 & 0 & 1 \\ 1 & 0 & 1 & 1 & 0 \\ 1 & 0 & 1 & 1 & 0 \end{pmatrix}, \quad B = \begin{pmatrix} 1 & 0 & 1 & 1 & 1 \\ 1 & 1 & 0 & 1 & 0 \\ 1 & 0 & 1 & 0 & 1 \\ 1 & 1 & 1 & 0 & 0 \\ 1 & 1 & 0 & 1 & 0 \end{pmatrix} \in \mathcal{U}(\mathcal{R}, \mathcal{S}).
$$

In graph theory, the matrices in $\mathcal{U}(\mathcal{R}, \mathcal{S})$ are the incidence matrices of the bipartite simple graphs of $BG(R, S)$. For more details, we direct the interested reader to Brualdi's expository paper [26].

Here are a few additional problems for your practice.

1. Let

$$
A = \begin{pmatrix} a & b \\ c & d \end{pmatrix}
$$

be invertible, and let $x = \dfrac{a+d}{2\sqrt{\det(A)}}$. Show that

$$
A^n = \det(A)^{(n-1)/2} U_{n-1}(x) A - \det(A)^{n/2} U_{n-2}(x) I,
$$

where $U_n(x)$ is the Chebyshev polynomial of the second kind.

2. **Problem 12025** (Proposed by A. Dzhumadil'daev, 125(2), 2018). The Chebyshev polynomials of the second kind are defined by the recurrence relation

$$U_0(x) = 1, \ U_1(x) = 2x, \text{ and } U_n(x) = 2xU_{n-1}(x) - U_{n-2}(x) \text{ for } n \geq 2.$$

For an integer n with $n \geq 2$, prove

$$\det \begin{pmatrix} 0 & 1 & 1 & \cdots & 1 & 1 \\ x & 0 & 1 & \cdots & 1 & 1 \\ x^2 & x & 0 & \cdots & 1 & 1 \\ \vdots & \vdots & \vdots & \ddots & \vdots & \vdots \\ x^{n-2} & x^{n-3} & x^{n-4} & \cdots & 0 & 1 \\ x^{n-1} & x^{n-2} & x^{n-3} & \cdots & x & 0 \end{pmatrix} = (-1)^{n-1}x^{n/2}U_{n-2}(\sqrt{x}).$$

3. Let

$$A = \begin{pmatrix} a & b & 0 & \cdots & 0 & 0 \\ b & a & b & \cdots & 0 & 0 \\ 0 & b & a & \cdots & 0 & 0 \\ \vdots & \vdots & \vdots & \ddots & \vdots & \vdots \\ 0 & 0 & 0 & \cdots & a & b \\ 0 & 0 & 0 & \cdots & b & a \end{pmatrix} \quad \text{and}$$

$$B = \begin{pmatrix} 0 & 0 & \cdots & 0 & b & a \\ 0 & 0 & \cdots & -b & -a & -b \\ 0 & 0 & \cdots & a & b & 0 \\ \vdots & \vdots & \vdots & \ddots & \vdots & \vdots \\ (-1)^{n-2}b & (-1)^{n-2}a & \cdots & 0 & 0 & 0 \\ (-1)^{n-1}a & (-1)^{n-1}b & \cdots & 0 & 0 & 0 \end{pmatrix}.$$

Determine the eigenvalues of A and B.

4. **SIAM Problem 61-5** (Proposed by C. Sealander). Let $A = (a_{ij})_{i,j=1}^{n-1}$ with

$$a_{ij} = \sin \frac{i\pi}{n} \sin \frac{j\pi}{n} \cos \frac{(i-j)\pi}{n}.$$

Find the eigenvalues of A.

5. **Problem 11415** (Proposed by F. Holland, 116(2), 2009). Let $(A_1, \ldots, A_n)$ be a list of n positive-definite 2×2 matrices of complex numbers. Let G be the group of all unitary 2×2 complex matrices, and define the function F on the Cartesian product G^n by

$$F(U) = F(U_1, U_2, \ldots, U_n) = \det \left(\sum_{k=1}^{n} U_k^* A_k U_k \right).$$

Show that

$$\min_{U \in G^n} F(U) = \sum_{k=1}^{n} \sigma_1(A_k) \cdot \sum_{k=1}^{n} \sigma_2(A_k),$$

where $\sigma_1(A_j)$ and $\sigma_2(A_j)$ denote the greatest and the least eigenvalues of A_j, respectively.

6. **Problem 12100** (Proposed by F. Holland, T. Laffey, and R. Smyth, 126(3), 2019). For a positive integer n, let A_n be the n-by-n tridiagonal matrix whose i, j-entry is given by

$$a_{ij} = \begin{cases} -2j(n - j + 1) & \text{if } j = i; \\ j(n - j + 1) & \text{if } j = i \pm 1; \\ 0 & \text{if } |i - j| > 1. \end{cases}$$

Determine the eigenvalues of A_n.

5.6 A matrix of secants

Problem 11969 (Proposed by A. Dzhumadil'daev, 124(3), 2017). Let $x_1, \ldots, x_n$ be indeterminates, and let A be the $n \times n$ matrix with (i, j) entry $\sec(x_i - x_j)$. Prove

$$\det(A) = (-1)^{\binom{n}{2}} \prod_{1 \le i < j \le n} \tan^2(x_i - x_j).$$

Discussion.
We begin by checking out a few special cases. When $n = 2$, the proposed equality follows from

$$\det(A) = \begin{vmatrix} 1 & \sec(x_1 - x_2) \\ \sec(x_2 - x_1) & 1 \end{vmatrix} = 1 - \sec^2(x_1 - x_2) = -\tan^2(x_1 - x_2).$$

When $n = 3$,

$$\det(A) = \begin{vmatrix} 1 & \sec(x_1 - x_2) & \sec(x_1 - x_3) \\ \sec(x_2 - x_1) & 1 & \sec(x_2 - x_3) \\ \sec(x_3 - x_1) & \sec(x_3 - x_2) & 1 \end{vmatrix}.$$

Multiplying the last row by $-\sec(x_i - x_3)$, then adding to the ith row for $i = 1, 2$, we have

$$\det(A) = \begin{vmatrix} 1 - \sec(x_3 - x_1)\sec(x_1 - x_3) & \sec(x_1 - x_2) - \sec(x_3 - x_2)\sec(x_1 - x_3) & 0 \\ \sec(x_2 - x_1) - \sec(x_3 - x_1)\sec(x_2 - x_3) & 1 - \sec(x_3 - x_2)\sec(x_2 - x_3) & 0 \\ \sec(x_3 - x_1) & \sec(x_3 - x_2) & 1 \end{vmatrix}.$$

Using the trigonometric identity

$$\sec(x_i - x_j) - \sec(x_i - x_3)\sec(x_3 - x_j) = -\tan(x_i - x_3)\tan(x_j - x_3)\sec(x_i - x_j),$$

then applying the cofactor expansion, as desired, we find that

$$\det(A) = \begin{vmatrix} -\tan^2(x_1 - x_3) & -\tan(x_1 - x_3)\tan(x_2 - x_3)\sec(x_1 - x_2) & 0 \\ -\tan(x_1 - x_3)\tan(x_2 - x_3)\sec(x_2 - x_1) & -\tan^2(x_2 - x_3) & 0 \\ \sec(x_3 - x_1) & \sec(x_3 - x_2) & 1 \end{vmatrix}$$

$$= \begin{vmatrix} -\tan^2(x_1 - x_3) & -\tan(x_1 - x_3)\tan(x_2 - x_3)\sec(x_1 - x_2) \\ -\tan(x_1 - x_3)\tan(x_2 - x_3)\sec(x_2 - x_1) & -\tan^2(x_2 - x_3) \end{vmatrix}$$

$$= \tan^2(x_1 - x_3)\tan^2(x_2 - x_3)\begin{vmatrix} 1 & \sec(x_1 - x_2) \\ \sec(x_2 - x_1) & 1 \end{vmatrix}$$

$$= -\tan^2(x_1 - x_3)\tan^2(x_2 - x_3)\tan^2(x_1 - x_2).$$

Here we not only have verified the special cases $n = 2, 3$, but also have found the idea to proceed for the general case. In fact, a proof based on the above idea is a nice application of induction.

Another approach is to use the *Cauchy matrix*. Let $a_i = b_i = \exp(2\tau x_i)$ with $\tau = \sqrt{-1}, 1 \leq i \leq n$. Since

$$\cos(x_i - x_j) = \frac{\exp(\tau(x_i - x_j)) + \exp(-\tau(x_i - x_j))}{2} = \frac{a_i + a_j}{2\exp(\tau x_i)\exp(\tau x_j)},$$

we have

$$\det(A) = \det\left(\left[\frac{2\exp(\tau x_i)\exp(\tau x_j)}{a_i + a_j}\right]_{i,j=1}^n\right)$$

$$= 2^n \left(\prod_{i=1}^n \exp(\tau x_i)\right)^2 \det\left(\left[\frac{1}{a_i + a_j}\right]_{i,j=1}^n\right)$$

$$= \prod_{i=1}^n (2a_i) \det\left(\left[\frac{1}{a_i + a_j}\right]_{i,j=1}^n\right).$$

In this form, the proposed equality directly follows from the *Cauchy determinant* ([68], p. 348):

$$\det\left(\left[\frac{1}{a_i + b_j}\right]_{i,j=1}^n\right) = \frac{\prod_{1 \leq i < j \leq n}(a_i - a_j)(b_i - b_j)}{\prod_{1 \leq i,\, j \leq n}(a_i + b_j)}. \tag{5.9}$$

Solution I.
We use induction on n for $n \geq 2$. Denote the n-by-n matrix A by A_n. We have proved the case $n = 2$ in the discussion above. For $n \geq 3$ we proceed as we did in the case $n = 3$. Multiplying the last row by $-\sec(x_i - x_n)$, then adding to the ith row for $1 \leq i < n$, we have

$$\det(A_n) = \begin{vmatrix} \genfrac{}{}{0pt}{}{1-\sec(x_n-x_1)\sec(x_1-x_n)}{\sec(x_2-x_1)-\sec(x_n-x_1)\sec(x_2-x_n)} & \genfrac{}{}{0pt}{}{\sec(x_1-x_2)-\sec(x_n-x_2)\sec(x_1-x_n)}{1-\sec(x_n-x_2)\sec(x_2-x_n)} & \cdots & 0 \\ \vdots & \vdots & \ddots & \vdots \\ \genfrac{}{}{0pt}{}{\sec(x_{n-1}-x_1)-\sec(x_n-x_1)\sec(x_{n-1}-x_n)}{\sec(x_n-x_1)} & \genfrac{}{}{0pt}{}{\sec(x_{n-1}-x_2)-\sec(x_n-x_2)\sec(x_{n-1}-x_n)}{\sec(x_n-x_2)} & \cdots & 1 \end{vmatrix}.$$

Since

$$\sec(x_i-x_j)-\sec(x_i-x_n)\sec(x_n-x_j) = -\tan(x_i-x_n)\tan(x_j-x_n)\sec(x_i-x_j),$$

we find that

$$\det(A_n) = \begin{vmatrix} \genfrac{}{}{0pt}{}{-\tan^2(x_1-x_n)}{-\tan(x_1-x_n)\tan(x_2-x_n)\sec(x_2-x_1)} & \genfrac{}{}{0pt}{}{-\tan(x_1-x_n)\tan(x_2-x_n)\sec(x_1-x_2)}{-\tan^2(x_2-x_n)} & \cdots & 0 \\ \vdots & \vdots & \ddots & \vdots \\ \sec(x_n-x_1) & \sec(x_n-x_2) & \cdots & 1 \end{vmatrix}.$$

In the above determinant, for $1 \le i, j < n$, there is a common factor $\tan(x_i - x_n)$ in ith row and a common factor $\tan(x_j - x_n)$ in jth column. After extracting these factors, then expanding the determinant along the last column yields

$$\det(A_n) = (-1)^{n-1} \prod_{i=1}^{n-1} \tan^2(x_i - x_n) \cdot \det(A_{n-1}).$$

Invoking the induction hypothesis, we obtain

$$\begin{aligned} \det(A_n) &= (-1)^{n-1} \prod_{i=1}^{n-1} \tan^2(x_i - x_n) \cdot (-1)^{n-2} \\ &\quad \prod_{i=1}^{n-2} \tan^2(x_i - x_{n-1}) \cdot \det(A_{n-2}) \\ &= (-1)^{n-1} \prod_{i=1}^{n-1} \tan^2(x_i - x_n) \cdot (-1)^{n-2} \\ &\quad \prod_{i=1}^{n-2} \tan^2(x_i - x_{n-1}) \cdots (-1)\tan^2(x_1 - x_2) \\ &= (-1)^{(n-1)+(n-2)+\cdots+1} \prod_{1\le i<j\le n} \tan^2(x_i - x_j) \\ &= (-1)^{\binom{n}{2}} \prod_{1\le i<j\le n} \tan^2(x_i - x_j) \end{aligned}$$

as desired. □

Solution II.

Since the Cauchy determinant (5.9) is not necessarily well-known, for the convenience of the reader, we start by presenting an inductive proof. Let

$$C_n = \det\left(\left[\frac{1}{a_i + b_j}\right]_{i,j=1}^n\right).$$

Define $a_{ij} = 1/(a_i + b_j)$. Since

$$a_{ij} - a_{i1} = \frac{1}{a_i + b_j} - \frac{1}{a_i + b_1} = \frac{b_1 - b_j}{(a_i + b_j)(a_i + b_1)},$$

subtracting column 1 from the jth column for $2 \leq j \leq n$ gives

$$\det(C_n) = \begin{vmatrix} \frac{1}{a_1+b_1} & \frac{b_1-b_2}{(a_1+b_1)(a_1+b_2)} & \frac{b_1-b_3}{(a_1+b_1)(a_1+b_3)} & \cdots & \frac{b_1-b_n}{(a_1+b_1)(a_1+b_n)} \\ \frac{1}{a_2+b_1} & \frac{b_1-b_2}{(a_2+b_1)(a_2+b_2)} & \frac{b_1-b_3}{(a_2+b_1)(a_2+b_3)} & \cdots & \frac{b_1-b_n}{(a_2+b_1)(a_2+b_n)} \\ \vdots & \vdots & \vdots & \ddots & \vdots \\ \frac{1}{a_n+b_1} & \frac{b_1-b_2}{(a_n+b_1)(a_n+b_2)} & \frac{b_1-b_3}{(a_n+b_1)(a_n+b_3)} & \cdots & \frac{b_1-b_n}{(a_n+b_1)(a_n+b_n)} \end{vmatrix}.$$

Factoring $1/(a_i + b_1)$ from each row and $b_1 - b_j$ from columns 2 to n yields

$$\det(C_n) = \frac{\prod_{j=2}^{n}(b_1 - b_j)}{\prod_{i=1}^{n}(a_i + b_1)} \begin{vmatrix} 1 & \frac{1}{a_1+b_2} & \frac{1}{a_1+b_3} & \cdots & \frac{1}{a_1+b_n} \\ 1 & \frac{1}{a_2+b_2} & \frac{1}{a_2+b_3} & \cdots & \frac{1}{a_2+b_n} \\ \vdots & \vdots & \vdots & \ddots & \vdots \\ 1 & \frac{1}{a_n+b_2} & \frac{1}{a_n+b_3} & \cdots & \frac{1}{a_n+b_n} \end{vmatrix}.$$

Similarly, subtracting row 1 from the ith row for $2 \leq i \leq n$ gives

$$\det(C_n) = \frac{\prod_{j=2}^{n}(b_1 - b_j)}{\prod_{i=1}^{n}(a_i + b_1)} \cdot \frac{\prod_{i=2}^{n}(a_1 - a_i)}{\prod_{j=1}^{n}(a_1 + b_j)} \begin{vmatrix} 1 & 1 & 1 & \cdots & 1 \\ 0 & \frac{1}{a_2+b_2} & \frac{1}{a_2+b_3} & \cdots & \frac{1}{a_2+b_n} \\ \vdots & \vdots & \vdots & \ddots & \vdots \\ 0 & \frac{1}{a_n+b_2} & \frac{1}{a_n+b_3} & \cdots & \frac{1}{a_n+b_n} \end{vmatrix}.$$

Condensing the products and expanding the determinant along the first column yields

$$\det(C_n) = \frac{\prod_{i=2}^{n}(a_1 - a_i)(b_1 - b_i)}{\prod_{1 \leq i, j \leq n}(a_i + b_1)(a_1 + b_j)} \det(C_{n-1}).$$

In view of the induction hypothesis, this proves (5.9).

To prove the proposed equality, let $a_i = b_i = \exp(2\tau x_i)$ with $\tau = \sqrt{-1}$. Then

$$\tan(x_i - x_j) = \frac{a_i - a_j}{\tau(a_i + a_j)}.$$

Finally, applying (5.9), we obtain

$$\begin{aligned} \det(A) &= \prod_{i=1}^{n}(2a_i) \det\left(\left[\frac{1}{a_i + a_j}\right]_{i,j=1}^{n}\right) \\ &= \prod_{i=1}^{n}(2a_i) \cdot \frac{\prod_{1 \leq i < j \leq n}(a_i - a_j)^2}{\prod_{1 \leq i, j \leq n}(a_i + a_j)^2} = \prod_{1 \leq i < j \leq n}\left(\frac{a_i - a_i}{a_i + a_j}\right)^2 \\ &= \prod_{1 \leq i < j \leq n}(\tau \tan(x_i - x_j))^2 = (-1)^{\binom{n}{2}} \prod_{1 \leq i < j \leq n} \tan^2(x_i - x_j). \end{aligned}$$

□

Remark. Eliminating all denominators on the left-hand side of (5.9) leads to

$$\det\left(\left[\frac{1}{a_i + b_j}\right]_{i,j=1}^{n}\right) = \frac{1}{\prod_{1\le i,j\le n}(a_i + b_j)}\det\left([A_{ij}]_{i,j=1}^{n}\right),$$

where

$$A_{ij} = (a_i + b_1)\cdots(a_i + b_{j-1})(a_i + b_{j+1})\cdots(a_i + b_n) = \prod_{k\ne j}(a_i + b_k).$$

From (5.9) we recover *Krattenthaler's formula* (1990)

$$\det\left([A_{ij}]_{i,j=1}^{n}\right) = \prod_{1\le i<j\le n}(a_i - a_j)(b_i - b_j). \tag{5.10}$$

This formula is viewed as the extension of the *Vandermonde determinant*

$$V(x_1, x_2, \cdots, x_n) = \begin{vmatrix} 1 & 1 & \cdots & 1 \\ x_1 & x_2 & \cdots & x_n \\ \vdots & \vdots & \ddots & \vdots \\ x_1^{n-1} & x_2^{n-1} & \cdots & x_n^{n-1} \end{vmatrix} = \prod_{1\le i<j\le n}(x_i - x_j)$$

and has played an important role in the proof of the *alternating sign matrix conjecture*.

An *alternating sign matrix* is a square matrix of 0s, 1s, and −1s for which

- the sum of the entries in each row and in each column is 1,

- the nonzero entries of each row and of each column alternate in sign.

An example of such matrix is

$$\begin{pmatrix} 0 & 1 & 0 \\ 1 & -1 & 1 \\ 0 & 1 & 0 \end{pmatrix}.$$

In 1866, Charles Dodgson, better known under his pen name of Lewis Carroll, discovered a *condensation algorithm* for calculating determinants via only 2×2 determinants: For any $n \times n$ matrix A, let $A_r(k, l)$ be the $r \times r$ connected submatrix whose upper leftmost corner is the entry a_{kl}, then

$$\det(A) = \frac{\det(A_{n-1}(1,1))\det(A_{n-1}(2,2)) - \det(A_{n-1}(1,2))\det(A_{n-1}(2,1))}{\det(A_{n-2}(2,2))}.$$

For example,

$$\det\begin{pmatrix} 1 & -2 & -1 \\ 2 & 1 & -1 \\ -1 & -2 & 1 \end{pmatrix} = \begin{vmatrix} 1 & -2 \\ 2 & 1 \end{vmatrix}\cdot\begin{vmatrix} 1 & -1 \\ -2 & 1 \end{vmatrix} - \begin{vmatrix} -2 & -1 \\ 1 & -1 \end{vmatrix}\cdot\begin{vmatrix} 2 & 1 \\ -1 & -2 \end{vmatrix}.$$

In contrast to expressing the determinant of an $n \times n$ matrix as a sum over the permutations of n letters, by generalizing Dodgson condensation, Robbins and Rumsey (early 1980s) proved that the determinant is a sum over alternating sign matrices. We know that the number of $n \times n$ permutation matrices is $n!$. Naturally, Robbins and Rumsey wanted to known the number of $n \times n$ alternating sign matrices. Based on the pattern obtained by ALTRAN (a computer algebra package), they made the *Alternating sign matrix conjecture*: The total number of $n \times n$ alternating sign matrices is

$$A_n := \prod_{k=0}^{n-1} \frac{(3k+1)!}{(n+k)!} = \frac{1!\, 4!\, 7! \cdots (3n-2)!}{n!\, (n+1)! \cdots (2n-1)!}. \tag{5.11}$$

Zeilberger (1995) proved (5.11) by showing that alternating sign matrices are equinumerous with totally symmetric, self-complementary plane partitions. (5.10) was used to determine the generating function for plane partitions. For the reader interested in learning more on alternating sign matrices and the determinant formulas such as (5.9) and (5.10), please refer to David Bressoud's beautiful expository book [25].

We now end this section by compiling some problems for additional practice.

1. Let $n \in \mathbb{N}$ and
$$A = \left(\binom{2(i+j-1)}{i+j-1} \right)_{i,j=1}^{n}.$$
Show that $\det(A) = 2^n$.

2. **Problem 10387** (Proposed by S. Rabinowitz and P. J. Costa, 101(5), 1994). Let $T_n = (t_{i,j})$ be the $n \times n$ matrix with $t_{i,j} = \tan(i+j-1)x$, i.e.,

$$\begin{pmatrix} \tan x & \tan 2x & \tan 3x & \cdots & \tan nx \\ \tan 2x & \tan 3x & \tan 4x & \cdots & \tan(n+1)x \\ \vdots & \vdots & \vdots & \ddots & \vdots \\ \tan nx & \tan(n+1)x & \tan(n+2)x & \cdots & \tan(2n-1)x \end{pmatrix}.$$

Computer experiments suggest that

$$\det(T_n) = (-1)^{\lfloor n/2 \rfloor} \sec^n nx \prod_{k=1}^{n-1} (\sin^2(n-k)x \sec kx \sec(2n-k)x)^k$$

$$\times \begin{cases} \sin n^2 x & \text{if } n \text{ odd,} \\ \cos n^2 x & \text{if } n \text{ even.} \end{cases}$$

Prove or disprove this conjecture.

Comment. The featured solution by Robbins actually evaluated a more general determinant:

$$\det \left(\frac{ax_i + by_j}{x_i + y_j} \right).$$

Haeringen offered another generalization: Let $D_n(t) = \det\left([\tan(t + i + j - 1)x]_{i,j=1}^n\right)$. Then

$$D_{n+1}(t-1)D_{n-1}(t+1) = D_n(t+1)D_n(t-1) - D_n^2(t).$$

It is interesting to see that $A_n(t) = \det\left([(t + i + j - 1)^{-1}]_{i,j=1}^n\right)$ satisfies the same recurrence.

3. **Problem 11475** (Proposed by Ömer Egecioglu, 117(1), 2010). Let $H_k = \sum_{j=1}^k 1/j$, and let D_n be the determinant of the $(n+1)\times(n+1)$ Hankel matrix with (i,j) entry H_{i+j+1} for $1 \le i,j \le n$. (Thus, $D_1 = -5/12$ and $D_2 = 1/216$.) Show that for $n \ge 1$,

$$D_n = \frac{\prod_{i=1}^n i!^4}{\prod_{i=1}^{2n+1} i!} \cdot \sum_{j=0}^n \frac{(-1)^j (n+j+1)! H_{j+1}}{j!(j+1)!(n-j)!}.$$

Hint. Let H be the $(n+1)\times(n+1)$ Hilbert matrix with (i,j) entry $1/(i+j+1)$ for $0 \le i,j \le n$. Show that

$$\det(H) = \frac{\prod_{i=1}^n i!^4}{\prod_{i=1}^{2n+1} i!}.$$

4. **Problem 11463** (Proposed by X. Chang, 116(9), 2009). Let A be a positive-definite $n \times n$ Hermitian matrix with minimum eigenvalue λ and maximum eigenvalue Λ. Show that

$$\left(\frac{n}{\mathrm{tr}((A + \lambda I)^{-1})} - \lambda\right)^n \le \det(A) \le \left(\frac{n}{\mathrm{tr}((A + \Lambda I)^{-1})} - \Lambda\right)^n.$$

Hint: Notice that $\mathrm{tr}((A + \lambda I)^{-1}) = \sum_{k=1}^n 1/(\lambda + \lambda_k)$, where $\lambda_1, \ldots, \lambda_n$ denote the eigenvalue of A.

5. **Problem 11471** (Proposed by F. Holland, 116(10), 2009). Let A be an $r \times r$ matrix with distinct eigenvalues $\lambda_1, \ldots, \lambda_n$. For $n \ge 0$, let $a(n)$ be the trace of A^n. Let $H(n)$ be the $r \times r$ Hankel matrix with (i,j) entry $a(i+j+n-2)$. Show that

$$\lim_{n\to\infty} |\det H(n)|^{1/n} = \prod_{k=1}^r |\lambda_k|.$$

Hint: Show that $\det H(0) \ne 0$ and $\det H(n+1) = (\prod_{k=1}^r \lambda_k) \cdot \det H(n)$.

6. **Problem 11293** (Proposed by S. Sadov, 114(5), 2007). Let $S_n(q)$

be the $n \times n$ matrix in which the entries are q through q^{n^2}, spiraling inwards with q in the $(1, 1)$ entry, so that, for instance,

$$S_2 = \begin{pmatrix} q & q^2 \\ q^4 & q^3 \end{pmatrix}, \qquad S_3 = \begin{pmatrix} q & q^2 & q^3 \\ q^8 & q^9 & q^4 \\ q^7 & q^6 & q^5 \end{pmatrix}.$$

Show that for $n \geq 2$

$$\det S_n = (-1)^{(n-2)(n-1)/2} q^{(2n^3 - 6n^2 + 13n - 6)/3} \prod_{k=0}^{n-2} \left(1 - q^{2+4k}\right).$$

7. (G. Kuperberg and J. Propp) Prove that

(a) $\det \left(\left[\binom{i+j}{i} \binom{2n-i-j}{n-i} \right]_{i,j=0}^{n} \right) = \dfrac{[(2n+1)!]^{n+1}}{\prod_{k=0}^{2n+1} k!}.$

(b) $\det \left(\left[\dfrac{1 - s^{i+j-1}}{1 - t^{i+j-1}} \right]_{i,j=1}^{n} \right) = t^{n^3/3 - n^2/2 + n/6} \prod_{1 \leq i < j \leq n} (1 - t^{j-i})^2$
$\prod_{i,j=1}^{n} \dfrac{1 - st^{j-i}}{1 - t^{i+j-1}}.$

8. (T. Amdeberhan and D. Zeilberger) Prove that

(a) $\det \left(\left[\dfrac{1}{a_i + b_j + t a_i b_j} \right]_{i,j=1}^{n} \right) = \dfrac{\prod_{1 \leq i < j \leq n} (a_i - a_j)(b_i - b_j)}{\prod_{1 \leq i,\, j \leq n} (a_i + b_j + t a_i b_j)}.$

(b) $\det \left(\left[\dfrac{1}{a_i + b_j} - \dfrac{1}{1 + a_i b_j} \right]_{i,j=1}^{n} \right) = \dfrac{\prod_{1 \leq i < j \leq n} (1 - a_i a_j)(1 - b_i b_j)(a_i - a_j)(b_i - b_j)}{\prod_{1 \leq i,\, j \leq n} (a_i + b_j)(1 + a_i b_j)}.$
$\prod_{i=1}^{n} (1 - x_i)(1 - y_i).$

(c) $\det \left(\left[\dfrac{s a_i + t b_j}{a_i + b_j} \right]_{i,j=1}^{n} \right) = (t - s)^{n-1}$
$\dfrac{(t \prod_{j=1}^{n} b_j + (-1)^{n-1} s \prod_{i=1}^{n} a_i) \prod_{1 \leq i < j \leq n} (a_i - a_j)(b_i - b_j)}{\prod_{1 \leq i,\, j \leq n} (a_i + b_j)}.$

9. Show that

$$\det \begin{pmatrix} 1 + x_1^{2n-2} & x_1 + x_1^{2n-3} & \cdots & x_1^{n-1} + x_1^{n-1} \\ 1 + x_2^{2n-2} & x_2 + x_2^{2n-3} & \cdots & x_2^{n-1} + x_2^{n-1} \\ \vdots & \vdots & \ddots & \vdots \\ 1 + x_n^{2n-2} & x_n + x_n^{2n-3} & \cdots & x_n^{n-1} + x_n^{n-1} \end{pmatrix}$$

$$= 2 \left(\prod_{1 \leq i < j \leq n} (x_i - x_j)(x_i x_j - 1) \right).$$

10. Let $P(x_1, x_2, \ldots, x_n)$ be an alternating polynomial (i.e., P changes sign when any two variables are transposed) of degree d. Show that

$$\dfrac{P(x_1, x_2, \ldots, x_n)}{\prod_{1 \leq i < j \leq n} (x_i - x_j)}$$

is a symmetric polynomial of degree $d - \binom{n}{2}$. Use this result to evaluate

$$\det\left(\left[\frac{1}{1 - a_i b_j}\right]^n_{i,j=1}\right).$$

Hint: Use the fact that $\prod(1 - a_i b_j) \det(1/(1 - a_i b_j))$ is an alternating polynomial in a_i and in b_j with degree $n - 1$, respectively.

11. **Problem 12066** (Proposed by X. Chang, 125(8), 2018). Let n and k be integers greater than 1, and let A be an n-by-n positive definite Hermitian matrix. Prove

$$(\det A)^{1/n} \leq \left(\frac{\text{trace}^k(A) - \text{trace}(A^k)}{n^k - n}\right)^{1/k}.$$

12. **Independent Study** — Chapman's evil determinant problem. Let p be an odd prime and define $k = (p + 1)/2$. Let A_p be the $k \times k$ matrix with (i, j) entry $\left(\frac{j-i}{p}\right)$ (where $\left(\frac{-}{p}\right)$ denotes the Legendre symbol modulo p). For example

$$A_7 = \begin{pmatrix} 0 & 1 & 1 & -1 \\ -1 & 0 & 1 & 1 \\ -1 & -1 & 0 & 1 \\ 1 & -1 & -1 & 0 \end{pmatrix}.$$

Evaluate $\det(A_p)$. For $p \equiv 3 \pmod 4$, *Mathematica* shows that $\det(A_p) = 1$ for $p < 100$. For $p \equiv 1 \pmod 4$, *Mathematica* shows that $\det(A_p)$ is always negative and even. You may use these facts to make your own conjectures.

5.7 A gcd-weighted trigonometric sum

Problem 12003 (Proposed by N. Osipov, 124(8), 2017). Given an odd positive integer n, compute

$$\sum_{k=1}^n \frac{\gcd(k, n)}{\cos^2(k\pi/n)}.$$

Discussion.

It is known that, for any odd positive integers m,

$$\sum_{k=1}^m \frac{1}{\cos^2(k\pi/m)} = m^2. \tag{5.12}$$

Thus, in the early stages of analysis, it is natural to think about how to convert the weighted sum into unweighted sums as (5.12). Here we investigate the special case $n = 9$ and hope to shed some light on proceeding with the general problem. Notice that

$$\{\gcd(k, 9)_{k=1}^{9}\} = \{1, 1, 3, 1, 1, 3, 1, 1, 9\}.$$

This contains three different numbers $\{1, 3, 9\}$, which are all factors of 9 exactly. Observe that

$$
\begin{aligned}
\gcd(1, 9) &= 1 = \phi(1); \\
\gcd(3, 9) &= 3 = \phi(1) + \phi(3); \\
\gcd(9, 9) &= 9 = \phi(1) + \phi(3) + \phi(9),
\end{aligned}
$$

where $\phi(m)$ is the Euler totient function, which is defined as the number of positive integers $\leq m$ that are relative prime to m. Thus, we can regroup the weighted sum as the sums with the same weights of $\phi(1), \phi(3)$, and $\phi(9)$, respectively. i.e.,

$$\sum_{k=1}^{9} \frac{\gcd(k, 9)}{\cos^2(k\pi/9)} = \phi(1) \sum_{k=1}^{9} \frac{1}{\cos^2(k\pi/9)} + \phi(3) \sum_{k=1}^{3} \frac{1}{\cos^2(3k\pi/9)}$$
$$+ \phi(9) \frac{1}{\cos^2(9\pi/9)}.$$

Using (5.12), we find that

$$\sum_{k=1}^{9} \frac{\gcd(k, 9)}{\cos^2(k\pi/9)} = 9^2 + 2 \cdot 3^2 + 6 \cdot 1 = 105.$$

We proceed to the general case using the same argument.

Solution.
Let $m = \gcd(k, n) = \sum_{d|m} \phi(d)$ and $r = k/d$. Then

$$
\begin{aligned}
\sum_{k=1}^{n} \frac{\gcd(k, n)}{\cos^2(k\pi/n)} &= \sum_{k=1}^{n} \left(\sum_{d|\gcd(k,n)} \frac{\phi(d)}{\cos^2(k\pi/n)} \right) \\
&= \sum_{d|n} \phi(d) \left(\sum_{r=1}^{n/d} \frac{1}{\cos^2(r\pi/(n/d))} \right)
\end{aligned}
$$
$$\text{(exchange the sum order)}$$
$$= \sum_{d|n} \phi(d) \left(\frac{n}{d} \right)^2 \quad \text{(use (5.12)).}$$

To simplify the last sum above, first let $n = p^r$, where p is prime and r is a positive integer. Since $d = p^i (0 \leq i \leq r)$ and $\phi(p^i) = p^i - p^{i-1}$ for $i \geq 1$, we find

$$\sum_{d|n} \phi(d) \left(\frac{n}{d} \right)^2 = p^{2r} + \sum_{i=1}^{r} (p^i - p^{i-1}) p^{2(r-i)} = p^{2r} + p^{2r-1} - p^{r-1}. \quad (5.13)$$

Recall that f is *multiplicative* if $f(x \cdot y) = f(x) \cdot f(y)$ for x and y relatively prime. It is well-known that ϕ is multiplicative. When n_1 and n_2 are relative prime, the divisors of $n_1 n_2$ are the products of the divisors of n_1 and n_2. Thus, n/d is multiplicative, and so $n^2 \phi(d)/d^2$ is a multiplicative function of n.

Finally, let $n = \prod_{i=1}^{l} p_i^{r_i}$ be its prime factorization. Using the multiplicative property and (5.13), we obtain

$$\sum_{k=1}^{n} \frac{\gcd(k, n)}{\cos^2(k\pi/n)} = \prod_{i=1}^{l} (p_i^{2r_i} + p_i^{2r_i - 1} - p_i^{r_i - 1}). \tag{5.14}$$

For example, let $n = 1125 = 3^2 5^3$. Then

$$\sum_{k=1}^{n} \frac{\gcd(k, n)}{\cos^2(k\pi/n)} = 105 \cdot 18725 = 1966125.$$

$\square$

Remark. For completeness, we give an elementary proof of (5.12). Recall that if $r_1, \ldots, r_n$ are roots of a nth degree polynomial $P_n(x)$, then

$$\sum_{k=1}^{n} \frac{1}{x - r_k} = \frac{P_n'(x)}{P_n(x)}.$$

In particular, if all $r_k \neq 0$ for $1 \leq k \leq n$, we have

$$\sum_{k=1}^{n} \frac{1}{r_k} = -\frac{P_n'(0)}{P_n(0)}. \tag{5.15}$$

Thus, it suffices to find an nth degree polynomial which has zeros $\cos^2(k\pi/n)$ $(1 \leq k \leq n)$. To this end, let $T_n(x)$ be the nth Chebyshev polynomial of the first kind. Since

$$T_n(\cos\theta) = \cos n\theta,$$

this implies that $\cos(2k\pi/n)$ $(1 \leq k \leq n)$ are zeros of $T_n(x) - 1$. Using $\cos^2(k\pi/n) = (1 + \cos(2k\pi/n))/2$ and n is odd, we see that $\cos^2(k\pi/n) \neq 0$ $(1 \leq k \leq n)$ are zeros of

$$P_n(x) := T_n(2x - 1) - 1.$$

As desired, (5.12) now follows from (5.15) by

$$\sum_{k=1}^{n} \frac{1}{\cos^2(k\pi/n)} = -\frac{P_n'(0)}{P_n(0)} = -\frac{2T_n'(-1)}{T_n(-1) - 1} = -\frac{2n^2}{-2} = n^2.$$

In general, when m is a positive integer, the formulas for the power sums

$\sum_{k=1}^{n} 1/\cos^{-2m}(k\pi/n)$ and their extensions can be obtained by the generating function method (For example, see Chapter 14 in [29]).

The restriction to odd n is necessary to ensure the neat answer form of (5.14). In fact, if n is even, we have

$$\sum_{\substack{k \neq n/2}}^{n} \frac{1}{\cos^2(k\pi/n)} = \frac{1}{3}(n^2 - 1),$$

which is not a multiplicative function of n. The same argument above then only offers us

$$\sum_{\substack{k \neq n/2}}^{n} \frac{\gcd(k,n)}{\cos^2(k\pi/n)} = \sum_{d|n, d\neq 2} \phi(d) \frac{1}{3}\left(\left(\frac{n}{d}\right)^2 - 1\right),$$

which cannot be simplified as neatly as (5.14).

Another extension of the proposed problem is to consider sums of the form:

$$\sum_{k=1}^{n} \frac{\gcd(k,n)}{\cos^{2m}(k\pi/n)},$$

where m is a positive integer. For example, notice that

$$\sum_{k=1}^{n} \frac{1}{\cos^4(k\pi/n)} = \frac{n^2}{3}(n^2 + 2).$$

Along the same lines, we find that

$$\sum_{k=1}^{n} \frac{\gcd(k,n)}{\cos^4(k\pi/n)} = \sum_{d|n} \phi(d) \left(\sum_{r=1}^{n/d} \frac{1}{\cos^4(r\pi/(n/d))}\right)$$

$$= \sum_{d|n} \frac{1}{3}\phi(d) \left(\frac{n}{d}\right)^2 \left(\left(\frac{n}{d}\right)^2 + 2\right).$$

Since $x^2 + 2$ is not multiplicative, again, we can't expect a simple formula like (5.14).

Here we provide a few more problems for additional practice.

1. Let n be a positive integer. Compute

$$\sum_{k=1}^{n-1} \frac{\gcd(k,n)}{\sin^2(k\pi/n)}.$$

2. For an odd positive integer n, let

$$S_m(n) = \sum_{k=1}^{n} \frac{1}{\cos^{2m}(k\pi/n)} \quad \text{and} \quad G(n,t) = \sum_{m=1}^{\infty} S_m(n)t^{2m}.$$

Show that
$$G(n,t) = \frac{nt}{\sqrt{1-t^2}} \tan(n \arcsin t).$$

3. **Problem 12099** (Proposed by M. Bataille, 126(3), 2019). Let m and n be integers with $0 \le m \le n-1$. Evaluate
$$\sum_{k=0,\, k \ne m}^{n-1} \cot^2\left(\frac{(m-k)\pi}{n}\right).$$

Hint: Show the sum is same as $\sum_{k=1}^{n-1} \cot^2\left(\frac{k\pi}{n}\right)$ first.

5.8 A lcm-sum weighted Dirichlet series

Problem 12114 (Proposed by Z. Franco, 126(5), 2019). Let n be a positive integer, and let
$$A_n = \{1/n, 2/n, \dots, n/n\}.$$
Let a_n be the sum of the numerators in A_n when these fractions are expressed in lowest terms. For example, $A_6 = \{1/6, 1/3, 1/2, 2/3, 5/6, 1/1\}$, so $a_6 = 1+1+1+2+5+1 = 11$. Find $\sum_{n=1}^{\infty} a_n/n^4$.

Discussion.
Recall that a *Dirichlet series* is a series of the form: $F(s) = \sum_{n=1}^{\infty} f(n)/n^s$, where $f(n)$ is a number-theoretic function. The sum of the series $F(s)$ is called the generating function of $f(n)$. For example, it is well-known that [48]
$$\sum_{n=1}^{\infty} \frac{\mu(n)}{n^s} = \frac{1}{\zeta(s)}, \quad (s > 1) \tag{5.16}$$
and
$$\sum_{n=1}^{\infty} \frac{\sigma(n)}{n^s} = \zeta(s)\zeta(s-1), \quad (s > 2), \tag{5.17}$$
where $\mu(n)$ is the Möbius function and $\sigma(n)$ is the sum of the divisors of n.

Motivated by such results, we expect to represent the a_n in terms of some number-theoretic functions. To search for the answer, we input the first several terms of a_n into *the On-Line Encyclopedia of Integer Sequence* (OEIS, https://oeis.org) and obtain the following output:

```
A057661        a(n) = Sum_{k=1..n} lcm(n,k)/n.
1, 2, 4,  6,  11,  11,  22,  22,  31,  32,  56,  39,  79,  65,  74,  86,  137,
92, 172, 116, 151, 167,  254, 151, 261, 236, 274, 237, 407,
```

221, 466, 342, 389, 410, 452, 336, 667, 515, 550,
452, 821, 452, 904, 611, 641, 761, 1082, 599, 1051,
  782, 956, 864, 1379, 821, 1166,

Here $\mathrm{lcm}(n, k)$ is the least common multiple of n and k. Once we know the answer, the verification becomes routine.

Solution.
We establish the following generalization: For $s > 3$,

$$\sum_{n=1}^{\infty} \frac{a_n}{n^s} = \frac{1}{2}\zeta(s)\left(1 + \frac{\zeta(s-2)}{\zeta(s-1)}\right),$$

where ζ is the Riemann zeta function. In particular, the value of the proposed series is $\frac{1}{2}\zeta(4)(1 + \zeta(2)/\zeta(3))$.

Let q be the numerator of k/n when this fraction is in the lowest terms. Then

$$q = \frac{k}{\gcd(n, k)} = \frac{nk}{n\gcd(n, k)} = \frac{\gcd(n, k)\mathrm{lcm}(n, k)}{n\gcd(n, k)} = \frac{\mathrm{lcm}(k, n)}{n},$$

and so

$$a_n = \frac{1}{n}\sum_{k=1}^{n} \mathrm{lcm}(n, k).$$

Notice that, for $1 \leq k < n$,

$$\mathrm{lcm}(n, k) + \mathrm{lcm}(n, n-k) = \frac{nk}{\gcd(n, k)} + \frac{n(n-k)}{\gcd(n, n-k)} = \frac{n^2}{\gcd(n, k)}.$$

Hence,

$$2\sum_{k=1}^{n-1} \mathrm{lcm}(n, k) = \sum_{k=1}^{n-1} (\mathrm{lcm}(n, k) + \mathrm{lcm}(n, n-k)) = n\sum_{k=1}^{n-1} \frac{n}{\gcd(n, k)}.$$

Let $d = \gcd(n, k)$. Then d appears in $\sum_{k=1}^{n} n/\gcd(n, k)$ exactly $\phi(d)$ times, where ϕ is the Euler totient function. Thus,

$$\sum_{k=1}^{n} \frac{n}{\gcd(n, k)} = \sum_{d|n} d \cdot \phi(d). \tag{5.18}$$

Since $\mathrm{lcm}(n, n) = \gcd(n, n) = n$, we have

$$a_n = \frac{1}{2}\left(1 + \sum_{d|n} d \cdot \phi(d)\right).$$

Next, let $b_n = \sum_{d|n} d \cdot \phi(d)$. To determine the generating function of b_n defined by

$$D(s) := \sum_{n=1}^{\infty} \frac{b_n}{n^s},$$

we apply the *Dirichlet convolution theorem* [48]: If

$$F(s) = \sum_{n=1}^{\infty} \frac{f(n)}{n^s} \quad \text{and} \quad G(s) = \sum_{n=1}^{\infty} \frac{g(n)}{n^s},$$

then

$$F(s)G(s) = \sum_{n=1}^{\infty} \frac{(f*g)(n)}{n^s},$$

where

$$(f*g)(n) = \sum_{d|n} f(d)g(n/d).$$

Let $f(n) = n\phi(n), g(n) = 1$. We obtain

$$D(s) = \left(\sum_{n=1}^{\infty} \frac{\phi(n)}{n^{s-1}} \right) \left(\sum_{n=1}^{\infty} \frac{1}{n^s} \right) = \zeta(s) \sum_{n=1}^{\infty} \frac{\phi(n)}{n^{s-1}}.$$

Since

$$\phi(n) = \sum_{d|n} \mu(d) \frac{n}{d},$$

where $\mu(n)$ is the Möbius function, using the Dirichlet convolution theorem with $f(n) = \mu(n), g(n) = n$ again, we have

$$D(s) = \zeta(s) \left(\sum_{n=1}^{\infty} \frac{\mu(n)}{n^{s-1}} \right) \left(\sum_{n=1}^{\infty} \frac{n}{n^{s-1}} \right) = \zeta(s)\zeta(s-2) \sum_{n=1}^{\infty} \frac{\mu(n)}{n^{s-1}}.$$

In view of (5.16), we finally find that

$$D(s) = \frac{\zeta(s)\zeta(s-2)}{\zeta(s-1)}.$$

In summary, we obtain

$$\sum_{n=1}^{\infty} \frac{a_n}{n^s} = \frac{1}{2} \left(\sum_{n=1}^{\infty} \frac{1}{n^s} + D(s) \right) = \frac{1}{2}\zeta(s) \left(1 + \frac{\zeta(s-2)}{\zeta(s-1)} \right),$$

as claimed.   □

Remark. When you encounter an integer sequence, you may check it with the OEIS first. As the largest database of integer sequences, OEIS records information on integer sequences of interest to both professional mathematicians and amateurs, and is widely cited. As of June 2019 it contains more than 320,000 sequences. Each sequence in the OEIS contains the leading terms of the sequence, keywords, mathematical motivations, literature links, and more, including the option to generate a graph or play a musical representation of the sequence. This database is searchable by keyword and by subsequence. If the sequence you studied is not in the record, you may become the author once your submission is approved.

In view of the solution above, it is interesting to see that (5.18) also offers a solution to the following Monthly problem:

Problem 10829 (Proposed by W. Janous, 107(8), 2000). For a positive integer m, let $f(m) = \sum_{r=1}^{m} m/\gcd(m, r)$. Evaluate $f(m)$ in terms of the canonical factorization of m into a product of powers of distinct primes.

Indeed, since ϕ is multiplicative, by (5.18), $f(m)$ is multiplicative as well. Thus, it suffices to evaluate f at prime powers. Since $\phi(p^a) = p^a - p^{a-1}$ when p is prime and $a \in \mathbb{N}$, we have

$$f(p^a) = 1 + \sum_{k=1}^{a} p^k(p^k - p^{k-1}) = \frac{1 + p^{2a+1}}{1 + p},$$

and so

$$f\left(\prod_{i=1}^{k} p_i^{a_i}\right) = \prod_{i=1}^{k} f(p_i^{a_i}) = \prod_{i=1}^{k} \left(\frac{1 + p_i^{2a+1}}{1 + p_i}\right).$$

This implies that if $n = \prod_{i=1}^{k} p_i^{a_i}$ then

$$\sum_{i=1}^{n} \operatorname{lcm}(n, i) = \frac{1}{2}\left(1 + \prod_{i=1}^{k}\left(\frac{1 + p_i^{2a+1}}{1 + p_i}\right)\right).$$

Similarly, we have

$$\sum_{i=1}^{n} \gcd(n, i) = \sum_{d|n} d\,\phi(n/d) = \prod_{i=1}^{k}\left((a_i + 1)p_i^{a_i} - a_i p_i^{a_i-1}\right).$$

The gcd-sum arises in deriving asymptotic estimates for a lattice point counting problem, for integer coordinate points under the square root curve. Similar to the lcm-sum, its Dirichlet series also has a closed representation in terms of the Riemann zeta function (See the Problem 3 below). Gould and Shonhiwa [48] provided an excellent overview of the commonly known number-theoretic functions together with their corresponding Dirichlet series.

However, deriving the asymptotic forms for gcd-sum/lcm-sum and their partial sums of the Dirichlet series is much more challenging. Finding the asymptotic expressions like $\sum_{i,j=1}^{n} \gcd(i, j)$ is even harder because the function fails to be multiplicative. We now compile a few problems for additional practice. The partial results related to deriving the asymptotic expressions are summarized in the independent study.

1. Let $\mu(n)$ be the Möbius function. Prove that

$$\sum_{n=1}^{\infty} \frac{|\mu(n)|}{n^s} = \frac{\zeta(s)}{\zeta(2s)}, \qquad s > 1.$$

2. Let $\tau(n)$ be the number of divisors of n. Prove that, for $s > 1$,

   (a) $\sum_{n=1}^{\infty} \frac{\tau(n)}{n^s} = \zeta^2(s)$.

   (b) $\sum_{n=1}^{\infty} \frac{\tau(n^2)}{n^s} = \frac{\zeta^3(s)}{\zeta(2s)}$.

   (c) $\sum_{n=1}^{\infty} \frac{\tau^2(n)}{n^s} = \frac{\zeta^4(s)}{\zeta(2s)}$.

3. Let $g(n) = \sum_{k=1}^{n} \gcd(n, k)$. Let the generating function of $g(n)$ be $G(s)$. For $s > 2$, prove that

$$G(s) = \sum_{n=1}^{\infty} \frac{g(n)}{n^s} = \frac{\zeta^2(s-1)}{\zeta(s)}.$$

4. Let f be any number-theoretic function. Prove that, for $s > 2$,

$$\sum_{n=1}^{\infty} \frac{\sum_{k=1}^{n} f(\gcd(n, k))}{n^s} = \frac{\zeta(s-1)}{\zeta(s)} \sum_{n=1}^{\infty} \frac{f(n)}{n^s}.$$

5. Let

$$\Omega(n) = \sum_{p^i | n} 1 = \text{the total number of prime factors of } n$$

   counting repetitions of prime.

   For example, if $n = \prod_{i=1}^{k} p_i^{a_i}$, then $\Omega(n) = \sum_{i=1}^{k} a_i$. Define *Liouville's function* $\lambda(n)$ as $\lambda(1) = 1$ and $\lambda(n) = (-1)^r$ for $n \neq 1$, where $r = \Omega(n)$. Prove that, for $s > 1$,

$$\sum_{n=1}^{\infty} \frac{\lambda(n)}{n^s} = \frac{\zeta(2s)}{\zeta(s)} \quad \text{and} \quad \sum_{n=1}^{\infty} \frac{\lambda(n)\tau^2(n)}{n^s} = \frac{\zeta^3(2s)}{\zeta^4(s)}.$$

6. Let

$$\gamma(n) = \begin{cases} 1, & \text{if } n = 1, \\ p_1 p_2 \cdots p_k, & \text{if } n = \prod_{i=1}^{k} p_i^{a_i}. \end{cases}$$

   Prove that, for $s > 1$,

$$\sum_{n=1}^{\infty} \frac{\gamma(n)}{n^s} = \zeta(s) \prod_{p} \left(\frac{p^s + p - 1}{p^s} \right).$$

7. Let k be a positive integer. If $s > 2k + 1$, show that

$$\sum_{n=1}^{\infty} \frac{\sum_{i=1}^{n} \operatorname{lcm}^k(n, i)}{n^s}$$

$$= \zeta(s - k) \left(1 + \frac{1}{(k+1)\zeta(s - 2k)} \sum_{i=1}^{k+1} \binom{k+1}{i} B_{k+1-i}\zeta(s - k - i) \right),$$

where B_n is the nth Bernoulli's number.

8. **Problem 10797** (Proposed by P. Bateman and J. Kalb, 107(4), 2000). Let h and k be integers with $k > 0, h+k > 0$, and $\gcd(h, k) = 1$. For $n \geq 1$, let $L(n)$ be the least common multiple of the n numbers $h + k, h + 2k, \ldots, h + nk$. Prove that

$$\lim_{n \to \infty} \frac{\ln(L(n))}{n} = \frac{k}{\phi(k)} \sum_{1 \leq m \leq k, \gcd(m,k)=1} \frac{1}{m}.$$

9. **Problem 6615** (Proposed by K. Lau, 96(9), 1989). Prove that as n tends to infinity

(a) $\sum_{i=1}^{n} \sum_{j=1}^{n} \gcd(i, j) \sim \frac{n^2 \ln n}{\zeta(2)}$,

(b) $\sum_{i=1}^{n} \sum_{j=1}^{n} \operatorname{lcm}(i, j) \sim \frac{\zeta(3)n^4}{4\zeta(2)}$,

where ζ denotes the Riemann zeta function.
Hint. You may use the following known asymptotic formulas: for $x, n \geq 2$,

$$\sum_{n \leq x} \frac{\phi(n)}{n} = \frac{x}{\zeta(2)} + O(\ln x),$$

$$\sum_{k \leq n} \ln(n/k) = n \ln n - \ln(n!) = n + O(\ln n).$$

10. (**Independent Study**) We study the asymptotic expressions related to the gcd-sum and the lcm-sum. Let

$$g(n) := \sum_{k=1}^{n} \gcd(n, k) \quad \text{and} \quad l(n) := \sum_{k=1}^{n} \operatorname{lcm}(n, k).$$

(a) Let $r \in \mathbb{N}$ and $x > e$. Show that

    (i) $\sum_{n \leq x} \phi(n) = \frac{x^2}{2\zeta(2)} + O(x(\ln x)^{2/3}(\ln(\ln x))^{2/3})$,

    (ii) $\sum_{n \leq x} n^r \phi(n) = \frac{x^{r+2}}{(r+2)\zeta(2)} + O(x^{r+1}(\ln x)^{2/3}(\ln(\ln x))^{2/3})$.

(b) As $x \to \infty$, show that

    (i) $\sum_{n \leq x} g(n) = \frac{x^2 \ln x}{2\zeta(2)} + O(x^2)$;

    (ii) $\sum_{n \leq x} \frac{g(n)}{n^s} = \frac{x^{2-s} \ln x}{(2-s)\zeta(2)} + O(x^{2-s})$ for $s < 2$;

(iii) $\sum_{n\le x} \frac{g(n)}{n^2} = \frac{\ln^2 x}{2\zeta(2)} + O(\ln x)$;

(iv) $\sum_{n\le x} \frac{g(n)}{n^s} = \frac{x^{2-s}\ln x}{(2-s)\zeta(2)} + \frac{\zeta^2(s-1)}{\zeta(s)} + O(x^{2-s})$ for $s > 2$.

(c) As $x \to \infty$, show that

(i) $\sum_{n\le x} l(n) = \frac{\zeta(3)}{8\zeta(2)} x^4 + O(x^3(\ln x)^{2/3}(\ln(\ln x))^{4/3})$;

(ii) $\sum_{n\le x} \sum_{k=1}^{n} \text{lcm}^a(n,k) = \frac{\zeta(a+2)}{2(a+1)^2\zeta(2)} x^{2a+2} + O(x^{2a+1}(\ln x)^{2/3} (\ln(\ln x))^{4/3})$;

(iii) $\sum_{n\le x} \sum_{k=1}^{n} \frac{1}{\text{lcm}(n,k)} = \frac{\ln^3 x}{6\zeta(2)} + \frac{\ln^2 x}{2\zeta(2)}(\gamma + \ln(A^{12}/2\pi)) + O(\ln x)$, where A is the Glaisher-Kinkelin constant.

(d) Let $s \ge 2$. Prove that

$$\sum_{n=1}^{\infty} \sum_{k=1}^{n} \frac{1}{\text{lcm}^s(n,k)} = \frac{\zeta(s)}{2}\left(1 + \frac{\zeta^s}{\zeta(2s)}\right).$$

5.9   An infinite sum-product identity

Problem 11883 (Proposed by H. Ohtsuka, 123(1), 2016). For $|q| > 1$, prove that

$$\sum_{k=0}^{\infty} \frac{1}{(q^{2^0} + q)(q^{2^1} + q)\cdots(q^{2^k} + q)} = \frac{1}{q-1} \prod_{i=0}^{\infty} \frac{1}{q^{1-2^i} + 1}.$$

Discussion.
Setting $x = 1/q$ converts the proposed identity into

$$\sum_{k=0}^{\infty} \prod_{j=0}^{k} \frac{x^{2^j}}{1 + x^{2^j-1}} = \frac{x}{1-x} \prod_{i=0}^{\infty} \frac{1}{1 + x^{2^i-1}}. \tag{5.19}$$

We naturally consider the finite partial sums of the left-hand series in (5.19)

$$S_n := \sum_{k=0}^{n} \prod_{j=0}^{k} \frac{x^{2^j}}{1 + x^{2^j-1}}.$$

For $n = 1$ and 2, we have

$$S_1 = \frac{x}{1+1} + \frac{x}{1+1}\cdot\frac{x^2}{1+x} = \frac{x + x^2 + x^3}{(1+1)(1+x)} = \frac{x - x^4}{1-x}\cdot\frac{1}{(1+1)(1+x)}$$

$$S_2 = \frac{x}{1+1} + \frac{x}{1+1}\cdot\frac{x^2}{1+x} + \frac{x}{1+1}\cdot\frac{x^2}{1+x}\cdot\frac{x^4}{1+x^3}$$

$$= \frac{x + x^2 + \cdots + x^7}{(1+1)(1+x)(1+x^3)} = \frac{x - x^8}{1-x}\cdot\frac{1}{(1+1)(1+x)(1+x^3)}.$$

Based on these facts, we conjecture the general identity as

$$S_n = \frac{x - x^{2^{n+1}}}{1 - x} \prod_{i=0}^{n} \frac{1}{1 + x^{2^i - 1}}, \tag{5.20}$$

that can be confirmed by induction.

Solution.
We now prove (5.20) by induction. The base case $n = 1$ has been verified in the discussion. Assume (5.20) is true for $n = N$. Then

$$
\begin{aligned}
S_{N+1} &= \sum_{k=0}^{N} \prod_{j=0}^{k} \frac{x^{2^j}}{1 + x^{2^j - 1}} + \prod_{j=0}^{N+1} \frac{x^{2^j}}{1 + x^{2^j - 1}} \\
&= S_N + \frac{\prod_{j=0}^{N+1} x^{2^j}}{1 + x^{2^{N+1} - 1}} \prod_{i=0}^{N} \frac{1}{1 + x^{2^i - 1}} \\
&= \frac{x - x^{2^{N+1}}}{1 - x} \prod_{i=0}^{N} \frac{1}{1 + x^{2^i - 1}} + \frac{x^{2^{N+2} - 1}}{1 + x^{2^{N+1} - 1}} \prod_{i=0}^{N} \frac{1}{1 + x^{2^i - 1}} \\
&= \frac{1}{1 - x} \left[(x - x^{2^{N+1}})(1 + x^{2^{N+1} - 1}) + (1 - x)x^{2^{N+2} - 1} \right] \prod_{i=0}^{N+1} \frac{1}{1 + x^{2^i - 1}} \\
&= \frac{x - x^{2^{N+2}}}{1 - x} \prod_{i=0}^{N+1} \frac{1}{1 + x^{2^i - 1}}.
\end{aligned}
$$

Here the induction hypothesis is used in the second equality and

$$\prod_{j=0}^{N+1} x^{2^j} = x^{\sum_{j=0}^{N+1} 2^j} = x^{2^{N+2} - 1} \tag{5.21}$$

in the third equality. Thus, by induction, (5.20) is true for all $n \in \mathbb{N}$. Since $|x| < 1$, letting $n \to \infty$ in (5.20) yields the desired identity (5.19), which is equivalent to the proposed identity. $\qquad\square$

Remark. In view of (5.21), the identity (5.19) can be rewritten as

$$\sum_{k=0}^{\infty} x^{2^{k+1} - 1} \prod_{j=k+1}^{\infty} (1 + x^{2^j - 1}) = \frac{x}{1 - x}.$$

The featured solution by GCHQ Problem Solving Group described this as a nearly binary expansion. Indeed, as a formal power series, this is the statement that every positive integer has a unique expression as a sum of distinct numbers of the form $2^j - 1$ for $j \geq 1$, except that the smallest number used (expressed as $2^{k+1} - 1$) can appear once or twice. Using induction on k, they confirmed this result by partitioning the positive integers into blocks of the form $[2^k - 1, 2^{k+1} - 2]$ for $k \geq 1$.

The study of infinite sum-product identities can be traced back to Euler. To determine an efficient way for computing $p(n)$, the number of partitions of n (expressing n as a sum of distinct integers from the set $\{1, 2, \ldots, m\}$), Euler discovered the following *Pentagonal number theorem*:

$$\prod_{k=1}^{\infty} (1 - q^k) = 1 + \sum_{n=1}^{\infty} (-1)^n \left(q^{n(3n-1)/2} + q^{n(3n+1)/2} \right),$$

and finally established the generating function for $p(n)$:

$$\sum_{n=0}^{\infty} p(n)q^n = \prod_{k=1}^{\infty} \frac{1}{1 - q^k}.$$

Inspired by Euler's results and (5.19), we try to find the infinite series of $\prod_{n=0}^{\infty} (1 + xq^{2^n})$, where x is a parameter. Direct computation gives

$$\prod_{n=0}^{\infty} (1+xq^{2^n}) = 1+xq+xq^2+x^2q^3+xq^4+x^2q^5+x^2q^6+x^3q^7+xq^8+x^2q^9+x^2q^{10}+\cdots.$$

The exponents of x fall into a sequence:

$$0, 1, 1, 2, 1, 2, 2, 3, 1, 2, 2, 3, 2, 3, 3, 4, \cdots$$

OEIS reveals this is the 1's counting sequence $s_2(n)(A000120)$, which is defined as the number of 1's in the binary expansion of n. For example, $s_2(7) = 3$ since $7 = 111_2$. By induction, we have

$$\prod_{k=0}^{n} (1 + xq^{2^k}) = \sum_{n=0}^{2^{n+1}-1} x^{s_2(n)}q^n.$$

In particular, if $x = 1$ and $|q| < 1$, we obtain

$$\prod_{k=0}^{\infty} (1 + q^{2^k}) = \sum_{n=0}^{\infty} q^n = \frac{1}{1 - q}.$$

A further study indicates that $s_2(n)$ appears to be ubiquitous in Number Theory and Combinatorics. For example, $s_2(n)$ is the largest integer such that $2^{s_2(n)} | \binom{2n}{n}$; and the nth row of Pascal's triangle has $2^{s_2(n)-1}$ distinct odd binomial coefficients. $s_2(n)$ also appeared in the solution of the following Monthly problem:

Problem E2692 (Proposed by D. Woods, 85(1), 1978) Determine the limit of the sequence

$$\frac{x}{x+1}, \left(\frac{x}{x+1}\right) \Big/ \left(\frac{x+2}{x+3}\right), \frac{\left(\frac{x}{x+1}\right) \Big/ \left(\frac{x+2}{x+3}\right)}{\left(\frac{x+4}{x+5}\right) \Big/ \left(\frac{x+6}{x+7}\right)}, \cdots.$$

David Robbins' beautiful solution [78] presented a recurrence for this sequence: $f_0(x) = x$ and $f_{n+1}(x) = f_n(x)/f_n(x + 2^n)$ for $n \geq 0$ with

$$f_n(x) = \prod_{k=0}^{2^n-1} (x + k)^{\theta(k)},$$

where $\theta(k) = (-1)^{s_2(k)}$ is called *the Thue-Morse sequence* (A106400). In particular, letting $x = 1$ yields a surprising identity:

$$\prod_{n=0}^{\infty} \left(\frac{2n + 1}{2n + 2} \right)^{\theta(n)} = \frac{\sqrt{2}}{2},$$

which is often refereed to as *the Woods-Robbins identity* now. Since then, many results have been obtained in this direction, such as

$$\prod_{n=1}^{\infty} \left(\frac{1 + 1/n}{1 + 1/(n+1)} \frac{1 + 1/(2n+2)}{1 + 1/(2n)} \right)^{s_2(n)} = \frac{\pi}{2}.$$

Notice that $s_2(n)$ satisfies

$$s_2(2n) = s_2(n); \qquad s_2(2n + 1) = s_2(n) + 1.$$

This suggests that the series of the form

$$\sum_n s_2(n) f(n) \qquad \text{or} \qquad \sum_n (-1)^{s_2(n)} f(n)$$

can be computed explicitly for some appropriate sequences $\{f(n)\}$. For example, for $f(n) = \frac{1}{n(n+1)}$, we have

$$
\begin{aligned}
\sum_{n=1}^{\infty} \frac{s_2(n)}{n(n+1)} &= \sum_{k=0}^{\infty} \frac{s_2(2k+1)}{(2k+1)(2k+2)} + \sum_{k=1}^{\infty} \frac{s_2(2k)}{2k(2k+1)} \\
&= \sum_{k=0}^{\infty} \frac{s_2(k) + 1}{(2k+1)(2k+2)} + \sum_{k=1}^{\infty} \frac{s_2(k)}{2k(2k+1)} \\
&= \sum_{k=0}^{\infty} \frac{1}{(2k+1)(2k+2)} \\
&\quad + \sum_{k=1}^{\infty} s_2(k) \left(\frac{1}{2k(2k+1)} + \frac{1}{(2k+1)(2k+2)} \right) \\
&= \ln 2 + \frac{1}{2} \sum_{k=1}^{\infty} \frac{s_2(k)}{k(k+1)}.
\end{aligned}
$$

This implies that

$$\sum_{n=1}^{\infty} \frac{s_2(n)}{n(n+1)} = 2 \ln 2. \tag{5.22}$$

We now end this section with problems for additional practice. Some problems are generalizations of the Woods-Robbins identity and (5.22).

1. **Putnam Problem A4, 1977.** For $0 < x < 1$, express

$$\sum_{n=0}^{\infty} \frac{x^{2^n}}{1 - x^{2^{n+1}}}$$

   as a rational function of x.

2. (Rogers-Ramanujan identities). Show that, for $|q| < 1$,

   (i) $\sum_{k=0}^{\infty} \frac{q^{k^2}}{\prod_{i=1}^{k}(1-q^i)} = \prod_{n=0}^{\infty} \frac{1}{(1-q^{5n+1})(1-q^{5n+4})}$.

   (ii) $\sum_{k=0}^{\infty} \frac{q^{k^2+k}}{\prod_{i=1}^{k}(1-q^i)} = \prod_{n=0}^{\infty} \frac{1}{(1-q^{5n+2})(1-q^{5n+3})}$.

3. Prove that

$$\sum_{n=1}^{\infty} s_2(n) \frac{2n+1}{n^2(n+1)^2} = \frac{\pi^2}{9}.$$

4. For $p > 1$, show that

$$\sum_{n=1}^{\infty} s_2(n) \left(\frac{1}{n^p} - \frac{1}{(n+1)^p} \right) = \frac{1 - 2^{1-p}}{1 - 2^{-p}} \zeta(p).$$

5. Show that the finite generating function of $s_2(n)$ can be expressed as the following finite *Lambert series*:

$$\sum_{k=1}^{2^n-1} s_2(k)x^k = \frac{1-x^{2^n}}{1-x} \sum_{i=0}^{n-1} \frac{x^{2^i}}{1+x^{2^i}}.$$

6. Let $u(n)$ be the number of digit blocks of 11 in the binary expansion of n. For example, $u(14) = 2$ since $14 = 1110_2$. Show that

   . (i) $\sum_{n=1}^{\infty} \frac{u(n)}{n(n+1)} = \frac{3}{2} \ln 2 - \frac{\pi}{4}$.

   (ii) $\prod_{n=0}^{\infty} \left(\frac{(2n+1)^2}{(n+1)(4n+1)} \right)^{(-1)^{u(n)}} = \frac{\sqrt{2}}{2}$.

7. Let $s_b(n)$ be the sum of all digits in the expansion of n in base b. Prove that

   (i) $\sum_{n=1}^{\infty} \frac{s_b(n)}{n(n+1)} = \frac{b}{b-1} \ln b$.

   (ii) $\prod_{n=0}^{\infty} \prod_{k=1,3,\dots}^{b-1} \left(\frac{nb+k}{nb+k+1} \right)^{(-1)^{s_b(n)}} = \frac{\sqrt{b}}{b}$ (Sondow).

   Comment. The digit sum sequence $s_b(n)$ can be implemented by *Mathematica* as

```
Digitsum[n_, b_: 10] := Total[IntegerDigits[n, b]]
```

8. **Problem 11685** (Proposed by D. Knuth, 120(1), 2013). Prove that

$$\prod_{k=0}^{\infty}\left(1+\frac{1}{2^{2^k}-1}\right) = \frac{1}{2}+\sum_{k=0}^{\infty}\frac{1}{\prod_{j=0}^{k-1}\left(2^{2^j}-1\right)}.$$

9. **Problem 11762** (Proposed by R. Stanley, 121(3), 2014). Let $f(n)$ be the least number of strokes needed to draw the Young diagrams of all the partitions of n. Let

$$F(x) = \sum_{n=1}^{\infty} f(n)x^n = x + 2x^2 + 5x^3 + 12x^4 + 21x^5 + 40x^6 + \cdots.$$

Find the coefficients $g(n)$ of the power series $G(x) = \sum_{n=1}^{\infty} g(n)x^n$ satisfying

$$F(x) = 1 + x + \frac{G(x)}{\prod_{i=1}^{\infty}(1-x^i)}.$$

10. **Problem 11828** (Proposed by R. Tauraso, 122(3), 2015). Let n be a positive integer, and let z be a complex number that is not a kth root of unity for any k with $1 \le k \le n$. Let S be the set of al lists $(a_1, \ldots, a_n)$ of n nonnegative integers such that $\sum_{k=1}^{n} ka_k = n$. Prove that

$$\sum_{a\in S}\prod_{k=1}^{n}\frac{1}{a_k! k^{a_k}(1-z^k)^{a_k}} = \prod_{k=1}^{n}\frac{1}{1-z^k}.$$

For example, for $n = 3$ we have

$$\frac{1}{6(1-z)^3}+\frac{1}{2(1-z)(1-z^2)}+\frac{1}{3(1-z^3)} = \frac{1}{(1-z)(1-z^2)(1-z^3)}.$$

11. **Problem 12113** (Proposed by R. Stanley, 126(5), 2019). Define $f(n)$ and $g(n)$ for $n \ge 0$ by

$$\sum_{n\ge 0} f(n)x^n = \sum_{j\ge 0}x^{2^j}\prod_{k=0}^{j-1}(1+x^{2^k}+x^{3\cdot 2^k})$$

and

$$\sum_{n\ge 0} g(n)x^n = \prod_{i\ge 0}(1+x^{2^i}+x^{3\cdot 2^i}).$$

Find all values of n for which $f(n) = g(n)$, and find $f(n)$ for these values.

12. **Independent Study.** Recall that an L-function is defined as $\sum_{n\geq 1} \frac{\chi(n)}{n^s}$, where χ is a Dirichlet character and satisfies $\chi(nk+r) = \chi(r)$, which resembles $s_b(nb+r) = s_b(n) + r$. Thus, the methods used to study the L-function could potentially lead to the study of Dirichlet series of $s_b(n)$.

(a) Let the Hurwitz function be defined by

$$\zeta(s,x) := \sum_{n=0}^{\infty} \frac{1}{(x+n)^s}.$$

For $s > 0, s \neq 1$ and $x \geq 0$, prove that

$$\sum_{k=1}^{b^n-1} s_b(n) \left(\frac{1}{(x+n)^s} - \frac{1}{(x+n+1)^s} \right)$$

$$= \sum_{i=0}^{n-1} \frac{1}{b^{si}} \left[\zeta(s, 1+x/b^i) - \zeta(s, 1+(x+b^n)/b^i) \right]$$

$$- \sum_{i=1}^{n} \frac{b}{b^{si}} \left[\zeta(s, 1+x/b^i) - \zeta(s, 1+(x+b^n)/b^i) \right].$$

(b) Let the Barnes zeta function be defined by

$$\zeta_2(s, x; (a, b)) := \sum_{n_1, n_2 \geq 0} \frac{1}{(x + an_1 + bn_2)^s}.$$

Let $s_b(n)$ be the sum of all digits in the expansion of n in base b. Prove that

$$\sum_{k=1}^{b^n-1} \frac{s_b(k)}{(n+x)^s}$$

$$= \sum_{i=0}^{n-1} \left[\zeta_2(s, x+b^i; (1, b^i)) - \zeta_2(s, x+b^i + b^n; (1, b^i)) \right]$$

$$- b \sum_{i=1}^{n} \left[\zeta_2(s, x+b^i; (1, b^i)) - \zeta_2(s, x+b^i + b^n; (1, b^i)) \right].$$

5.10   A polynomial zero identity

Problem 12077 (Proposed by M. A. Alekseyev, 125(10), 2018). Let $f(x)$ be a monic polynomial of degree n with distinct zeros $a_1, \ldots, a_n$. Prove

$$\sum_{i=1}^{n} \frac{a_i^{n-1}}{f'(a_i)} = 1.$$

Discussion.

By the assumption, let

$$f(x) = \prod_{i=1}^{n} (x - a_i).$$

Then

$$f'(x) = \sum_{i=1}^{n} \prod_{j \neq i} (x - a_j).$$

In particular, $f'(a_i) = \prod_{j \neq i}(a_i - a_j)$. Recall the Lagrange polynomials with nodal points $a_1, \ldots, a_n$

$$L_i(x) = \prod_{j \neq i} \frac{x - a_j}{a_i - a_j} = \frac{1}{f'(a_i)} \prod_{j \neq i} (x - a_j), \quad (i = 1, 2, \ldots, n).$$

In view of the left-hand side of the proposed identity, this suggests that we apply the Lagrange interpolation for some selected polynomial.

Solution.

Let $f(x)$ and $L_i(x)\,(1 \leq i \leq n)$ be as defined in the discussion above. Let

$$P(x) = x^n - f(x),$$

which is a polynomial with degree not exceeding $n-1$. Applying the Lagrange interpolation formula yields

$$P(x) = \sum_{i=1}^{n} L_i(x) P(a_i) = \sum_{i=1}^{n} L_i(x) a_i^n. \tag{5.23}$$

We now consider two cases.

(I) If all $a_i \neq 0$, letting $x = 0$ in (5.23) gives

$$P(0) = (-1)^{n+1} a_1 a_2 \ldots a_n = \sum_{i=1}^{n} L_i(0) a_i^n. \tag{5.24}$$

Since

$$L_i(0) = (-1)^{n-1} \frac{1}{f'(a_i)} \prod_{i \neq j} a_j,$$

from (5.24) it follows that

$$a_1 a_2 \ldots a_n = a_1 a_2 \ldots a_n \left(\sum_{i=1}^{n} \frac{a_i^{n-1}}{f'(a_i)} \right).$$

This proves the desired identity by cancelling the common factor $a_1 \ldots a_n$.
(II) If one of $a_i = 0$, without loss of generality, we assume that $a_1 = 0$.
Removing the common factor x in (5.23) then applying the above process
yields

$$a_2 \ldots a_n = a_2 \ldots a_n \left(\sum_{i=2}^{n} \frac{a_i^{n-1}}{f'(a_i)} \right),$$

which is equivalent to the proposed identity.

Remark. Based on the idea of the proof above, we can shorten the proof by
directly establishing

$$x^{n-1} = \sum_{i=1}^{n} \frac{a_i^{n-1}}{f'(a_i)} \frac{f(x)}{x - a_i}.$$

The proposed equality follows from matching the coefficients of x^{n-1}. Similarly, the reader can try to prove the following generalization:

$$\sum_{i=1}^{n} \frac{a_i^k}{f'(a_i)} = \begin{cases} 0, & \text{if } 0 \le k < n - 1; \\ 1, & \text{if } k = n - 1; \\ \sum_{i=1}^{n} a_i, & \text{if } k = n. \end{cases}$$

We now end this section by offering a few more problems for additional practice.

1. **Problem 10697** (Proposed by J. L. Diaz-Barrero, 105(10), 1998).
   Given n distinct nonzero complex numbers $z_1, z_2, \ldots, z_n$, show that

   $$\sum_{k=1}^{n} \frac{1}{z_k} \prod_{j=1, j \ne k}^{n} \frac{1}{z_k - z_j} = \frac{(-1)^{n+1}}{z_1 z_2 \cdots z_n}.$$

2. **Problem 11008** (Proposed by J. L. Diaz-Barrero and J. Eqozcue,
   110(4), 2003). Let $A(z) = \sum_{k=0}^{n} a_k z^k$ be a monic polynomial with
   complex coefficients and with zeros $z_1, z_2, \ldots, z_n$. show that

   $$\frac{1}{n} \sum_{k=1}^{n} |z_k|^2 < 1 + \max_{1 \le k \le n} |a_{n-k}|^2.$$

3. **Problem 11012** (Proposed by C. Popescu, 110(5), 2008). Given a
   positive integer n, find the minimum value of

   $$\frac{x_1^3 + \cdots + x_n^3}{x_1 + \cdots + x_n}$$

   subject to the condition that $x_1, \ldots, x_n$ be distinct positive integers.

4. **Problem 11098** (Proposed by C. Hillar and D. Rhea, 111(7), 2004). Let

$$f(n) = \sum_{i=1}^{n} \frac{(-1)^{i+1}}{2^i - 1} \binom{n}{i}.$$

   Prove that there are constants c and c' such that $c \leq f(n)/\ln n \leq c'$ for sufficiently large n (that is, $f(n) = \Theta(\ln n)$).

5. **Problem 11403** (Proposed by Y. Yu, 115(10), 2008). Let n be an integer greater than 1, and let f_n be the polynomial given by

$$\sum_{i=0}^{n} \binom{n}{i} (-x)^{n-i} \prod_{j=0}^{i-1} (x + j).$$

   Find the degree of f_n.

6. **Problem 11354** (Proposed by M. Beck and A. Berkovich, 115(3), 2008). Find a polynomial f in two variables such that for all pairs (s, t) of relatively prime positive integers,

$$\sum_{m=1}^{s-1} \sum_{n=1}^{t-1} |mt - ns| = f(s, t).$$

7. **Problem 11577** (Proposed by P. P. Dalyay, 120(5), 2013). Let n be a positive even integer and let p be prime. Show that the polynomial f given by $f(z) = p + \sum_{k=1}^{n} z^k$ is irreducible over $\mathbb{Q}$.

8. **Problem 11798** (Proposed by F. Holland, 121(10), 2014). For positive integer n, let f_n be the polynomial given by

$$f_n(x) = \sum_{k=0}^{n} \binom{n}{k} x^{\lfloor k/2 \rfloor}.$$

   (a) Prove that if $n + 1$ is prime, then f_n is irreducible over $\mathbb{Q}$.

   (b) Prove for all n,

$$f_n(1 + x) = \sum_{k=0}^{\lfloor n/2 \rfloor} \binom{n-k}{k} 2^{n-2k} x^k.$$

9. **Problem 11736** (Proposed by M. Merca, 120(9), 2013). For $n \geq 1$, let f be the symmetric polynomial in variables $x_1, \ldots, x_n$, given by

$$f(x_1, \ldots, x_n) = \sum_{k=0}^{n-1} (-1)^{k+1} e_k(x - 1 + x_1^2, x_2 + x_2^2, \ldots, x_n + x_n^2),$$

where e_k is the kth elementary polynomial in n variables. Also, let ω be a primitive nth root of unity. Prove that

$$f(1, \omega, \omega^2, \ldots, \omega^{n-1}) = L_n - L_0,$$

where L_k is the kth Lucas number.

10. **Problem 11720** (Proposed by I. Gessel, 120(7), 2013). Let $E_n(t)$ be the Eulerian polynomial defined by

$$\sum_{k=0}^{\infty} (k+1)^n t^k = \frac{E_n(t)}{(1-t)^{n+1}},$$

and let B_n be the nth Bernoulli number. Show that

$$(E_{n+1}(t) - (1-t)^n) B_n$$

is a polynomial with integer coefficients. *Hint.* Use that

$$E_n(t) = \sum_{k=0}^{n} \left(\sum_{i=0}^{k} (-1)^i \binom{n+1}{i} (k+1-i)^n \right) t^k.$$

11. **Problem 11947** (Proposed by G. Stoica, 123(10), 2016). Let n be a positive integer, and let $z_1, \ldots, z_n$ be the zeros of $z^n + 1$. For $a > 0$, prove

$$\frac{1}{n} \sum_{k=1}^{n} \frac{1}{|z_k - a|^2} = \frac{\sum_{k=0}^{n-1} a^{2k}}{(1+a^n)^2}.$$

12. **Problem 12022** (Proposed by M. Merca, 125(2), 2018). Let n be a positive integer, and let x be a real number not equal to -1 or 1. Prove

$$\sum_{k=0}^{n-1} \frac{(1-x^n)(1-x^{n-1}) \cdots (1-x^{n-k})}{1 - x^{k+1}} = n$$

and

$$\sum_{k=0}^{n-1} (-1)^k \frac{(1-x^n)(1-x^{n-1}) \cdots (1-x^{n-k})}{1 - x^{k+1}} x^{\binom{n-k-1}{2}} = n x^{\binom{n}{2}}.$$

A

List of Problems

In this list, bold face denotes the Monthly problem featured solution in each section.
P = Putnam Problem, M = Mathematics Magazine, C = The College Mathematics Journal.

Problems from *The American Mathematical Monthly, Mathematics Magazine, and The College Mathematics Journal* ©Mathematical Association of America, 2020. All rights reserved.

11559; 118(3), 2011
E3356; 98(4), 1991
11528; 117(9), 2010
11659; 119(7), 2012
11995; 124(7), 2017
E3034; 91(4), 1984
1966-A3 (P)
11773; 121(4), 2014
E1557; 70(9), 1963
6376; 89(1), 1982
12079; 125(10), 2018
2087(M); 93(1), 2020
11976; 124(4), 2017
11941; 123(9), 2016
E1245; 63(10), 1956
4828; 66(2), 1959
11225; 113(5), 2006
11611; 118(10), 2011
12120; 126(6), 2019
12153; 127(1), 2020
2097 (M); 93(3), 2020
11535; 117(9), 2010
11438; 116(5), 2009
11853; 122(7), 2015
11930; 123(8), 2016
12090; 126(2), 2019
12101; 126(3), 2019
12118; 126(6), 2019
11505; 117(5), 2010

11367; 115(5), 2008
12063; 125(8), 2018
1966-A6 (P)
11967; 124(3), 2017
11592; 118(8), 2011
11494; 117(3), 2010
11612; 118(10), 2011
12029; 125(3), 2018
11677; 119(10), 2012
11821; 122(2) , 2015
11206; 113(2), 2006
11637; 119(4), 2012
12031; 125(3), 2018
12181; 127(5), 2020
11837; 124(1), 2017
2012-B4 (P)
11068; 111(3), 2004
4552; 60(7), 1953
4305; 55(7), 1948
4946; 68(1), 1961
11885; 123(1), 2016
11810; 122(1), 2015
4431; 58(2), 1951
4564; 62(2), 1955
10635; 105(1), 1998
10754; 106(8), 1999
12102; 126(3), 2019
11765; 121(3), 2014
11509; 117(6), 2010

11164; 112(6), 2005
11499; 117(2), 2010
11873; 122(10), 2015
12189; 127(6), 2020
12060; 125(7), 2018
11302; 114(6), 2007
11633; 119(3), 2012
11802; 121(8), 2014
11921; 123(6), 2016
11400; 115(10), 2008
11333; 114(10), 2007
11793; 121(7), 2014
11755; 121(2), 2014
11519; 117(7), 2010
11682; 119(10), 2012
12134; 126(8), 2019
12194; 127(6), 2020
11829; 122(3), 2015
11865; 122(9), 2015
12084; 126(1), 2019
12012; 124(10), 2017
11739; 120(9), 2013
11685; 120(1), 2013
11883; 123(1), 2016
11423; 116(3), 2009
2020 (M), 90(2), 2017
5529; 74(8), 1967
11329; 114(10), 2007
11426; 116(4), 2009

B

Glossary

Abel's Limit Theorem. Let $f(x) = \sum_{n=0}^{\infty} a_n x^n$ be a power series which converges for $|x| < R$. If $\sum_{n=0}^{\infty} a_n R^n$ converges, then

$$\lim_{x \to R^-} f(x) = \sum_{n=0}^{\infty} a_n R^n.$$

Abel's Summation Formula. Let $a_1, a_2, \ldots, a_n$ and $b_1, b_2, \ldots, b_n$ be real or complex numbers with $A_k = \sum_{i=1}^{k} a_i$ for $1 \leq k \leq n$. Then

$$\sum_{i=1}^{n} a_i b_i = A_n b_{n+1} + \sum_{i=1}^{n} A_i (b_i - b_{i+1}).$$

The limit version of this formula is

$$\sum_{i=1}^{\infty} a_i b_i = \lim_{n \to \infty} (A_n b_{n+1}) + \sum_{i=1}^{\infty} A_i (b_i - b_{i+1}).$$

Arithmetic-Geometric-Harmonic Mean Inequality (AM-GM-HM Inequality). Let $a_1, a_2, \ldots, a_n$ be positive real numbers. Then

$$\frac{n}{\frac{1}{a_1} + \frac{1}{a_2} + \cdots + \frac{1}{a_n}} \leq \sqrt[n]{a_1 a_2 \cdots a_n} \leq \frac{a_1 + a_2 + \cdots + a_n}{n},$$

with equality if and only if $a_1 = a_2 = \cdots = a_n$.

The weighted version assumes that, for any $p_i > 0 \, (1 \leq i \leq n)$ with $\sum_{i=1}^{n} p_i = 1$,

$$\frac{1}{\frac{p_1}{a_1} + \frac{p_2}{a_2} + \cdots + \frac{p_n}{a_n}} \leq a_1^{p_1} a_2^{p_2} \cdots a_n^{p_n} \leq p_1 a_1 + p_2 a_2 + \cdots + p_n a_n.$$

Bernoulli's Inequality. Let $\alpha > 0$. Then

$$(1 + x)^\alpha \geq 1 + \alpha x \quad \text{for all } x \geq -1.$$

Bernoulli Numbers. The Bernoulli numbers B_n are defined by the generating function

$$\frac{x}{e^x - 1} = \sum_{n=0}^{\infty} \frac{B_n}{n!} x^n = 1 - \frac{x}{2} + \sum_{n=1}^{\infty} \frac{B_{2n}}{(2n)!} x^{2n}.$$

The first few Bernoulli numbers are $B_1 = -1/2$, $B_2 = 1/6$, $B_4 = -1/30$, and $B_6 = 1/42$. They satisfy the recurrence relation

$$B_0 = 1, \quad B_n = -\frac{1}{n+1} \sum_{k=0}^{n-1} \binom{n+1}{k} B_k, \quad n \geq 1.$$

This implies that all the Bernoulli numbers are rational. Asymptotically, as $n \to \infty$,

$$B_{2n} \sim (-1)^{n+1} 4\sqrt{\pi n} \left(\frac{n}{\pi e}\right)^{2n}.$$

Bernoulli Polynomials. The Bernoulli polynomials $B_n(x)$ are defined by the generating function

$$\frac{z e^{xz}}{e^z - 1} = \sum_{n=0}^{\infty} \frac{B_n(x)}{n!} z^n.$$

Here $B_n(0) = B_n$, the nth Bernoulli number for all $n \geq 1$, and

$$B_n(x) = \sum_{k=0}^{n} \binom{n}{k} B_k x^{n-k}.$$

Beta Function. For $p, q > 0$, the beta function is defined by

$$B(p, q) = \int_0^1 x^{p-1} (1 - x)^{q-1} \, dx.$$

A change of variables leads two equivalent forms:

$$B(p, q) = \int_0^\infty \frac{x^{p-1}}{(1 + x)^{p+q}} \, dx$$

$$= 2 \int_0^{\pi/2} \sin^{2p-1} x \cos^{2q-1} x \, dx.$$

In general, if $a < b, p, q > 0$, then

$$\int_a^b (x - a)^{p-1} (b - x)^{q-1} \, dx = (b - a)^{p+q-1} B(p, q).$$

In terms of the gamma function, we have

$$B(p, q) = \frac{\Gamma(p)\Gamma(q)}{\Gamma(p + q)}.$$

Asymptotically, as $p \to \infty$ and $q \to \infty$,

$$B(p, q) \sim \sqrt{2\pi} \frac{p^{p-1/2} q^{q-1/2}}{(p + q)^{p+q-1/2}}.$$

Catalan Numbers. The Catalan numbers are defined by

$$C_n = \frac{1}{n+1}\binom{2n}{n} \quad \text{for } n \geq 0,$$

or alternatively by the recurrence relation

$$C_{n+1} = \sum_{k=0}^{n} C_k C_{n-k}, \quad C_0 = 1.$$

The generating function of $\{C_n\}$ is given by

$$C(x) := \frac{1 - \sqrt{1 - 4x}}{2x}.$$

Carleman's Inequality. Let $a_1, a_2, \ldots, a_n, \ldots$ be a positive real number sequence. Then

$$\sum_{n=1}^{\infty} (a_1 a_2 \cdots a_n)^{1/n} \leq e \sum_{n=1}^{\infty} a_n.$$

Cauchy-Schwarz Inequality (Discrete version). Let $a_1, a_2, \ldots, a_n$ and $b_1, b_2, \ldots, b_n$ be real numbers. Then

$$\left(\sum_{i=1}^{n} a_i b_i\right)^2 \leq \left(\sum_{i=1}^{n} a_i^2\right)\left(\sum_{i=1}^{n} b_i^2\right).$$

Equality holds if and only if a_i and b_i are proportional for all $1 \leq i \leq n$.
Cauchy-Schwarz Inequality (Integral version). Let $f, g : [a, b] \to \mathbb{R}$ be nonnegative and integrable. Then

$$\left(\int_a^b f(x)g(x)\, dx\right)^2 \leq \left(\int_a^b f^2(x)\, dx\right)\left(\int_a^b g^2(x)\, dx\right).$$

If both f and g are continuous, equality holds if and only if f and g are proportional.
Chebyshev Polynomials. The Chebyshev polynomials of the *first kind* are defined by

$$T_n(\cos\theta) = \cos n\theta,$$

or alternatively by the *generating function*:

$$\sum_{n=0}^{\infty} T_n(x)t^n = \frac{1 - xt}{1 - 2xt + t^2}.$$

The first few Chebyshev polynomials of the first kind are

$$T_0(x) = 1, \ T_1(x) = x, \ T_2(x) = 2x^2 - 1, \ T_3(x) = 4x^3 - 3x, \ T_4(x) = 8x^4 - 8x^2 + 1.$$

They also satisfy the recurrence relation

$$T_{n+1}(x) = 2xT_n(x) - T_{n-1}(x).$$

In general,

$$T_n(x) = \sum_{k=0}^{\lfloor n/2 \rfloor} \binom{n}{2k} x^{n-2k}(x^2-1)^k = 2^{n-1} \prod_{k=1}^{n} \left[x - \cos\left(\frac{(2k-1)\pi}{2n}\right)\right].$$

The Chebyshev polynomials of the *second kind* are defined by

$$U_n(\cos\theta) = \frac{\sin(n+1)\theta}{\sin\theta},$$

or alternatively by the *generating function*:

$$\sum_{n=0}^{\infty} U_n(x)t^n = \frac{1}{1 - 2xt + t^2}.$$

The first few Chebyshev polynomials of the second kind are

$$U_0(x) = 1, \ U_1(x) = 2x, \ U_2(x) = 4x^2 - 1, \ U_3(x) = 8x^3 - 4x, \ U_4(x)$$
$$= 16x^4 - 12x^2 + 1.$$

They also satisfy the recurrence relation

$$U_{n+1}(x) = 2xU_n(x) - U_{n-1}(x).$$

In general,

$$U_n(x) = \sum_{k=0}^{\lfloor n/2 \rfloor} \binom{n+1}{2k+1} x^{n-2k}(x^2-1)^k = 2^n \prod_{k=1}^{n} \left[x - \cos\left(\frac{k\pi}{n+1}\right)\right].$$

Digamma Function. The digamma function $\psi(z)$ (also called psi function) is define by

$$\psi(z) = (\ln\Gamma(z))' = \frac{\Gamma'(z)}{\Gamma(z)} = -\gamma + \sum_{k=1}^{\infty} \left(\frac{1}{k} - \frac{1}{z+k-1}\right).$$

It has a series representation

$$\psi(z) = \frac{1}{z} - \gamma + \sum_{n=2}^{\infty} (-1)^n \zeta(n) z^{n-1},$$

where γ is Euler's constant and ζ is the Riemann zeta function.

Some useful formulas of $\psi(z)$ with $\text{Re}(z) > 0$:

$$\psi(z+n) = \sum_{k=0}^{n-1} \frac{1}{z+k} + \psi(z) \quad \text{for } n \geq 1;$$

$$\psi(z) = -\int_0^1 \left(\frac{1}{\ln t} + \frac{t^{z-1}}{1-t} \right) dt;$$

$$\psi(z) = \ln z - \frac{1}{2z} - \int_0^\infty \frac{2t\,dt}{(z^2+t^2)(e^{2\pi t}-1)}.$$

In particular, when $z = p/q, 0 < p < q$, a rational number, Gauss formula assumes that

$$\psi\left(\frac{p}{q}\right) = -\gamma - \frac{\pi}{2}\cot\left(\frac{p\pi}{q}\right) - \ln q + 2\sum_{n=1}^{\lfloor q/2 \rfloor} \cos\left(\frac{2np\pi}{q}\right) \ln\left(2\sin\frac{n\pi}{q}\right).$$

Asymptotically, as $x \to \infty$,

$$\psi(x) \sim \ln x - \frac{1}{2x} - \frac{1}{12x^2} + \frac{1}{120x^4} - \frac{1}{256x^6} + O(x^{-8}).$$

Euler's Constant. Euler's constant γ is defined by

$$\gamma = \lim_{n\to\infty}\left(1 + \frac{1}{2} + \cdots + \frac{1}{n} - \ln n\right) = 0.5772156649\ldots.$$

Euler-Maclaurin Summation Formula. Let B_n be nth Bernoulli number. Then

$$\sum_{k=m}^{n} f(k) = \int_m^n f(x)\,dx + \frac{1}{2}(f(m) + f(n))$$

$$+ \sum_{k=1}^{\lfloor p/2 \rfloor} \frac{B_{2k}}{(2k)!}\left(f^{(2k-1)}(n) - f^{(2k-1)}(m)\right) + R_p.$$

Euler's Product Formula for the Riemann Zeta Function. Let $\zeta(s)$ be the Riemann zeta function. Then, for all $s > 1$,

$$\zeta(s) = \sum_{n=1}^{\infty} \frac{1}{n^s} = \prod_{\text{primes } p} \frac{1}{1 - p^{-s}}.$$

Euler's Product Formula for Sine:

$$\frac{\sin \pi x}{\pi x} = \prod_{n=1}^{\infty}\left(1 - \frac{x^2}{n^2}\right).$$

Logarithmic differentiation yields the following partial fraction expansions:

$$\pi \cot \pi x \;=\; \frac{1}{x} - \sum_{n=1}^{\infty} \left(\frac{1}{x+n} + \frac{1}{x-n} \right),$$

$$\frac{\pi}{\sin \pi x} \;=\; \frac{1}{x} + 2x \sum_{n=1}^{\infty} \frac{(-1)^n}{x^2 - n^2},$$

$$\pi \tan \pi x \;=\; \sum_{n=-\infty}^{\infty} \frac{1}{n - x + 1/2} = \lim_{n \to \infty} \sum_{k=-n}^{n} \frac{1}{k - x + 1/2},$$

$$\pi \sec \pi x \;=\; \sum_{n=-\infty}^{\infty} \frac{1}{n + x + 1/2} = \lim_{n \to \infty} \sum_{k=-n}^{n} \frac{1}{k + x + 1/2},$$

$$\frac{\pi^2}{\sin^2 \pi x} \;=\; \sum_{n=-\infty}^{\infty} \frac{1}{(n+x)^2} = \lim_{n \to \infty} \sum_{k=-n}^{n} \frac{1}{(k+x)^2}.$$

Faà di Bruno's Formula. This formula gives an explicit expansion for the nth derivative of the composition function $(f \circ g)(x)$. Let both f and g be n times differentiable. Then

$$[(f \circ g)(x)]^{(n)} = \sum \frac{n!}{k_1! \cdots k_n!} (D^k f)(g(x)) \left(\frac{Dg(x)}{1!} \right)^{k_1} \cdots \left(\frac{D^n g(x)}{n!} \right)^{k_n},$$

where $k = k_1 + \cdots + k_n$ and the sum is over all partition of n, i.e., values of $k_1, \ldots, k_n$ such that

$$k_1 + 2k_2 + \cdots + nk_n = n.$$

For example, when $n = 3$,

$$[(f \circ g)(x)]^{(3)} = 3g'(x)f''(g(x))g''(x) + (g'(x))^3 f^{(3)}(g(x)) + f'(g(x))g^{(3)}(x).$$

Fibonacci Numbers. The Fibonacci numbers F_n are defined recursively by $F_1 = F_2 = 1$ and

$$F_n = F_{n-1} + F_{n-2} \quad \text{for all } n \geq 3.$$

The generating function of $\{F_n\}$ is

$$\sum_{n=1}^{\infty} F_n x^n = \frac{x}{1 - x - x^2}.$$

Let

$$\phi = \frac{1 + \sqrt{5}}{2} \quad \text{and} \quad \hat{\phi} = \frac{1 - \sqrt{5}}{2},$$

where ϕ is called the golden ratio. The closed form of F_n is given by Binet's formula

$$F_n = \frac{1}{\sqrt{5}} (\phi^n - \hat{\phi}^n).$$

Famous identities include

- Cesàro identity: $\sum_{k=0}^{n} \binom{n}{k} F_k = F_{2n}$.

- Catalan's identity: $F_n^2 - F_{n+m}F_{n-m} = (-1)^{n-m}F_m^2$.

Frullani Integrals. If f is continuous on $[0, \infty)$ and $f(\infty) = \lim_{x \to \infty} f(x)$ exists, then

$$\int_0^\infty \frac{f(ax) - f(bx)}{x}\, dx = (f(0) - f(\infty)) \ln \frac{b}{a}.$$

Gamma Function. The gamma function is defined by

$$\Gamma(x) = \int_0^\infty t^{x-1}e^{-t}\, dt \qquad \text{for all } x > 0.$$

A product representation due to Weierstrass asserts that

$$\frac{1}{\Gamma(x)} = xe^{\gamma x} \prod_{n=1}^{\infty} \left\{ \left(1 + \frac{x}{n}\right) e^{-x/n} \right\},$$

where γ is Euler's constant.

The gamma function satisfies many interesting identities:

- Euler reflection formula: $\Gamma(x)\Gamma(1 - x) = \frac{\pi}{\sin \pi x}$.

- Legendre duplication formula: $\Gamma(2x)\Gamma(1/2) = 2^{2x-1}\,\Gamma(x)\Gamma\left(x + \frac{1}{2}\right)$.

- Gauss multiplication formula: $\prod_{k=0}^{n-1} \Gamma\left(x + \frac{k}{n}\right) = (2\pi)^{(n-1)/2}n^{1/2-nx}\Gamma(nx)$.

The gamma function can be defined for all complex numbers except the non-positive integers. It is related to the *Riemann zeta function* $\zeta(s)$ by

$$\zeta(s)\Gamma(s) = \int_0^\infty \frac{t^{s-1}}{e^t - 1}\, dt;$$

$$\Gamma\left(\frac{s}{2}\right) \pi^{-s/2}\zeta(s) = \Gamma\left(\frac{1-s}{2}\right) \pi^{-(1-s)/2}\zeta(1 - s).$$

Asymptotically, as $x \to \infty$,

$$\Gamma(x) \sim \sqrt{2\pi}\, x^{x-1/2}e^{-x}.$$

Equivalently,

$$\ln \Gamma(x) \sim \left(x - \frac{1}{2}\right) \ln x - x + \frac{1}{2} \ln(2\pi) + O(1/x).$$

Hardy's Inequality (Discrete version). Let a_n be a nonnegative real sequences. For $p > 1$,

$$\sum_{i=1}^{\infty} \left(\frac{1}{n} \sum_{k=1}^{n} a_k\right)^p \le \left(\frac{p}{p-1}\right)^p \sum_{i=1}^{\infty} a_n^p.$$

Equality holds if and only if $a_n = 0$ for all $n \geq 1$. Moreover, the constant $\left(\frac{p}{p-1}\right)^p$ is the best possible.

Hardy's Inequality (Integral version). Let $f : [0, \infty] \to [0, \infty)$ be continuous such that $f \in L^p$ with $p > 1$. Then

$$\int_0^\infty \left(\frac{1}{x}\int_0^x f(t)\,dt\right)^p \leq \left(\frac{p}{p-1}\right)^p \int_0^\infty f^p(x)\,dx.$$

Equality holds if and only if $f = 0$ identically. Moreover, the constant $\left(\frac{p}{p-1}\right)^p$ is the best possible.

Harmonic Numbers. The nth harmonic numbers, H_n, is defined by

$$H_n = 1 + \frac{1}{2} + \cdots + \frac{1}{n} = \sum_{k=1}^n \frac{1}{k},$$

which has the generating function:

$$\sum_{n=1}^\infty H_n x^n = -\frac{\ln(1-x)}{1-x}.$$

H_n satisfies the elementary inequality

$$\frac{1}{24(n+1)^2} < H_n - \ln(n+1/2) - \gamma < \frac{1}{24n^2}.$$

Asymptotically, as $n \to \infty$,

$$H_n \sim \ln n + \gamma + \frac{1}{2n} - \sum_{k=1}^\infty \frac{B_{2k}}{2kn^{2k}}$$
$$= \ln n + \gamma + \frac{1}{2n} - \frac{1}{12n^2} + \frac{1}{120n^4} - \frac{1}{252n^6} + \cdots.$$

where B_{2k} is the $(2k)$th Bernoulli numbers.

Hermite-Hadamard Inequality. Let $f : [a, b] \to \mathbb{R}$ be a convex function. Then

$$f\left(\frac{a+b}{2}\right) \leq \frac{1}{b-a}\int_a^b f(x)\,dx \leq \frac{f(a)+f(b)}{2}.$$

Hölder's Inequality (Discrete version). Let $a_1, a_2, \ldots, a_n$ and $b_1, b_2, \ldots, b_n$ be positive numbers. If $p, q > 1$ and $1/p + 1/q = 1$, then

$$\sum_{i=1}^n a_i b_i \leq \left(\sum_{i=1}^n a_i^p\right)^{1/p} \left(\sum_{i=1}^n b_i^q\right)^{1/q}.$$

Equality holds if and only if a_i and b_i are proportional for all $1 \leq i \leq n$.

Hölder's Inequality (Integral version). Let f and g be nonnegative function. If $f \in L^p[a,b], g \in L^q[a,b]$ with $p, q > 1$ and $1/p + 1/q = 1$, then

$$\int_a^b f(x)g(x)\,dx \le \left(\int_a^b f^p(x)\,dx \right)^{1/p} \left(\int_a^b g^q(x)\,dx \right)^{1/q}.$$

L'Hôpital's Monotone Rule. Let $f, g : [a, b] \to \mathbb{R}$ be continuous functions that are differentiable on (a, b) with $g'(x) \ne 0$ on (a, b). If $f'(x)/g'(x)$ is increasing (decreasing) on (a, b), then the functions

$$\frac{f(x) - f(a)}{g(x) - g(a)} \quad \text{and} \quad \frac{f(x) - f(b)}{g(x) - g(b)}$$

are likewise increasing (decreasing) on (a, b).

Hurwitz Zeta Function. Hurwitz zeta function is defined by

$$\zeta(x, a) = \sum_{n=0}^{\infty} \frac{1}{(n + a)^x}.$$

For $x > 1$, the integral representation is

$$\zeta(x, a) = \frac{1}{\Gamma(x)} \int_0^\infty \frac{t^{x-1} e^{-at}}{1 - e^{-t}}\,dt.$$

In particular, if $x = k \in \mathbb{N}, a \in (0, 1)$, then

$$\zeta(k, a) - (-1)^{k-1} \zeta(k, 1 - a) = \frac{(-1)^{k-1} \pi}{(k - 1)!} \frac{d^{k-1}}{dt^{k-1}} \left(\cot(\pi t) \right)|_{t=a}.$$

Jensen's Inequality. Let $f : [a, b] \to \mathbb{R}$ be a convex function. For any $p_i > 0, x_i \in [a, b] \, (1 \le i \le n)$ with $\sum_{i=1}^n p_i = 1$,

$$f(p_1 x_1 + p_2 x_2 + \cdots + p_n x_n) \le p_1 f(x_1) + p_2 f(x_2) + \cdots + p_n f(x_n).$$

Mathieu's Inequality. For $c \ne 0$,

$$\frac{1}{c^2 + 1/2} < \sum_{n=1}^{\infty} \frac{2n}{(n^2 + c^2)^2} < \frac{1}{c^2}.$$

Alzer refined this inequality as

$$\frac{1}{c^2 + 1/(2\zeta(3))} < \sum_{n=1}^{\infty} \frac{2n}{(n^2 + c^2)^2} < \frac{1}{c^2 + 1/6}.$$

Parserval's Identity. Let $f \in L^2[-\pi, \pi]$ and

$$f(x) = \frac{a_0}{2} + \sum_{n=1}^{\infty} (a_n \cos nx + b_n \sin nx).$$

Then

$$\frac{1}{\pi} \int_0^\pi [f(x)]^2 \, dx = \frac{a_0^2}{2} + \sum_{n=1}^\infty (a_n^2 + b_n^2).$$

For a Fourier sine series on $[0, \pi]$, i.e., $f(x) = \sum_{n=1}^\infty b_n \sin nx$, Parserval's identity assumes the form

$$\frac{2}{\pi} \int_0^\pi [f(x)]^2 \, dx = \sum_{n=1}^\infty b_n^2.$$

Polylogarithm Function. For $|z| \le 1$, the polylogarithm Li_n is defined by $\text{Li}_1(z) = -\ln(1 - z)$ and

$$\text{Li}_n(z) := \sum_{k=1}^\infty \frac{z^k}{k^n} = \int_0^z \frac{\text{Li}_{n-1}(t)}{t} \, dt \quad \text{for } n \ge 2.$$

It has an integral representation:

$$\text{Li}_s(z) = \frac{1}{\Gamma(s)} \int_0^\infty \frac{t^{s-1}}{e^t/z - 1} \, dt.$$

In particular, the dilogarithm $\text{Li}_2(z)$ satisfies

$$\text{Li}_2(z) = \int_z^0 \frac{\ln(1 - t)}{t} \, dt = -\int_0^1 \frac{\ln(1 - zt)}{t} \, dt$$

and the functional equations:

$$\text{Li}_2(z) + \text{Li}_2(-z) = \frac{1}{2} \text{Li}_2(z^2);$$

$$\text{Li}_2(z) + \text{Li}_2(1 - z) = \zeta(2) - \ln z \ln(1 - z).$$

Riemann Zeta Function. The Riemann zeta function is defined by

$$\zeta(s) = \sum_{n=1}^\infty \frac{1}{n^s}, \quad \text{for all } s > 1.$$

It has the integral representation

$$\zeta(s) = \frac{1}{\Gamma(s)} \int_0^\infty \frac{x^{s-1}}{e^x - 1} \, dx.$$

It is well-known that $\zeta(2) = \pi^2/6, \zeta(4) = \pi^4/90$. In general,

$$\zeta(2k) = \frac{(-1)^{k-1} 2^{2k-1} B_{2k}}{(2k)!} \pi^{2k},$$

where B_{2k} is the $(2k)$th Bernoulli number. Since

$$\lim_{s \to 1} \left(\zeta(s) - \frac{1}{s-1} \right) = \gamma,$$

$\zeta(s)$ has the following *Laurent series expansion*:

$$\zeta(s) = \frac{1}{s-1} + \sum_{n=0}^{\infty} \frac{(-1)^n \gamma_n}{n!} (s-1)^n,$$

where γ_n are the so-called *Stieltjes constants*. These constants are recursively defined by $\gamma_0 = \gamma$ and

$$\gamma_n = \lim_{m \to \infty} \left(\sum_{k=1}^{m} \frac{\ln^n k}{k} - \frac{\ln^{n+1} m}{n+1} \right), \quad \text{for any } n \geq 1.$$

The reciprocal of the $\zeta(s)$ can be represented by a Dirichlet series:

$$\frac{1}{\zeta(s)} = \sum_{n=1}^{\infty} \frac{\mu(n)}{n^s},$$

where $\mu(n)$ is the *Möbius function* in the Number Theory.

Series Tests:

Guass's Test. Let a_n be a positive sequence such that

$$\frac{a_{n+1}}{a_n} = 1 - \frac{r}{n} + O(1/n^p)$$

for some constant r and $p > 1$. Then the series $\sum_{n=1}^{\infty} a_n$ converges if $r > 1$ and diverges if $r \leq 1$.

Kummer's Test. Let a_n be a positive sequences.

(a) $\sum_{n=1}^{\infty} a_n$ converges iff there is a positive series $\sum_{n=1}^{\infty} b_n$ and a constant $c > 0$ such that

$$c_n = \frac{a_n}{a_{n+1}} b_n - b_{n+1} \geq c \quad \text{for all } n.$$

(b) $\sum_{n=1}^{\infty} a_n$ diverges iff there is a positive series $\sum_{n=1}^{\infty} b_n$ such that $\sum_{n=1}^{\infty} 1/b_n$ diverges and

$$c_n = \frac{a_n}{a_{n+1}} b_n - b_{n+1} \leq 0 \quad \text{for all } n.$$

Remark. Many other series tests, for example, Ratio test, Raabe's test, Bertrand's test and Guass's test above are all special cases of Kummer's test obtained by choosing specific parameter series $\sum_{n=1}^{\infty} b_n$.

Stirling's Formula. As $n \to \infty$,

$$n! \sim \sqrt{2\pi n} \left(\frac{n}{e} \right)^n.$$

More accurately, we have

$$n! = \sqrt{2\pi n} \left(\frac{n}{e}\right)^n \left(1 + \frac{1}{12n} + \frac{1}{288n^2} - \frac{139}{51840n^3} + O(1/n^4)\right).$$

The corresponding logarithmic form becomes

$$\ln n! = \left(n + \frac{1}{2}\right)\ln n - n + \frac{1}{2}\ln(2\pi) + \frac{1}{12n} - \frac{1}{360n^3} + O(1/n^5).$$

Stolz-Cesàro Theorem. Let a_n and b_n be two real sequences.

(a) Assume that $a_n \to 0$, $b_n \to 0$ as $n \to \infty$ and b_n is decreasing. If $\lim_{n\to\infty}(a_{n+1} - a_n)/(b_{n+1} - b_n)$ exists, then

$$\lim_{n\to\infty} \frac{a_n}{b_n} = \lim_{n\to\infty} \frac{a_{n+1} - a_n}{b_{n+1} - b_n}.$$

(b) Assume that $b_n \to \infty$ as $n \to \infty$ and b_n is increasing. If $\lim_{n\to\infty}(a_{n+1} - a_n)/(b_{n+1} - b_n)$ exists, then

$$\lim_{n\to\infty} \frac{a_n}{b_n} = \lim_{n\to\infty} \frac{a_{n+1} - a_n}{b_{n+1} - b_n}.$$

Wallis Product Formula. Let $(2n)!! = 2n \cdot (2n - 2) \cdots 2$; $(2n - 1)!! = (2n - 1) \cdot (2n - 3) \cdots 1$. Then

$$\lim_{n\to\infty} \frac{1}{2n + 1} \frac{[(2n)!!]^2}{[(2n - 1)!!]^2} = \prod_{n=1}^{\infty} \left(\frac{2n}{2n - 1} \cdot \frac{2n}{2n + 1}\right) = \frac{\pi}{2}.$$

Some Useful Inequalities.

1. $y^{1/n} - x^{1/n} \le (y - a)^{1/n} - (x - a)^{1/n}$ for $y \ge x \ge a \ge 0, n \in \mathbb{N}$.

2. $\sqrt{xy} \le \frac{x-y}{\ln x - \ln y} \le \left(\frac{\sqrt{x}+\sqrt{y}}{2}\right)^2 \le \frac{x+y}{2}$ for $x, y > 0$.

3. $\sqrt{xy} \le \frac{x^{1-\alpha}y^{\alpha} + x^{\alpha}y^{1-\alpha}}{2} \le \frac{x+y}{2}$ for $x, y > 0, \alpha \in [0, 1]$.

4. $\frac{2x}{2+x} \le \ln(1 + x) \le \frac{x}{\sqrt{x+1}}$ for $x \ge 0$.

5. $\frac{\ln x}{x-1} \le \frac{1}{\sqrt{x}}$ for $x > 0, x \ne 1$.

6. $x - \frac{1}{2}x^3 \le x\cos x \le x - \frac{1}{6}x^3 \le \sin x \le \frac{1}{3}(x\cos x + 2x)$.

7. $\max\left\{\frac{2}{\pi}, \frac{\pi^2-x^2}{\pi^2+x^2}\right\} \le \frac{\sin x}{x} \le 1 + \frac{1}{3}x^2 \le \frac{\tan x}{x}$ for $0 \le x \le \pi/2$.

8. $\frac{4}{\pi}\frac{x}{1-x^2} < \tan\left(\frac{\pi}{2}x\right) < \frac{\pi}{2}\frac{x}{1-x^2}$ for $x \in (0, 1)$.

9. $e\left(1 - \frac{1}{2n}\right) \le \left(1 + \frac{1}{n}\right)^n < e\left(1 - \frac{1}{2(n+1)}\right)$ for $n \in \mathbb{N}$.

10. $e\left(\frac{n}{e}\right)^n \le \sqrt{2\pi n}\left(\frac{n}{e}\right)^n e^{1/(12n+1)} \le n! \le \sqrt{2\pi n}\left(\frac{n}{e}\right)^n e^{1/12n} \le en\left(\frac{n}{e}\right)^n$.

11. $\frac{4^n}{\sqrt{n\pi}}\left(1-\frac{1}{8n}\right) \le \binom{2n}{n} \le \frac{4^n}{\sqrt{n\pi}}\left(1-\frac{1}{9n}\right)$.

12. $\frac{1}{2n+2/5} < H_n - \ln n - \gamma < \frac{1}{2n+1/3}$ for all $n \in \mathbb{N}$.

Some Useful Power Series. Let H_n be the nth harmonic number and B_n be the nth Bernoulli number.

1. $\frac{1}{(1-x)^k} = \sum_{n=0}^{\infty}\binom{n+k-1}{n}x^n, \quad x \in (-1,1)$.

2. $-\frac{\ln(1-x)}{1-x} = \sum_{n=1}^{\infty} H_n x^n, \quad x \in (-1,1)$.

3. $\ln^2(1-x) = 2\sum_{n=2}^{\infty}\frac{H_{n-1}}{n}x^n, \quad x \in [-1,1)$.

4. $-\frac{\ln(1-x)}{(1-x)^k} = \sum_{n=1}^{\infty}\binom{n+k-1}{k-1}(H_{n+k-1} - H_{k-1})x^n$ for $k \ge 1, H_0 = 0, \quad x \in (-1,1)$.

5. $-\frac{1}{1-x^2}\ln\left(\frac{1+x}{1-x}\right) = \sum_{n=1}^{\infty}(2H_{2n} - H_n)x^{2n-1}, \quad x \in (-1,1)$.

6. $\frac{1}{2}\ln^2\left(\frac{1+x}{1-x}\right) = \sum_{n=1}^{\infty}\frac{2H_{2n}-H_n}{n}x^{2n}, \quad x \in (-1,1)$.

7. $\arctan^2 x = \frac{1}{2}\sum_{n=1}^{\infty}\frac{(-1)^{n-1}(2H_{2n}-H_n)}{n}x^{2n}, \quad x \in [-1,1]$.

8. $\frac{1}{2}\arctan x \ln(1+x^2) = \sum_{n=1}^{\infty}\frac{(-1)^{n-1}H_{2n}}{2n+1}x^{2n+1}, \quad x \in [-1,1]$.

9. $\sqrt{1+x} = \sum_{n=0}^{\infty}\binom{1/2}{n}x^n = 1 + \sum_{n=0}^{\infty}\frac{(-1)^n\binom{2n}{n}}{(n+1)2^{2n+1}}x^{n+1}, \quad x \in (-1,1)$.

10. $\frac{1}{\sqrt{1+x}} = \sum_{n=0}^{\infty}\frac{(-1)^n\binom{2n}{n}}{2^{2n}}x^n, \quad x \in (-1,1)$.

11. $\arcsin x = \sin^{-1} x = \sum_{n=0}^{\infty}\frac{\binom{2n}{n}}{2^{2n}(2n+1)}x^{2n+1}, \quad x \in [-1,1]$.

12. $\arcsin^2 x = \frac{1}{2}\sum_{n=1}^{\infty}\frac{2^{2n}}{n^2\binom{2n}{n}}x^{2n}, \quad x \in [-1,1]$.

13. $\text{arcsinh}\, x = \sinh^{-1}x = \ln(x+\sqrt{1+x^2}) = \sum_{n=0}^{\infty}\frac{(-1)^n\binom{2n}{n}}{2^{2n}(2n+1)}x^{2n+1}, \quad x \in [-1,1]$.

14. $\text{arcsinh}^2 x = \sum_{n=1}^{\infty}\frac{(-1)^{n-1}2^{2n-1}}{n^2\binom{2n}{n}}x^{2n}, \quad x \in [-1,1]$.

15. $\tan x = \sum_{n=0}^{\infty}\frac{(-1)^{n-1}2^{2n}(2^{2n}-1)B_{2n}}{(2n)!}x^{2n-1}, \quad x \in (-\pi/2, \pi/2)$.

16. $\tanh x = \sum_{n=0}^{\infty}\frac{2^{2n}(2^{2n}-1)B_{2n}}{(2n)!}x^{2n-1}, \quad x \in (-\pi/2, \pi/2)$.

17. $\pi x \cot(\pi x) = 1 - 2\sum_{n=1}^{\infty}\zeta(2n)x^{2n}, \quad x \in (-1,1)$.

18. $\ln\Gamma(1-x) = \gamma x + \sum_{n=2}^{\infty}\frac{\zeta(n)}{n}x^n, \quad x \in (-1,1)$.

19. $\sum_{n=0}^{\infty}\zeta(2n+2)x^{2n} = \sum_{n=1}^{\infty}\frac{1}{n^2-x^2} = 3\sum_{n=1}^{\infty}\frac{1}{(n^2-x^2)\binom{2n}{n}}\prod_{k=1}^{n-1}\frac{k^2-4x^2}{k^2-x^2}$.

20. $\sum_{n=0}^{\infty} \zeta(2n+3)x^{2n} = \sum_{n=1}^{\infty} \frac{1}{n(n^2-x^2)} = \frac{1}{2}\sum_{n=1}^{\infty} \frac{(-1)^{n-1}}{n^3\binom{2n}{n}} \frac{5n^2-x^2}{n^2-x^2}$
$\prod_{k=1}^{n-1} \frac{k^2-x^2}{k^2}$.

Some Fourier Series Expansions

1. $\frac{\pi-x}{2} = \sum_{n=1}^{\infty} \frac{1}{n}\sin nx$ for $x \in (0, 2\pi)$.

2. $\frac{x}{2} = \sum_{n=1}^{\infty} \frac{(-1)^{n-1}}{n}\sin nx$ for $x \in (-\pi, \pi)$.

3. $\ln(2\sin x) = -\sum_{n=1}^{\infty} \frac{1}{n}\cos 2nx$ for $x \in (0, \pi)$.

4. $\ln(2\cos x) = \sum_{n=1}^{\infty} \frac{(-1)^{n-1}}{n}\cos 2nx$ for $x \in (-\pi/2, \pi/2)$.

5. $\frac{1-r^2}{1-2r\cos x+r^2} = 1 + 2\sum_{n=1}^{\infty} r^n\cos nx$ for $|r| < 1$.

6. $\frac{r\sin x}{1-2r\cos x+r^2} = \sum_{n=1}^{\infty} r^n\sin nx$ for $|r| < 1$.

7. $B_{2n}(x) = \frac{(-1)^{n-1}2(2n)!}{(2\pi)^{2n}}\sum_{k=1}^{\infty} \frac{\cos 2k\pi x}{k^{2n}}$.

8. $B_{2n-1}(x) = \frac{(-1)^n 2(2n-1)!}{(2\pi)^{2n-1}}\sum_{k=1}^{\infty} \frac{\sin 2k\pi x}{k^{2n-1}}$.

9. $\left(\frac{1}{2}\cot(x/2)\right)^{1/2} = \sum_{n=1}^{\infty} a_n\sin nx = \sum_{n=0}^{\infty} a_n\cos nx$ for $0 < x < \pi$, where

$$a_{2n} = a_{2n+1} = \frac{(1/2)_n}{n!} = \frac{(1/2)(1/2+1)\cdots(1/2+n-1)}{n!}.$$

10. $\ln\frac{\Gamma(x)}{\sqrt{2\pi}} = -\frac{1}{2}\ln(2\sin\pi x) + \frac{1}{2}(\gamma+\ln(2\pi))(1-2x) + \frac{1}{\pi}\sum_{n=1}^{\infty} \frac{\ln n}{n}\sin(2n\pi x)$ for $x \in (0, 1)$.

Bibliography

[1] M. Abramowitz and I. A. Stegun (Eds), *Handbook of Mathematical Functions with Formulas, Graphs, and Mathematical Tables*, National Bureau of Standards, Applied Mathematics Series **55**, 9th printing, Washington, 1970. The latest update available at http://dlmf.nist.gov

[2] Z. Ahmed and P. M. Jarvis, Problem 10777, Problems and Solutions, *Amer. Math. Monthly*, **107**:956-957, 2000.

[3] M. Aigner and G. M. Ziegler, *Proofs from the Book*, 4th edition, Springer-Verlag, New York, 2010.

[4] G. L. Alexanderson,L. Klosinski and L. C. Larson (Eds), *The William Lowell Putnam Mathematical Competition Problems and Solutions: 1965-1984*, MAA Problem Books, The Mathematical Association of America, Washington, D.C., 1985.

[5] H. Alzer, On som inequalities for the incomplete gamma function, *Math. Comp.*, **66**:771-778, 1997.

[6] H. Alzer, J. L. Brenner and O. G. Ruehr, On Mathieu's inequality, *J. Math. Anal. Appl.*, **218**:607-610, 1998.

[7] T. Amdeberhan, O. Espinosa and V. H. Moll, The Laplace transform of the digamma function: An integral due to Glasser, Manna and Oloa, *Proc. Amer. Math. Soc.*, **136**:3211-3221, 2008.

[8] G. Andrews, R. Askey and R. Roy, *Special Functions*, Encyclopedia of Mathematics and Its Applications, **71**, Cambridge University Press, Cambridge, 1999.

[9] R. Apéry, Irrationalité de $\zeta(2)$ et $\zeta(3)$, *Astérisque*, **61**:11-13, 1979.

[10] D. H. Bailey, J. M. Borwein, V. Kapoor and E. W. Weisstein, Ten problems in experimental mathematics, *Amer. Math. Monthly*, **113** :481-509, 2006.

[11] G. Bennett, p-Free ℓ^p inequalities, *Amer. Math. Monthly*, **117**:334-351, 2010.

[12] B. Berndt, *Ramanujan's Notebooks, Part I*, Springer-Verlag, New York, 1985.

303

[13] M. G. Beumer, Some special integrals, *Amer. Math. Monthly*, **68** :645-647, 1961.

[14] G. Boros and V. H. Moll, An integral with three parameters, *SIAM Review*, **40**:972-980, 1998.

[15] G. Boros and V. H. Moll, *Irresistible Integrals*, Cambridge University Press, Cambridge, 2004.

[16] D. Borwein and J. M. Borwein, On an intriguing integral and some series related to $\zeta(4)$, *Proc. Amer. Math. Soc.*, **123**:1191-1198, 1995.

[17] D. Borwein, J. M Borwein and B. Sims, *Symmetry and the Monotonicity of Certain Riemann Sums*, In: D. Bailey et al. (Eds) From Analysis to Visualization. JBCC 2017. Springer Proceedings in Mathematics & Statistics, **313**, Springer, Cham, Switzerland, 2020.

[18] D. Borwein, J. M. Borwein and A. Straub, A sinc that sank, *Amer. Math. Monthly*, **119**:535-549, 2012.

[19] J. M. Borwein, D. Bailey and R. Girgensohn, *Experimentation in Mathematics–Computational Paths to Discovery*, A. K. Peters, Ltd., Wellesley, Massachusetts, 2004.

[20] J. M. Borwein and P. Borwein, *Pi and the AGM*, John Wiley, New York, 1987.

[21] J. M. Borwein and D. M. Bradley, Thirty-two Goldbach variations, *Inter. J. Number Theory*, **2**:65-103, 2006. Available at `http://carma.newcastle.edu.au/resources/jon/32goldbach.pdf`

[22] J. M. Borwein and R. M. Corless, Gamma and factorial in the Monthly, *Amer. Math. Monthly*, **125**:400-424, 2018.

[23] J. M. Borwein and A. Straub, Special values of generalized log-sine integrals, *Proceedings of ISSAC 2011 (International Symposium on Symbolic and Algebraic Computation)*, 43-50, San Jose, CA, USA, 2011.

[24] F. Brenti, Log-concave and unimodal sequences in algebra, combinatorics, and geometry: An update, *Contemporary Mathematics*, **178**:71-89, 1994.

[25] D. M. Bressoud, *Proofs and Confirmations–The Story of the Alternating Sign Matrix Conjecture*, The Mathematical Association of America, Washington, D.C., 1999.

[26] R. A. Brualdi, Matrices of zeros and ones with fixed row and column sum vectors, *Linear Algebra Appl.*, **33**:159-231, 1980.

[27] R. Chapman, A proof of Hadjicostas's conjecture, June 15, 2004. Available at `https://arxiv.org/abs/math/0405478`

[28] H. Chen, Evaluations of some variant Euler sums, *J. Integer Seq.*, **9**, Article 06.2.3, 2006. Available at https://cs.uwaterloo.ca/journals/JIS/VOL9/Chen/chen78.pdf

[29] H. Chen, *Excursions in Classical Analysis*, The Mathematical Association of America, Washington D.C., 2010.

[30] H. Chen, Interesting series associated with central binomial coefficients, Catalan numbers and harmonic numbers, *J. Integer Seq.*, **19**, Article 16.1.5, 2016. Available at https://cs.uwaterloo.ca/journals/JIS/VOL19/Chen/chen21.pdf

[31] H. Chen, On an infinite series for $(1 + 1/x)^x$ and its applications, *Int. J. Math. Math. Sci.*, **29**:675-680, 2002.

[32] W. Chu and L. D. Donno, Hypergeometric series and harmonic number identities, *Adv. Appl. Math.*, **34**:123-137, 2005.

[33] J. H. Conway and R. Guy, *The Book of Numbers*, Springer, New York, 2011.

[34] K. Dale, Solution to Problem 11369, *Amer. Math. Monthly*, **117**:377, 2010.

[35] A. Dixit, The Laplace transform of the psi function, *Proc. Amer. Math. Soc.*, **138**:593-603, 2010.

[36] W. Dunham, *Euler: The Master of Us All*, The Mathematical Association of America, Washington D. C., 1990.

[37] L. Euler, Meditationes circa singulare serierum genus, *Novi Comment. Acad. Sci. Petropolitanae*, **20**:140-186, 1776. Available at http://eulerarchive.maa.org/

[38] A. B. Farnell and J. B. Rosser, A definite integral–Solutions to Problem 4212, *Amer. Math. Monthly*, **54**:601-602, 1947.

[39] J. Fernández-Sánchez and W. Trutschnig, Nested square roots of 2 revisited, *Amer. Math. Monthly*, **127**:344-351, 2020.

[40] R. Feynman, *"Surely You're Joking, Mr. Feynman!": Adventures of a curious character*, Bantam Books, New York, 1985.

[41] P. Flajolet and B. Salvy, Euler sums and contour integral representations, *Experimental Math.*, **7**:15:35, 1998.

[42] T. M. Flett, A mean value theorem, *Math. Gazette*, **42**:38-39, 1958.

[43] O. Furdui, *Limits, Series, and Fractional Part Integrals*, Springer-Verlag, New York, 2013.

[44] O. Furdui and T. Trif, On the summation of certain iterated series, *J. Integer Seq.* **14**, Article 11.6.1, 2011. Available at https://cs.uwaterloo.ca/journals/JIS/VOL14/Furdui/furdui3.pdf

[45] C. Georghiou and A. N. Philippou, Harmonic sums and the zeta function, *Fibonacci Quart.*, **21**:29-36, 1983.

[46] M. L. Glasser and D. Manna, On the Laplace transform of the psi function, "Tapas in Experimental Mathematics" (T. Amdeberhan and V. H. Moll (Eds)), *Contemporary Mathematics*, **457**:205-214, Amer. Math. Soc., Providence, RI, 2008.

[47] A. M. Gleason, R. E. Greenwood and L. M. Kelly (Eds), *The William Lowell Putnam Mathematical Competition Problems and Solutions:1938-1964*, MAA Problem Books, The Mathematical Association of America, Washington, D.C., 1980.

[48] H. W. Gould and T. Shonhiwa, A catalog of interesting Dirichlet series, *Missouri J. Math. Sci.*, **20**:2-18, 2008.

[49] I. Gradshteyn and I. Ryzhik, *Table of Integrals, Series, and Product*, 5th edition, Edited by A. Jeffrey and D. Zwillinger, Academic Press, New York, 1994.

[50] R. Graham, D. Knuth and O. Patashnik, *Concrete Mathematics*, 2nd edition, Addison-Wesley, New York, 1994.

[51] M. Gromov and M. Taylor, Finite propagation speed, kernel estimates for functions of the Laplace operator, and the geometry of complete Riemannian manifolds, *J. Diff. Geom.*, **17**:15-53, 1982.

[52] J. Hadamard, *The Psychology of Invention in the Mathematical Field*, Dover Publications, Inc, New York, 1954.

[53] J. W. Hagood, Solution to Problem 11753, *Amer. Math. Monthly*, **122**:906, 2015.

[54] G. H. Hardy, J. E. Littlewood and G. Pólya, *Inequalities*, Cambridge University Press, Cambridge, 1967.

[55] W. P. Johnson, The Curious history of Faá di Bronu's formula, *Amer. Math. Monthly*, **109**:217-234, 2020.

[56] K. Kedlaya, B. Poonen and R. Vakil (Eds), *The William Lowell Putnam Mathematical Competition Problems and Solutions:1985-2000*, MAA Problem Books, The Mathematical Association of America, Washington, D.C., 2002.

[57] L. F. Klosinski, G. L. Alexanderson and M. Krusemeyer, The Seventy-Third William Lowell Putnam Mathematical Competition, *Amer. Math. Monthly*, **120**:686, 2013.

[58] M. Kneser, Problem 4305, Problems and Solutions, *Amer. Math. Monthly*, **57**:267-268, 1950.

[59] K. Knopp, Über Reihen mit positiven Gliedern, *J. London Math. Soc.*, **3**:205-211, 1928.

[60] O. Kouba, An inequality for bounded functions, *Math. Inequal. Appl.*, **17**:531-537, 2014.

[61] O. Kouba, Problem 11811, Problems and Solutions, *Amer. Math. Monthly*, **124**:837-838, 2017.

[62] J. Lagarias, An elementary problem equivalent to the Riemann Hypothesis, *Amer. Math. Monthly*, **109**: 534-543, 2002.

[63] L. C. Larson, *Problem-Solving through Problems*, Problem Books in Mathematics, Springer-Verlag, New York, 1983.

[64] P. D. Lax and L. Zalcman, *Complex Proofs of Real Theorems*, University Lecture Series, **58**, American Mathematics Society, Providence, Rhode Island, 2012.

[65] D. H. Lehmer, Interesting series involving the central binomial coefficients, *Amer. Math. Monthly*, **92**:449-457, 1985.

[66] R. G. Medhurst and J. H. Roberts. Evaluation of the integral $I_n(b) = \frac{2}{\pi} \int_0^\infty \left(\frac{\sin x}{x}\right)^n \cos(bx)\, dx$, *Mathematics of Computation*, **19**:113-117, 1965.

[67] D. S. Mitrinović, *Analytic Inequalities*, Springer-Verlag, New York, 1970.

[68] T. Muir, *A Treatise on the Theory of Determinants*, Dover, New York, 1960.

[69] H. Muzaffar, A new proof of a classical formula, *Amer. Math. Monthly*, **120**:355-358, 2013.

[70] I. Nemes, M. Petkovšek, H. S. Wilf, and D. Zeilberger, How to do Monthly problems with your computer, *Amer. Math. Monthly*, **104**:505-519, 1997.

[71] N. Nielsen, *Handbuch der Theorie der Gammafunktion*, Druck und Verlag·von B. G. Teubner, Leipzig, 1906. Reprinted by Chelsea Publishing Company, Bronx, New York, 1965.

[72] A. Nijenhuis, Small gamma products with simple values, *Amer. Math. Monthly*, **117**:733-737, 2010.

[73] O. Oloa, Some Euler-type integrals and a new rational series for Euler's constant, "Tapas in Experimental Mathematics" (T. Amdeberhan and V. H. Moll (Eds)), *Contemporary Mathematics*, **457**:253-264, Amer. Math. Soc., Providence, RI, 2008.

[74] M. Petkovsek, H. Wilf, and D. Zeilberger, $A = B$, A. K. Peters, Ltd., Wellesley, Massachusetts, 1996.

[75] G. Pólya and G. Szegö, *Problems and Theorems in Analysis (I)*, Classics in Mathematics, Springer-Verlag, New York, 1978.

[76] M. H. Protter and H. F. Weinberger, *Maximum Principles in Differential Equations*, Springer-Verlag, New York, 1984.

[77] T. T. Rădulescu, V. D. Rădulescu and T. Andreescu, *Problems in Real Analysis: Advanced Calculus on the Real Axis*, Springer-Verlag, New York, 2009.

[78] D. Robbins, Solution to Problem E2692, *Amer. Math. Monthly*, **86**:394-395, 1979.

[79] G. Robin, Grandes valeurs de la fonction somme des diviseurs et hypothése de Riemann, *J. Math. Pures Appl.*, **63**:187-213, 1984.

[80] J. B. Rosser and L. Schoenfeld, Approximate formulas for some functions of prime numbers, *Illiois J. Math.*, **6**:64-97, 1962.

[81] J. Sándor and L. Tóth, A remark on the gamma function, *Elem. Math.*, **44**:73-76, 1989.

[82] N. J. A. Sloane, The On-Line Encyclopedia of Integer Sequences®(OEIS®), Available at http://oeis.org

[83] J. Sondow, Double integrals for Euler's constant and $\ln 4/\pi$ and an analogue of Hadjicostas's formula, *Amer. Math. Monthly*, **112**:61-65, 2005.

[84] J. Sorel and H. Chen, Transcendental moments of the arctangent, *Math. Mag.*, **91**:157, 2018.

[85] H. M. Srivastava and J. Choi, *Series Associated with the Zeta and Related Functions*, Kluwer Academic Publishes, Dordrecht, 2001.

[86] R. P. Stanley, *Catalan Numbers*, Cambridge University Press, Cambridge, 2015.

[87] R. P. Stanley, Log-concave and unimodal sequences in algebra, combinatorics, and geometry, In *Graph theory and its applications: East and West (Jinan, 1986)*, *Ann. New York Acad. Sci.*, **576**:500-535, 1989.

[88] J. M. Steele, *The Cauchy-Schwarz Master Class*, Cambridge University Press, Cambridge, 2004.

[89] A. Stenger, Experimental math for math Monthly problems, *Amer. Math. Monthly*, **124**:116-131, 2017.

[90] J. S. Sumner, A. A. Jagers and J. Anglesio, Inequalities involving trigonometric functions, *Amer. Math. Monthly*, **98**:264-267, 1991.

[91] F. G. Tricomi and A. Erdélyi, The asymptotic expansion of a ratio of gamma functions, *Pac. J. Math.*, **1**:133-142, 1951.

[92] R. van der Veen and J. van de Craats, *The Riemann Hypothesis*, MAA Press, The Mathematical Association of America, Washington, D.C., 2015.

[93] J. G. Wendel, Note on the gamma function, *Amer. Math. Monthly*, **55**:563-564, 1948.

[94] E. Witten, On quantum gauge theories in two dimensions, *Comm. Math. Phys.*, **141**(1):153-209, 1991.

Index

Printed in the United States
by Baker & Taylor Publisher Services